VHDL

BASICS TO PROGRAMMING

VHDL
BASICS TO PROGRAMMING

Gaganpreet Kaur

Head of the Department
Electronics and Communication Engineering
Rayat-Bahra College of Engineering and Biotechnology for Women

PEARSON

Delhi • Chennai • Chandigarh

Assistant Acquisitions Editor: Sameer Gupta
Associate Production Editor: Amrita Naskar
Composition: Ace Pro India Pvt. Ltd
Printer: Rahul PrintoPack

ISBN 978-81-317-3211-3

First Impression

Published by Dorling Kindersley (India) Pvt. Ltd, licensees of Pearson Education in South Asia.

Head Office: 7th Floor, Knowledge Boulevard, A-8(A), Sector – 62, Noida, UP 201309, India.

Registered Office: 11 Community Centre, Panchsheel Park, New Delhi 110017, India.

Contents

Preface

Over the past few decades, the number of transistors in processors has grown at an exponential rate. Every 18 months, the number of transistors on a chip roughly doubles and circuit design problems become more complex. There is a great need for improvement in physical design tools to handle the increased design sizes. Since digital is ruling the world of technology today, digital design is going to become an important field in the future. The limitation of manual designing methods has paved the way for CAD tools to handle the complex digital designs and VHDL is a design tool to dictate the future design methodology. Infact, standardization of VHDL as IEEE standard itself justifies its capability to be the future digital design language and it becomes imperative for young engineers to be equipped with VHDL design language. VHDL is not just limited to digital design but can also revolutionize the analog design with VHDL-AMS.

With this view in mind and to equip the young designers with design tools of future, this book on VHDL has been written. Although numerous books on VHDL exist in the market, my idea behind writing this book was not to scare the young minds with complexities of design, but rather to enable them to have a smooth transition from budding engineers to exploring design engineers. This book presents an integrated view of a tutor and a designer on VHDL to ignite the spark of learning and innovation in engineers.

The approach taken for this book is to start from the basics, with emphasis on small examples right from the beginning to make the concepts clear. Care has been taken to address queries of beginners. This book is the result of my experience as the teacher of undergraduate engineering students and my knowledge of VHDL as a designer.

Organization of the Book

The first 13 chapters provide a comprehensive study on VHDL as a language and its usage in small designs. Chapters 14 and 15 deal with complex system designs, which will help students in building an approach for designing their projects in VHDL. A separate section on case studies presents design examples in VHDL. A gradual approach has been adopted to understand the significance of VHDL and its capabilities. Then VHDL syntax and features are introduced with suitable examples. Every effort has been taken to keep the language simple but at the same time convey the subject knowledge of VHDL. Even advanced programming concepts are explained with examples in VHDL. A chapter on PLDs has been included as present day complex design has to be implemented using PLDs and gate arrays. This book is intended for undergraduate and postgraduate students in electronics engineering, computer science engineering and electrical engineering. Apart from these, designers from industry too will find it useful, particularly with respect to issues on embedded system design and synthesis issues. Sufficient examples have been included to make the concepts clear to the students through VHDL programming. All the VHDL coding is done using Xilinx 9.2i and simulator used is Modelsim 6.2.

Acknowledgements

Over the entire duration of the writing and compilation of this book, many people have helped me. First and foremost, I would like to thank my students who put unfathomable faith in my proficiency and teaching and motivated me to start work on the book. I am grateful to ITM Gurgaon, particularly to Dr Swaran Ahuja, who guided me and helped me to go ahead with this book. Thanks to many experts who reviewed drafts of this book. Their positive and critical review has certainly helped to improve the content and presentation of the book.

I thank my family members and friends whose support and constant encouragement during the two years of writing the book made this effort worthwhile and without whose support, this book could never have been completed.

Above all, I am grateful to my mother who always stood by me and supported me in every possible way. This book is actually the result of her wishes, support and sacrifices made for me. My little son too shares credit, whose share of time and love I sacrificed to complete this book.

GAGANPREET KAUR

Lesson Plan

This lesson plan is for undergraduate/postgraduate engineering students (ECE/EE/CSE), based on the assumption that there will be three lectures per week and at least 12 weeks in a semester of 4 months

Lecture No.	Topics
L1	**CHAPTER 1** INTRODUCTION Design Flow EDA Tools for Design ROLE OF HDL's Introduction to HDLs Different HDLs
L2	**CHAPTER 2** HISTORY OF VHDL COMPARISON BETWEEN VHDL AND VERILOG FEATURES OF VHDL
L3	BASIC TERMINOLOGIES OF VHDL LEVELS OF HARDWARE ABSTRACTION Overview Entity Declaration Architecture Body Styles of Modelling Simple Example of VHDL
L4	DELAY MODELS IN VHDL Transport Delay Inertial Delay Delta Delay
L5	**CHAPTER 3** INTRODUCTION Identifiers DATA OBJECTS Signal Variable Constant
L6	DATA TYPES Predefined Types Scalar Types Composite Type Access Type File Type
L7	OPERATORS Arithmetic Logical Relational Shift Miscellaneous
L8	**CHAPTER 4** INTRODUCTION SEQUETIAL Vs CONCURRENT STATEMENTS PROCESS STATEMENT Sensitivity List
L9	VARIABLE ASSIGNMENT Variables Differences Between Signal and Variable SEQUENTIAL STATEMENTS IF CASE LOOP
L10	-DO- WAIT ASSERT EXIT NEXT NULL

Lecture No.	Topics	Lecture No.	Topics
L11	POSTPONED PROCESSES PASSIVE PROCESSES	L20	**CHAPTER 8** INTRODUCTION Need for Configurations
L12	**CHAPTER 5** INTRODUCTION CONDITIONAL ASSIGNMENT STATEMENTS When-Else With-When-Select	L21	WRITING CONFIGURATIONS Entity–Architecture Pair Configuration Component Entity Binding Configuration Specification
L13	BLOCK STATEMENT Using Block Statement Guarded Block Statement CONCURRENT ASSERT STATEMENT CONCEPT OF DELAY IN CONCURRENT LOGIC	L22	INTRODUCTION TO PACKAGES WRITING A PACKAGE USING A PACKAGE
L14	**CHAPTER 6** INTRODUCTION COMPONENT DECLARATION COMPONENT INSTANSIATION	L23	**CHAPTER 9** MULTIPLE DRIVERS Need for Resolution Function
L15	GENERICS Introduction Generic Map Generics with Behavioural Modelling	L24	ATTRIBUTES - Value Attributes, Function Attributes
L16	**CHAPTER 7** INTRODUCTION FUNCTIONS USING FUNCTIONS Functions Inside Main Code Functions Inside Package Scope of Functions	L25	- DO- - Array Attributes, Signal Attributes
L17	TYPES OF FUNCTIONS - Pure/Impure	L26	LIBRARIES ALIAS SIGNATURES
L18	PROCEDURES USING PROCEDURES DIFFERENCES BETWEEN FUNCTIONS AND PROCEDURES	L27	QUALIFIED EXPRESSIONS GENERATE STATEMENTS NULL TRANSACTIONS
L19	OVERLOADING Subprogram Overloading Overloading Operators	L28	**CHAPTER 10** INTRODUCTION Advantages of Testbenches

Lecture No.	Topics	Lecture No.	Topics
L29	WRITING A TESTBENCH	L35	**CHAPTER 13** INTRODUCTION Basic Components of Microcomputer VHDL CODE FOR EACH COMPONENT—PRACTICAL EXCERCISE
L30	WAVEFORM GENERATION	L36	**CHAPTER 14** WHAT ARE PLDs ADVANTAGES OF USING PROGRAMMABLE LOGIC
L31	**CHAPTER 11** ** INTRODUCTION DIFFERENCES BETWEEN SIMULATION AND SYNTHESIS	L37	PALs PLAs
L32	SYNTHESIS FLOW SYNTHESIS OF DATA TYPES	L38	CPLDs FPGAs
L33	SYNTHESIZING COMBINATIONAL LOGIC SYNTHESIZING SEQUENTIAL LOGIC	L39	DIFFERENCES BETWEEN FPGA AND CPLD IMPLEMENTATION USING FPGAs AND CPLDs
L34	MODELLING FSM IN VHDL		

** Note: Chapter 11 is generally not part of B.Tech curriculum and is included for better understanding of synthesis process . It may be treated as self study topics for students.

About the Author

Dr Gaganpreet Kaur is an Assistant Professor and Head of the Department, Electronics and Communication Engineering at Rayat-Bahra College of Engineering and Biotechnology for Women, Punjab. She started her career with Guru Nanak Dev Engineering College as graduate engineer and chose teaching as her career. During her career as a teacher, she has held a number of responsible administrative positions. She was involved as research pioneer in the field of VLSI design in college. She also taught at ITM Gurgaon for a brief period. She has published research papers in international and national journals and has presented papers at a number of international conferences. Presently, she teaches undergraduate and graduate students in the area of VLSI and Digital Design. She has conducted a number of training programmes for academicians and, most importantly, for students on 'Circuit design using VHDL'. Her research interests include evolutionary optimization techniques, VLSI partitioning and circuit design with VHDL.

DIGITAL DESIGN FLOW

1

CHAPTER OBJECTIVES

- Significance of CAD tools
- Design flow for circuit design
- Introduction to design tools
- Basics of HDLs

1.1 INTRODUCTION

With the growth in the integrated circuit technology, from large scale integration (LSI) to ultra large scale integration (ULSI), increasing the number of components on a chip with speed and power optimization, the designing of complex digital circuits has become a challenging task. Over the time, different methods of digital circuit design have come into picture like

(i) Manual paper–pen method, and
(ii) Computer-aided design (CAD) tools method.

The manual paper–pen method for circuit design is now obsolete because of its incompetence to capture the complex designs at the basic as well as at the detailed level of gates and flip-flops. It is not only tedious and time-consuming, but practically impossible due to complex, programmable logic devices (PLDs) such as FPGAs and CPLDs.

The CAD tools method, over the years, has assumed an important role in the complex circuit design, with a continuous scaling of semiconductor technology, thereby, leading to the automation of the design process. With reference to digital design, automation can be categorized into two phases namely,

(i) Schematic capture tools
(ii) Hardware description languages (HDLs)

The schematic capture tools for design automation were introduced around 1970s with tools such as Orcad, Pspice, Espresso, and so on. These tools were based on drag-and-drop approach, wherein large, complex circuits could simply be designed by choosing from a list of predefined components such as gates, which joined together to form bigger complex circuits. It was just an automated form of manual design process with the user being free from the responsibility of designing the basic digital components. These circuits could then be verified via simulation process to verify their functionality. It eased the design process but was not much of a breakthrough as the designer needed to have a draft of the circuit that is to be designed at hand to put in a schematic capture tool and then simulate it.

The next breakthrough in digital design automation came in the 1980s with the coming of HDLs for circuit design. HDLs allowed flexibility in the digital circuits to be designed at a higher level of functionality before getting converted in the form of basic gates and flip-flops. It also meant building of complex circuits from the basic components. This enabled the user to have a draft of the complete schematic program before the actual design's implementation. Further, a verification of circuits at both levels—simulation as well as synthesis—was allowed in HDLs without an actual hardware being used.

1.1.1 Design Flow

To see the application of HDLs, it is very important to understand the digital design flow. Before going into the complexities of design flow, let us get familiar with the basics of a digital system. A complex digital circuit is a collection of basic components, that is, gates and flip-flops interconnected to function as a programmable logic device or perform a particular function. The first step in designing any digital system is to take into account any information available about the circuit and determine at what level the system has to be described. Any digital system consists of a complex circuit spread on a single chip or multiple chips, for example, the motherboard of a computer is a CPU circuit designed on a single chip.

The system-level description is the highest level of description, wherein only the functionality of the circuit to be designed is of prime concern, irrespective of whatever components comprise the circuit (Figure 1.1). Next, it is determined if the circuit has to be designed on a single chip or multiple chips. If it is to be spread on multiple chips, then the modules of design are considered chipwise, verifying the functionality of each chip using CAD tools and then integrating them to form a system. On a chip, the most common and the highest level of description is at register–transfer logic level. The register–transfer level describes the circuit operation over several clock cycles with intermediate results stored in registers, that is, simply describing a circuit in terms of sequential registers and combinational circuitry. Now, this description of circuit in terms of sequential and combinational logic can further be broken into basic components to form the sequential and combinational circuits, that is, in terms of gates (combinational) and flip-flops (sequential). This is known as the gate-level description of the circuit. Now, each gate and storage element can further be broken down into simpler forms of circuit components, such as resistors, capacitors, and so on. Once the circuit-level description is made available, it is then implemented and fabricated in terms of silicon wafers—the lowest level of system description and, generally, a part of the back-end processing or physical design of circuits.

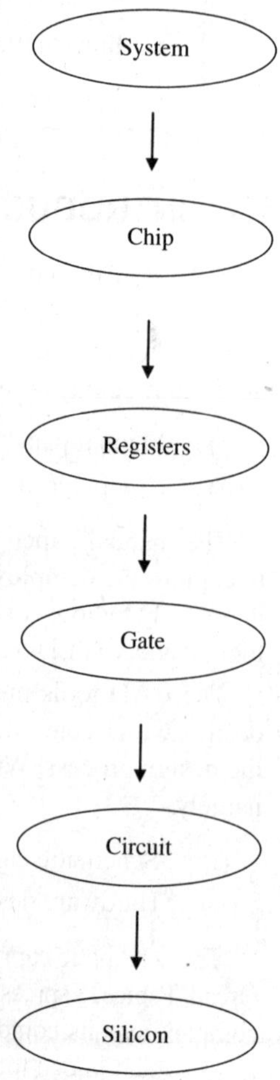

Figure 1.1 Levels of System Description

To start with the design process, the first step is to understand the system specifications, as to what may be the function of the circuit that is to be designed. For example, if adder is to be designed, designer should know what type of quantities will it add—integer number, bit values, or bit-vectors. Let's say if it adds, then how many bits will be input for addition and so on. Once

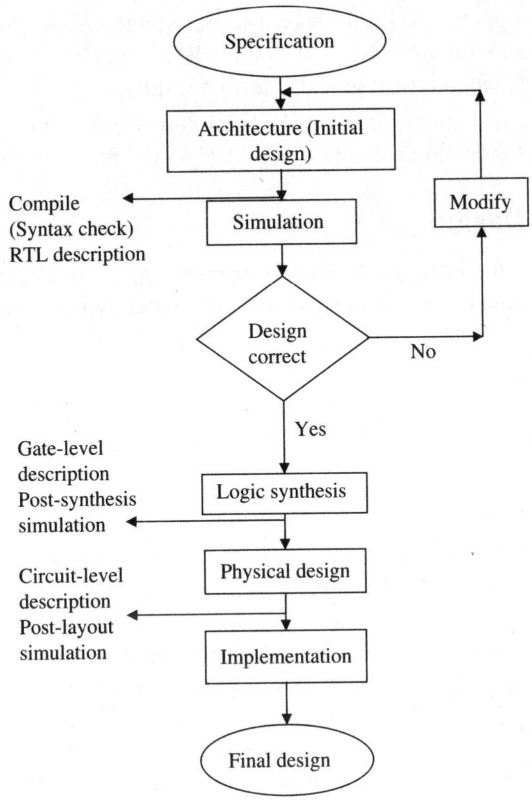

Figure 1.2 Design Flow

the specifications are clear, the designer is in a position to tell what type of and how many inputs are expected to be given to the circuit and what is the output expected. So, a study of specification is very important, as it determines the interface of the design. The next step is to define the functionality of the circuit in the form of its architecture. The logical relationship between the inputs to achieve the desired output is given in the form of architecture and forms the basic user input file in form of text or block diagram. Once the circuit is completely defined in terms of its functionality and interface, the next step is to verify if the circuit functions as expected, giving the desired output for a set of inputs. This functional verification of circuit without an actual hardware is known as simulation. So, the input to simulation is the design specification in the form of text/graphical input and the output is timing-waveform diagram showing values of output achieved for given inputs. Before simulating a circuit, a compilation check is done by the design language compiler to check for syntax errors, if any, in the design. The simulation makes use of an RTL description of the circuit, internally, to verify its correctness. If the design is correct, then the next step is carried out, else the required modifications are made to the design so as to yield the desired results. The next step in the row is logic synthesis, which describes the circuit in terms of its basic components, that is, gates and flip-flops in case of digital circuits. So, synthesis generates a gate-level description of the circuit and the netlist file. Generally, a post-synthesis simulation is carried out after this procedure to check for the circuit (component) delays in the circuit by using the target technology. If this is fine, then the next step is physical design, where the user-defined design file is mapped to the target technology and the outputs are checked.

After this step, it is very important to perform post-layout simulation to check for the design-rule violations and also cross-check for any interconnection delays in the target technology. The output is also checked with the target technology to give the desired results.

Once this is done, the final step is to implement the circuit which is more of a back-end activity, wherein silicon wafers and fabrication come into play to yield the final design.

1.1.2 EDA Tools for Design

With the use of CAD tools, the design process of complex digital circuits has been simplified. To design a digital circuit, a number of CAD tools are needed, which typically include tools for the following basic design tasks:

(i) Design entry
(ii) Simulation
(iii) Synthesis

1.1.2.1 Design Entry

The first and foremost step in designing a circuit using CAD tools is to know the system specifications and architecture (function) of the design, which are to be given to the CAD tools as an input. This is done using the two most popular design languages—very high speed integrated circuit (VHSIC) hardware description language (VHDL) and Verilog. So, the information about the circuit to be designed is given in the form of a text file written in VHDL or Verilog. The code giving the system specifications and the logical expressions to perform the desired function is written in HDL. The source code is usually a plain text explaining how a system works. The language compiler is used to check for any syntax errors. Usually, at this stage, the code is technology-independent and portable, giving the designer the flexibility to reuse the same code with different simulation and synthesis tools. Different vendors such as Aldec, Xilinx, Actel, and so on offer their design entry tools both with VHDL and Verilog editions.

1.1.2.2 Simulation

Once the design entry step is complete, presynthesis simulation is carried out to check if the circuit works as intended. The simulation tool gives its output in the form of timing waveforms, where the waveforms describe the output values for the given set of inputs. This step is very important as it lays the foundation of the whole design to perform the desired task, and also verifies the functionality of the design. ModelSim and Synplify Pro are the most popularly used simulation tools. VHDL supports the event-driven simulation—that is, whenever any change in the value of signal takes place, a simulation runs.

1.1.2.3 Synthesis

The synthesis of a circuit that is to be designed is basically the conversion of a design to gate-level schematics. It expresses the circuit in terms of basic components in accordance with the logical expressions that are used to describe the circuit design. Another important task of synthesis is to optimize the design with reference to component delays, critical path delay or area considerations. It also checks for any errors in the design violating any timing or voltage considerations. Netlist, log and schematic are the outputs of the synthesis. Synthesis is generally dependent on the target technology—say, for example, synthesis optimization differs for the design that is to be implemented in the Spartan family from that for the Vitrex family. This difference is due to the difference in speed and grade of each technology. Therefore, a synthesis determines how exactly the circuit will be realized in a specific hardware technology.

Once the synthesis is completed, a post-synthesis simulation is performed. It takes into account the timing considerations by evaluating the optimized result from the synthesis, for both component delay as well as interconnection delay in the circuit. Synplify Pro by Synopsys, Quatrus II by Altera and Leonardo spectrum by Mentor graphics are the popularly used synthesis tools.

Table 1.1 Commonly Used VHDL Tools

Vendor	Simulator	Synthesizer	RTL view	Gate view
Actel	Modelsim–Actel Edition	Synplify–Actel Edition	No	Yes
Aldec	Aldec Active HDL		Yes	Yes
Altera	Modelsim–Altera Edition	Altera Quartrus	Yes	Yes
Xilinx	ISE Simulator	XST	Yes	Yes
Mentor	Modelsim PE	Leonardo Spectrum	Yes	Yes
Lattice		Precision/Synplify Edition	No	Yes

1.2 ROLE OF HDL

After having got an insight into the design flow, let us examine the role of HDLs in the ever-growing semiconductor industry for circuit design. By now, it is clear that the HDLs have an important role to play as they are the CAD tools for the design entry. They are the means to provide a textual entry of circuit design that is to be processed and synthesized.

1.2.1 Introduction to HDLs

The hardware description language (HDL) is a language that is used for the design of electronic circuits, describing its interface, operation and verification of its functionality, without an actual hardware using simulation and synthesis. It is a text-based language to design a circuit, instead of conventional drag-and-drop circuit-design languages. It is similar to programming languages, but with explicit notations to describe the circuit syntax and the semantics of hardware needed to describe the circuit. Although it is possible to represent the hardware semantics using traditional-programming languages such as C++, these languages do not allow any explicit expression of time and concurrency.

Any hardware language should necessarily have the following features:

- It should have the capability to describe the hardware structure of the circuit to be designed.
- It should be able to capture the true abstract behaviour of the circuit.
- It should also have the flexibility in the levels of abstraction at which the circuit is described.

The first and the foremost characteristic of any HDL is to describe its hardware structure, thereby, providing the language ability to model its interface, that is, its external view without describing its functionality; whereas the characteristic to model the behaviour of the circuit gives HDL the capability to define the functionality of the circuit.

This characteristic to describe the internal structure determines the level of abstraction at which a circuit is modelled. The level of abstraction for any circuit design is determined by the objects that are used to design the circuit, the way these objects are manipulated and the operations that can be performed on these objects as an insight to circuit design.

1.2.2 Evolution of HDLs

The first HDLs were ISP, developed at Carnegie Mellon University, and KARL, developed at the University of Kaiserslautern, both around 1977. ISP was, however, more like a software-programming language that was used to describe the relations between the inputs and the outputs of the design; and, hence, could be used to simulate the design, but not to synthesize it. KARL included the design calculus language features supporting the VLSI chip floor planning and hardware design; and served as the basis for KARL's interactive graphic sister language, ABL, which was implemented in the early 1980s as the ABLED graphic VLSI design editor by the telecommunication research center, CSELT, at Torino, Italy. In the mid-1980s, a VLSI-design framework was implemented around KARL and ABL by an international consortium funded by the commission of the European Union. In 1983, Data-I/O introduced advanced boolean expression language (ABEL), which was targeted to describe the programmable logical devices, and was basically used to design the finite-state machines. The first modern HDL Verilog was introduced by Gateway Design Automation in 1985.

Cadence Design Systems later acquired the rights to Verilog-XL, the HDL simulator which became the standard for the next decade. But it was inadequate to describe the complex hardware designs. In 1987, a request from the US Department of Defence led to the development of very high speed integrated circuit hardware description language (VHDL). Its first version was VHDL 87, and was standardized by the IEEE. It was later upgraded to VHDL 93 after five years; and in 2002 was again upgraded to VHDL 1076 (1993–2002). IEEE 1164 was added to VHDL to introduce multi-valued logic.

Initially, both VHDL and Verilog were used to document and simulate circuit designs that were already captured and described in the form design file, but was later extended to work at the design level using HDL; and capabilities to synthesize complex circuits were added to generate circuit schematics without an actual hardware.

The introduction of logic synthesis for HDLs made HDL popular for front-end design tools. The synthesis tools compiled the HDL source files (written in a constrained format called register transfer logic (RTL)) into a gate/transistor-level netlist description.

Need to Know?

Netlist: A textual representation of circuit schematic in terms of a number of components, identifying each component and nets that are connecting the components.

RTL: A level of description of a digital circuit in which operation if spread over several clock-cycles describes circuit behaviour as a flow of data between registers.

Simulation: A process of functional verification of the circuit using a model of the circuit without an actual hardware in place under the given set of stimuli/conditions (test data), and the output is timing waveforms.

VHDL is an acronym for very high speed integrated circuit (VHSIC) hardware description language (HDL). VHDL was initially developed as a simulator and ultimately developed as a tool to make hardware designs as portable as possible (i.e., ASIC, FPGA, different micron-process technologies). It is a design-automation tool that is used to model digital systems at different levels of abstraction, verifying its functionality, generating test data, and map the verified circuit to hardware. VHDL finds its applications in field programmable gate arrays (FPGAs), complex programmable logic devices (CPLDs), and application specific integrated circuits (ASIC) design, making it a versatile language to design digital circuits, with levels of complexity varying from simple gates to VLSI circuits. VHDL was developed as vendor independent IEEE standard with IEEE 1076 standard as its first version in 1987 and IEEE 1164 standard as VHDL '93 version.

The VHDL model of a device or a circuit supports its external as well as its internal view, where the external view specifies the interface of the device with the user and the internal view captures the functionality of the device.

Although a number of HDLs have evolved over the time such as ABEL and bluespec, the most widely used are VHDL and Verilog.

1.2.3 Different HDLs

Before the coming of VHDL, different hardware languages existed in the market, such as AHPL, interactive design language (IDL), instruction set processor specification (ISPS), ABEL, Verilog and so on. But among all of these, advanced boolean equation language (ABEL) and Verilog were the only popular languages and are still used, whereas others had shortcomings which limited their scope only to be a part of possible digital circuits which could be designed. ABEL is used for PLD design and is not suitable for designing of simple circuits. ABEL hardware descriptive language allows one to describe the digital designs with equations, truth tables, state diagrams or any other combinations of the three, optimize and simulate the design without specifying a device or assigning pins. The ABEL-HDL permits to write hardware-independent programs; and after program verification and optimization, one can choose the appropriate PLD device. It helps the user to choose the best PLD in which the program fits in with optimal parameters.

Verilog is very similar to VHDL. In fact, VHDL, Verilog and ABEL are presently used in the market.

SUMMARY

- CAD tools are design automation tools used for designing of complex circuits.
- Use of CAD tools depends on the level of circuit to be designed.
- HDLs are important EDA tools that are used for circuit design even when circuit schematic is not known and only functionality of the circuit to be designed is known.
- Any EDA tool has three steps to follow: design entry, simulation and synthesis.
- Simulation is used for functional and timing verification, while synthesis is used for circuit generation and timing analysis, based on the family of components used.
- Any HDL should essentially have the capability to describe circuit functionality and the ability to abstract the true behaviour of a circuit.
- VHDL and Verilog are the two most popularly used HDLs.
- VHDL stands for very high speed integrated circuit (VHSIC) hardware description language.

QUESTIONS

Objective Questions

1. Netlist is a .... representation of a circuit.

   a. graphic   b. text-based   c. handwritten   d. pictorial

2. CAD stands for ....

   a. colour and design   b. computer and design
   c. compute and design   d. computer-aided design

3. VHDL is a .... tool for circuit design.
   a. design b. design automation c. simulation d. synthesis
4. Synthesis means a gate-level description of circuit. (True/False)
5. VHDL-designed circuits can be implemented using ....
   a. FPGAs b. CPLDs c. both d. none
6. RTL is the outcome of ....
   a. simulation b. synthesis c. physical design d. implementation
7. VHDL is a back-end tool. (True/False)
8. RTL stands for ....
   a. resistor–transfer logic b. register–transistor logic
   c. register–transfer logic d. none of these

Descriptive Questions

1. Discuss the steps for design and implementation of digital circuits.
2. What is the significance of HDLs?
3. Write a note on the features and capabilities of an HDL.
4. Discuss the different EDA tools for digital design.
5. Compare the schematic capture tools and HDLs.
6. Discuss the different levels of system design.
7. Give the differences between functional simulation and post-synthesis simulation.
8. VHDL is primarily a simulation language. Justify.
9. List the advantages of using VHDL for digital design over the traditional methods.

2

INTRODUCTION TO VHDL

CHAPTER OBJECTIVES

- Evolution of HDL
- Comparison between two popular HDLs—Verilog and VHDL
- Features of VHDL
- Introduction to the three styles of VHDL modelling
- Delay models

2.1 HISTORY OF VHDL

VHDL is an acronym for very high speed integrated circuit language (VHSIC) hardware description language (HDL). VHDL is used to describe the logical structure and functions of digital systems that are varying from simple gates to complex PLDs (FPGAs, ASICs). It facilitates design specification and simulation of digital systems. Even though initially it was not meant for synthesis, it can be used for synthesis within a limited domain. It was initiated by the US Department of Defense in the 1980s as a need for a standard HDL for their projects as different vendors were using different languages for IC design. In 1981, three companies—IBM, Texas Instruments and Intermetrics—were entrusted with the job of developing a common standard language for design, and VHDL was proposed as the high description language. In August 1985, first VHDL Version 7.2 was introduced publicly for Digital design and in 1986 VHDL was proposed IEEE standard as design language. VHDL 1076-1987 was introduced as the first IEEE standard and in 1988 as the ANSI standard too. In 1993, the previous standard was revised to IEEE 1164 to include multivalued logic. Later in 1995, Verilog IEEE standard 1364 was also introduced. Currently, IEEE 1076-2002 VHDL standard is being used. Also, a new version of VHDL has been introduced with new added features in the form of VHDL 2008. Presently, Verilog and VHDL are two design HDLs used widely.

2.2 COMPARISON BETWEEN VHDL AND VERILOG

Both VHDL and Verilog were widely accepted as industry as well as academia standards for digital design, but eventually VHDL attained more popularity due to its capabilities to capture complex digital designs better. In fact, VHDL-AMS has been introduced as IEEE VHDL 1076.1-1999, a superset of VHDL meant for analog system design.

Although both Verilog and VHDL are design languages, they have differences from data types to complexity level of circuits that are to be designed.

- VHDL allows both user-defined as well as predefined data types which makes it suitable as a design language for abstract hardware design. Whereas Verilog allows only predefined data types making it suitable as a design language for hardware-structured modelling.
- With respect to reusability of code, VHDL has much more to offer with features of packages and subprograms (functions and procedures). But as a contrast, Verilog has no provision for packages and only functions can be defined in the form of a file which can be reused.
- VHDL is a strongly typed language and includes advanced features like generics, configurations, generate statements and user-defined libraries. These features are absent in Verilog restricting to code reusability and code sharing.
- Overall advance features of VHDL make it suitable for large complex designs where as Verilog being a simple language is more suitable for primitive designs.

2.3 FEATURES OF VHDL

VHDL is a powerful design language with an ability to model digital systems at different levels of abstraction varying from gates to complex digital systems.

Features of VHDL are listed as follows:

1. VHDL is a hardware desiription language used for *design entry* and *simulation* of digital circuits.
2. VHDL is an *event-driven* language; that is, whenever an event occurs on signals in VHDL it triggers the execution of the statement. (event refers to the change in value).

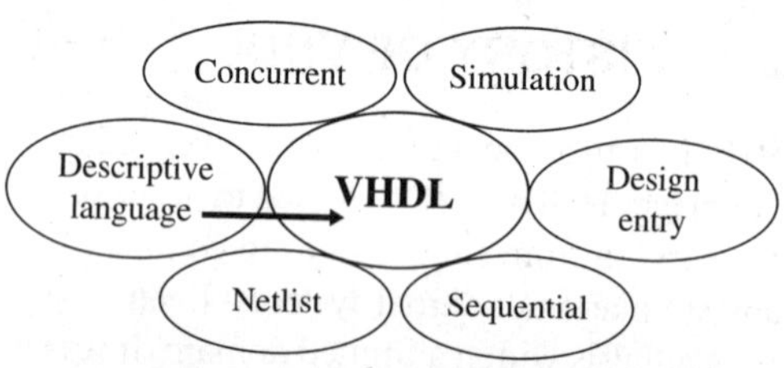

Figure 2.1 Features of VHDL

3. VHDL is *technology platform-independent language* and *portable* by allowing different vendors to use same design descriptions across different target technologies for implementation, even though it offers flexibility to work with specific technology too.
4. VHDL allows both *concurrent* as well as *sequential modelling* which are effective in capturing the features of real-life complex circuits consisting both concurrent combinational logic as well as synchronous sequential logic.
5. It gives user the flexibility to define the *data-types* that are specific to user needs apart from the predefined types.
6. It strongly supports *code reusability and code sharing* by including features of packages, user-defined libraries and subprograms.
7. It includes advance features of *configuration* allowing multiple architectures that are existing for the same design.
8. VHDL supports *flexible design methodologies: top-down approach,* in which the system is designed at a higher level and then refined to its schematic using simulation and synthesis; *bottom-up approach* in which the design is started at its gate level and the smaller modules are brought together and simulated to verify its functional behaviour as a system; or *mixed approach* which uses both top-down as well as bottom-up approach.

9. It is a *case-insensitive* language, that is, it does not differentiate between lowercase and uppercase letters. Internally, a compiler converts all characters to uppercase.
10. VHDL is also said to be *strongly typed language,* that is, it does not support implicit conversion between data types.
11. It allows *general purpose designs* with the help of generics and attributes.
12. It allows *design verification at three levels*: compilation (syntax check) and simulation, synthesis and post-synthesis simulation.

2.4 BASIC TERMINOLOGIES OF VHDL

Entity: Entity is referred to the circuit which is to be modelled. Every entity is defined by a unique name and encodes hardware abstraction of the actual hardware device. It represents the external interface of the circuit that is to be designed. An entity with one external view can have more than one internal view (functionality description). Since VHDL allows a hierarchical design, one entity when used in the description of another higher level entity is referred to as a component.

Architecture: All entities to be designed in VHDL have an architecture description. The architecture describes the functionality of the entity. A single entity can have multiple architectures.

Configuration: In case of entity with multiple architectures, a configuration is used to bind an entity to a particular architecture for simulation or bind a component instance to an entity—architecture pair.

Subprograms: Subprograms are user-defined pieces of code that are written to perform specific functions and can be stored and reused when needed. VHDL supports two types of subprograms—functions and procedures.

Package: A package is a collection of commonly used data types and subprograms that are used in a design allowing it for reusage by including the package in the design in a similar way as header files are included in C/C++.

Library: To create a design in VHDL the first step is to include libraries that are needed in design. A VHDL library is nothing but a collection of all in-built defined language constructs like predefined data types, conversion functions, language features and so on. Library stores all this information in the form of packages consisting of functions, type definitions and procedures. VHDL being an IEEE standard has a predefined library IEEE which contains packages such as std_logic_1164, std_logic_arith, std_logic_unsigned, and std_logic_signed; and library std containing package standard. All user-defined packages and programs are by default stored in LIBRARY 'WORK'.

2.5 LEVELS OF HARDWARE ABSTRACTION

A digital system can be represented at different levels of abstraction. Determining the level of system abstraction is very important in a digital design as it determines what level of complexity and representation is used for a system design. 'Abstraction' means 'relevant detail' in English. Therefore, abstraction in VHDL refers to at what level the design of system is represented. So the level of abstraction is determined by objects that can be manipulated and operations that can be performed on them.

2.5.1 Overview

In VHDL, a circuit can be designed at different levels of abstraction depending on the depth of circuit functionality and components available to the designer, and complexity of the circuit so as to make the circuit design manageable. There are three levels of system abstraction of digital design:

- Behavioural/Algorithmic
- Dataflow
- Structural/Physical

Behavioural is the highest level of abstraction that describes a system in terms of what it does or how it behaves rather than, in terms of its components and the interconnection between them. A behavioural description specifies the relationship between the input and output signals. Sequential statements are executed in the sequence in which they are specified. VHDL allows both concurrent as well as sequential signal assignments which will determine the manner in which they are executed. It is complex, needs complete understanding of system functionality and always preferred for functional and simulation analysis.

The **dataflow** representation describes how a data moves through the system. This is typically done in terms of data flow between registers through buses and interconnections. So this level of abstraction is close to register transfer level (RTL) of system representation. The data flow model makes use of concurrent statements that are executed in parallel as soon as the data arrives at the input. Both dataflow as well as the behavioural styles are based on the mapping relation between input(s) and output(s) with a difference that behavioural style is procedural following the sequence of steps to perform a desired function, where the steps may be Boolean expressions. Whereas dataflow on the other hand is concurrent, event-driven approach where logic equation is executed only when any of its input value changes.

The **structural** level, on the other hand, is the lowest level of abstraction. It describes a system as a collection of gates and components that are interconnected to perform a desired function. It is nothing but a physical description of the circuit. A structural description could be compared to a schematic of interconnected logic gates. It is a representation that is closer to the physical realization of a system.

2.5.2 Entity Declaration

VHDL is used to describe a model for digital hardware device. This model specifies the external view of the device through which it communicates with other models in its environment as well as the internal view of the device which specifies the functionality or structure. VHDL uses five design units for describing the circuit, namely:

- Entity
- Architecture
- Configuration
- Package declaration
- Package body

These units are classified as primary and secondary units as shown in Figure 2.2. The former can exist independently in a design whereas the latter are used in addition to the former. Of the five basic units, entity, configuration and package declaration are primary units while architecture and package body are secondary units.

A true hardware abstraction in a VHDL requires a VHDL entity and a VHDL architecture as follows:

- The entity defines an *external interface.*
- The architecture defines an *internal function.*

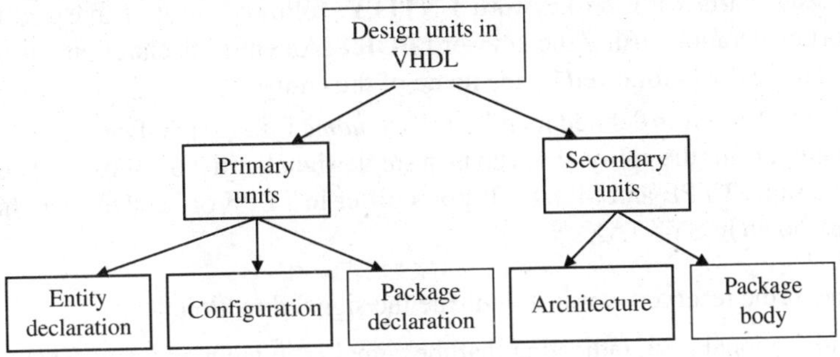

Figure 2.2 Design Units in VHDL

The complete entity description is as shown in Figure 2.3 comprising ENTITY DECLARATION and ARCHITECTURE.

Entity is specified by an entity declaration. The entity declaration specifies the name of the entity that is being modelled and lists the set of interface ports. It describes the external view of the circuit that is to be designed in terms of number of ports, their direction of dataflow, and acceptable data type.

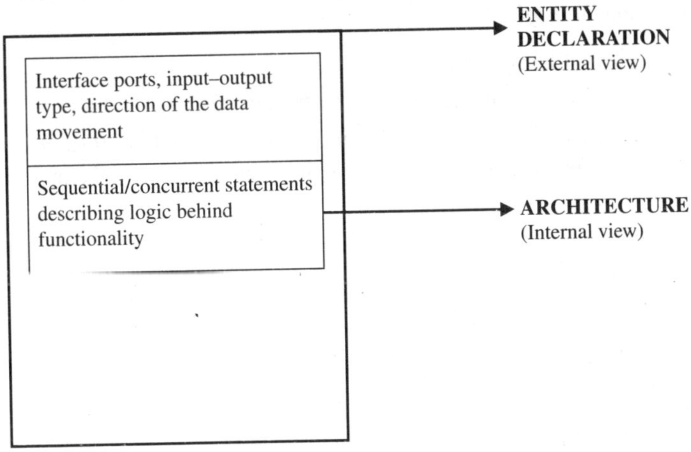

Figure 2.3 Entity Description

The entity declaration defines the name of the entity and lists the input and output ports, where the ports are the signals through which the entity communicates with the other models in its external environment.

The general form is as follows:

```
ENTITY name IS
      PORT (port_name1,[port_name2]: MODE type;
            port_name: MODE type;
                          :
            port_name: MODE type);
END [name];
```

Note: All keywords are written in bold and uppercase and [ ] is used to indicate the optional code.

An entity always starts with the keyword **ENTITY**, followed by its name and the keyword **IS**. Next are the port declarations using the keyword **PORT**. An entity declaration always ends with the keyword **END**, optionally [] followed by the name of the entity.

The *name* of entity is a user-defined identifier. *Port_name* is user-specified name of external signals (both input and output) for the entity. Port can be a single signal or list of signals separated by commas provided all have same TYPE and MODE. If ports differ in TYPE or MODE then they are separated by semicolon as shown in SYNTAX.

MODE: is one of the reserved words to indicate the signal direction:

- *in*—unidirectional port, indicating that the signal is an input and data can be written to
- *out*—Unidirectional port, indicating that the signal is an output of the entity whose value can only be read
- *buffer*—Bidirectional port, indicating that the signal is an output of the entity whose value can be read inside the entity's architecture
- *inout*—Bidirectional port, indicating that the signal can be an input or an output.

TYPE: a built-in or user-defined signal type. Examples of types are bit, bit_vector, Boolean, character, std_logic, and std_ulogic.

- *bit*—can have the values 0 and 1
- *bit_vector*—is a vector of bit values (e.g., bit_vector (0–7))
- *std_logic, std_ulogic, std_logic_vector, std_ulogic_vector*: can have 9 values to indicate the value and strength of a signal. Std_ulogic and std_logic are preferred over the bit or bit_vector types.
- *Boolean*—can have the values—TRUE and FALSE
- *integer*—can have a range of integer values
- *real*—can have a range of real values
- *character*—any printing character
- *time*—to indicate time.

2.5.3 Architecture Body

Once the entity is declared its external interface is defined. To completely define the entity, its functionality or operation too is required to be defined. This is done by defining the architecture of the entity under design. So, architecture describes the functional relationship between the input and output signals, giving an internal view of the entity. Though one entity can have multiple architectures, only one architecture body should be bound to an entity for execution. Every architecture of the entity created is identified by a unique user-defined name. The architecture body of an entity consists of two parts as follows:

- *Declarative* part: The declarative part of architecture includes necessary declarations such as type, signal, component and subprogram declarations. Anything declared inside the declarative part of the architecture body has its scope local to that particular architecture only and not to the whole entity itself.
- *Statement part*: The statement part is the actual executable part which includes statements describing the functionality of the entity in terms of logical and mathematical expressions. An architecture body may contain concurrent statements or sequential statements or both.

SYNTAX:

```
ARCHITECTURE architecture_name OF entity_name IS
–declarative_part
BEGIN
   Statement1;
   Statement2;
END [ architecture_name ] ;
```

An architecture name is a user-defined name of the architecture body. It can be the same as an entity name or different, but in case of multiple architectures of an entity all architectures should have unique different names.

Example 2.1 Example of entity–architecture for AND gate

```
  ENTITY and_gate IS
     PORT (a,b: IN BIT;
              c: OUT BIT);
  END and_gate;
ARCHITECTURE simpleOF and_gate IS
BEGIN
        c <= a AND b;
END simple;
```

2.5.4 Styles of Modelling

VHDL allows circuit design to represent at different levels of abstraction. For an entity to be designed, depending on the level of abstraction, architecture can be written in any of the permitted styles of modelling, though the entity declaration remains the same. Based on the three levels of abstraction the styles of modelling a circuit design in VHDL are listes as follows:

- Structural style
- Dataflow style
- Behavioural style
- Mixed style

2.5.4.1 Structural Style of Modelling

Structural style of modelling is the simplest style of modelling in VHDL and is based on the lowest level of abstraction of the entity that is to be designed. It is a representation of circuit or entity in terms of the interconnected components. The structural style of modelling corresponds to the most detailed level of representation of the circuit. Though it is very simple, it requires a designer to have a complete understanding of the circuit that is to be designed in terms of components comprising the circuit, which implies a draft of circuit should be known to the designer for a structural style to be used. This restricts the use of structural to more of simpler circuits. The simulation of structural design corresponds to gate-level simulation, closest to the physical design.

In the structural style of modelling, architecture body of entity that is to be designed essentially consists of

- component declaration
- component instantiation

A component declaration is very similar to an entity declaration, where each and every component to be used in the architecture description is declared with its unique name—identifier, its number, and type of input and output ports. Component declaration is done in the declarative section of the architecture body.

Component instantiation is nothing but a mapping between actual and formal ports of the entity and components, respectively. Component instantiation is the port-map statement written inside the executable section of the body of architecture.

Let us see example of half-adder using structural style of modelling half adder contains an AND gate and a XOR gate as shown in Figure 2.4.

Example 2.2 Design of half adder using the structural style of modelling

```
   ENTITY half_adder IS                                  -- entity declaration
        PORT (a,b: IN BIT;
                sum, carry: OUT BIT);
   END half_adder;
ARCHITECTURE struct OF half_adder IS                     -- declarative section of architecture
COMPONENT and_gate IS                                    -- component declaration
        PORT (x,y: IN BIT;
                z: OUT BIT);
   END and_gate;
COMPONENT xor_gate IS
        PORT (l,m: IN BIT;
                n: OUT BIT);
   END xor_gate;
BEGIN
X1:xor_gate PORT MAP(a,b,sum);                           -- component instantiation
A1:and_gate PORT MAP(a,b,carry);
END struct;
```

In Example 2.2 of half adder, we learnt that half adder comprises AND gate and XOR gate.

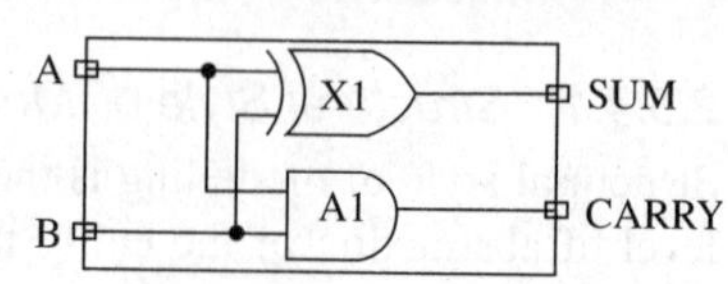

Figure 2.4 Half_adder

In the architecture body, the two components of half adder are AND gate and XOR gate as declared. The keyword **COMPONENT** is used followed by the name of the component, then the keyword PORT is used followed by the ports of the component, their mode and data type. After the component declaration, each component to be used is instantiated using PORT MAP statement to map the actual ports of the entity under design with the component ports. A component once declared in the architecture can be instantiated any number of times, that is, any number of instances of that component can be used.

Note that the structural style is the concurrent style of modelling, where the order of statements written inside the architecture body is not important.

2.5.4.2 Dataflow Modelling

Dataflow style of modelling expresses the functionality of a circuit as a flow of data through buses and interconnections. It does not use the structure or components but uses the logic expressions to express the flow of data from input to output through intermediate interconnections (signals). The dataflow style is also the concurrent style of modelling. It is less detailed than the structural style of modelling as no involvement of physical basic components is there. This style of modelling is closer to RTL level simulation. It corresponds to the middle level of abstraction varying between structural (lowest) and behavioural (highest) level of abstraction.

Example 2.3 Design of half adder using the dataflow style of modelling

```
  ENTITY half_adder IS                              -- entity declaration
    PORT (a,b: IN BIT;
          sum, carry: OUT BIT);
  END half_adder;
ARCHITECTURE dataflow OF half_adder IS              -- declarative section of architecture
BEGIN
  sum <= a XOR b;
  carry <= a AND b;
END dataflow;
```

In Example 2.3, the architecture of half adder is written using the dataflow style of modelling. It makes use of two simple concurrent statements, expressing the logic relation between the output ‘sum’ with inputs ‘a’ and ‘b’ as well between output ‘carry’ and inputs ‘a’ and ‘b’. Being concurrent style the order of statements is not important.

2.5.4.3 Behavioural Modelling

Behavioural modelling expresses the architecture of the entity under design at the highest level of abstraction. This style of modelling emphasizes on the behaviour of the circuit, that is, behaviour of the output with respect to changes in the input without going into details of how or using what components is the functionality carried out. It simply maps the relationship between the inputs and the outputs. This modelling is sequential style of modelling; so, the order in which the statements are written is particularly very important. It is also known as algorithmic style of modelling as approach in the behavioural modelling is to give steps for performing the required tasks in sequence (algorithm). This kind of modelling is most suitable for complex novel designs and is preferred for functional and simulation analysis of the circuit. In fact, being at the highest level of abstraction, it is used for faster simulation of the circuits and corresponds to the functional simulation of the circuit.

Example 2.4 Design of half adder using the behavioural style of modelling

```
  ENTITY half_adder IS                              -- entity declaration
    PORT (a,b: IN BIT;
          sum, carry: OUT BIT);
  END half_adder;
```

```
ARCHITECTURE behave OF half_adder IS                -- declarative section of architecture
  BEGIN
  PROCESS(a,b)                                      -- sequential construct
  BEGIN
     sum<= a XOR b;
     carry<= a AND b;
  END PROCESS;
END behave;
```

In Example 2.4, the half adder circuit is designed using behavioural modelling. It uses a special construct **PROCESS** within which everything written is treated as sequential. Inside the process, two statements expressing the functional relationship between the input and output are written. Note here the order of writing sum and carry statements is important.

2.5.4.4 Mixed Style of Modelling

Mixed style of modelling uses all three styles, namely, structural, dataflow and behavioural in a single architecture body. So, the concurrent and sequential statements can be a part of the same architecture. This gives the designer the flexibility to use different styles for forming complex designs.

Example 2.5 Design of half adder using the mixed style of modelling

```
  ENTITY half_adder IS                              -- entity declaration
     PORT (a,b: IN BIT;
            sum, carry: OUT BIT;
            clk:IN STD_LOGIC);
  END half_adder;
ARCHITECTURE behave OF half_adder IS                -- declarative section of architecture
 BEGIN
 carry<= a AND b;
 PROCESS (clk)                                      -- sequential construct
   BEGIN
   sum<= a XOR b;
  END PROCESS;
END behave;
```

In Example 2.5, the half adder is designed using the dataflow as well as the behavioural style modelling. Carry is generated by AND operation between the inputs 'a' and 'b'-dataflow statement, whereas the sum is generated only when there is a clock pulse. A signal 'clk' is introduced such that a change at the signal 'clk' triggers the process to generate 'sum' behavioural.

Since the behavioural coding is done inside the **PROCESS** construct and the process itself is a concurrent statement it does not matter if the process appears first or other concurrent statements appear first inside the architecture body as order is not important. Only the statements written inside the process are sequential and not the process itself.

2.5.5 A Simple Coding Example in VHDL

Example 2.6 Design of full adder using all three styles of modelling.

Figure 2.5 shows the architecture of a full adder circuit.

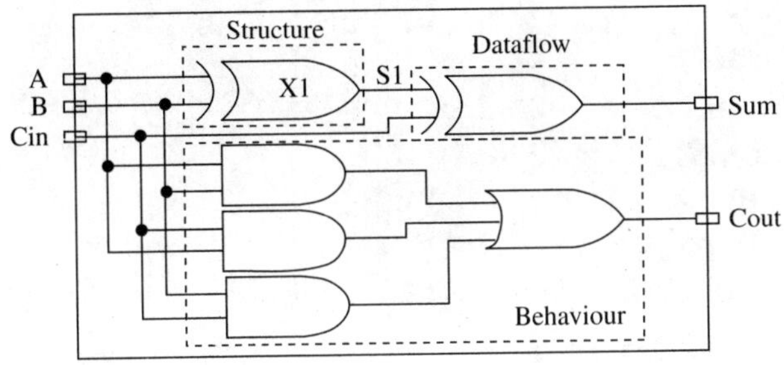

Figure 2.5 Full Adder Diagram

```
ENTITY full_adder IS                                    -- entity declaration
    PORT (a,b,c: IN BIT;
          sum, carry: OUT BIT);
END full_adder;
```

Dataflow architecture

```
ARCHITECTURE dataflow OF full_adder IS          -- declarative section of architecture
  BEGIN
    sum<= a XOR b XOR c ;
    carry<= (a OR b)AND (b OR c)AND (a OR c);
END dataflow;
```

Behavioural architecture

```
ARCHITECTURE behave OF full_adder IS            -- declarative section of architecture
  BEGIN
  PROCESS (a,b,c)
    sum<= a XOR b XOR c ;
    carry<= (a OR b)AND (b OR c)AND (a OR c);
  END PROCESS;
END behave;
```

Structural architecture

```
ARCHITECTURE struct OF full_adder IS            -- declarative section of architecture
COMPONENT and_gate IS                           -- component declaration
      PORT (x,y: IN BIT;
```

```
            z: OUT BIT);
    END and_gate;
COMPONENT xor_gate IS
        PORT (l,m: IN BIT;
              n: OUT BIT);
    END xor_gate;
COMPONENT or_gate IS
        PORT (p,q: IN BIT;
              r: OUT BIT);
    END or_gate;
    SIGNAL M1,M2,M3,M4,M5: BIT;                 -- intermediate signals

BEGIN                                           -- executable part of architecture
L1:xor_gate PORT MAP(a,b,M1);                   -- component instantiation
L2: xor_gate PORT MAP(M1,c,sum);
L3: or_gate PORT MAP(a,b,M2);
L4: or_gate PORT MAP(a,c,M3);
L5: or_gate PORT MAP(b,c,M4);
L6: and_gate PORT MAP(M2,M3,M5);
L7: and_gate PORT MAP(M5,M4,carry);
  END struct;
```

2.6 DELAY MODELS IN VHDL

Delay is the time gap between the giving value at the input and the time at which the change due to input is reflected at the output. By default, the propagation delay of the circuit is present but VHDL gives the user the flexibility to specify the delay to manage the correct updation of values in case of concurrent statements which are all executed in parallel.

There are three types of delay models in VHDL as follows:

- Inertial delay model
- Transport delay model
- Delta delay model

The delay model used affects the simulation results as handling of new events generated in simulation and scheduled events are all influenced by the type of delay model used.

2.6.1 Inertial Delay Model

Inertial delay model is the default delay model in VHDL design. It derives its meaning from the word 'inertia' which suggests that there has to be a minimal pulse length of the input for a circuit to respond. It represents the minimum length of time for which an input must be stable, to cause a change at the output. It takes into account the propagation delay of the circuit and suppresses the input pulses of duration less than that of the specified pulse width. These suppressed input pulses are not propagated to the output, thereby not affecting the output.

SYNTAX:

```
signal <= [ REJECT time_exp1] [INERTIAL] 'val' AFTER time_exp2;
example:
a <= REJECT 4 ns INERTIAL input AFTER 10 ns;          -- 'a' a BIT type signal
a <= input AFTER 10 ns;
b <= inp AFTER 5 ns
c<= 2 AFTER 15 ns, 3 AFTER 25ns,4 AFTER 35 ns;          -- 'c' a INTEGER type signal
```

In the inertial delay model, an optional **REJECT** expression is used to specify the minimum pulse length of the input pulse in time. So, **REJECT** *time_exp1* specifies that the input should be stable for at least the time equal to *time_exp1* for the input to change output if pulse length is less than *time_exp1*, pulse is rejected and no output change happens due to input. Even though the inertial delay model is a default delay model and there is no need to use keyword **INERTIAL** explicitly, still if **REJECT** clause is used then **INERTIAL** keyword has to be used explicitly. If **REJECT** clause is not to be used then **INERTIAL** keyword is implicit as it is default delay model. Inertial delay model uses **AFTER** clause to specify the time delay. If no delay time is specified using **AFTER** clause, for example, simple assignment statement as: a <= input, default delay is considered which is an infinitesimal small delay.

The waveforms for the above statement a <= **REJECT** 4 ns **INERTIAL** input **AFTER** 10 ns; are shown in Figure 2.6.

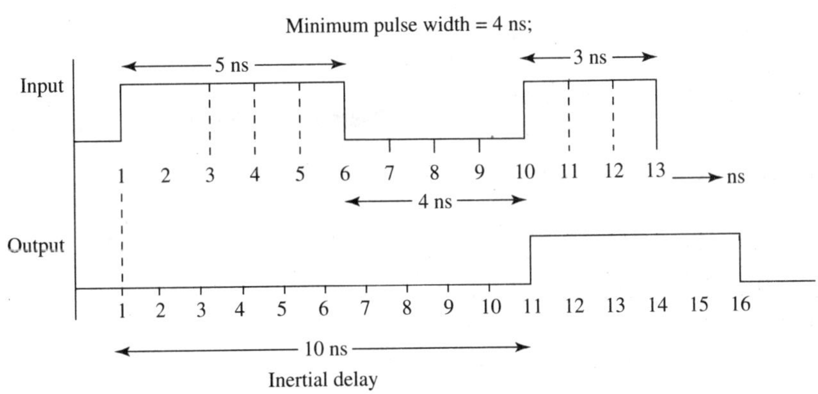

Figure 2.6 Inertial Delay Timing Waveform with Explicit Pulse Rejection Limit

In this statement, signal 'a' will be assigned the value of input, only if the input will be stable for a minimum time of 4 ns which is the pulse rejection limit. This value of pulse rejection is always a time expression and cannot be negative or greater than the inertial value. Next, **AFTER** clause specifies the time for the propagation delay of the input change that is to be reflected at the output. So, here, 'a' will change the value to input after 10 ns from the time instant input changes its value provided the input remains stable at that changed value for minimum 4 ns.

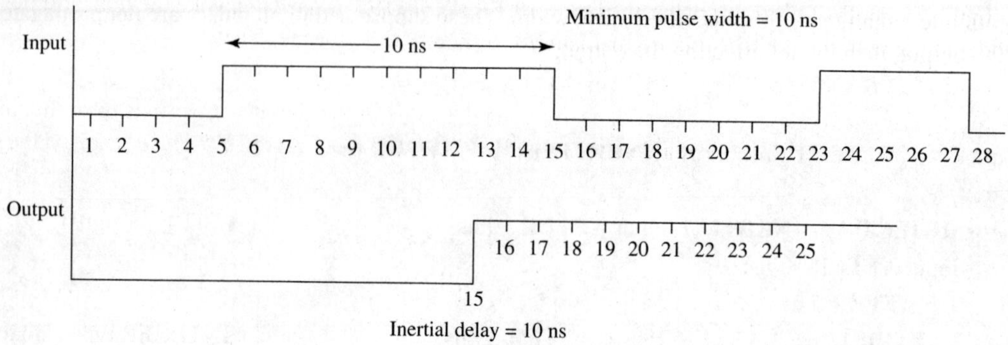

Figure 2.7 Inertial Delay Timing Waveform

For a <= input **AFTER** 10 ns;

In this statement, as **REJECT** clause is not used, there is no need of using **INERTIAL** keyword explicitly. The **AFTER** clause is only used to specify the propagation delay of 10 ns. In the absence of the **REJECT** clause, the propagation delay limit is considered as the pulse rejection limit by default. This means, the input will be propagated after 10 ns to output 'a' from time value at input changes provided the input remains stable at the changed value for a minimum time of 10 ns (pulse rejection) as shown in Figure 2.7.

Timing diagram

c <= 2 **AFTER** 15 ns, 3 **AFTER** 25 ns, 4 **AFTER** 35 ns; – 'c' a INTEGER type signal.

In this statement, signal 'c' of type integer takes value of 2 after 15 ns from the time the value changes to 2, the value changes to 3 after 25 ns from initial instant or after 10 ns from time the signal changes its value to 2, and further after 35 ns from initial timing instant value of c becomes 4. The timing waveform in Figure 2.8 shows the effect of this statement where each delay value is added to initial time t_1.

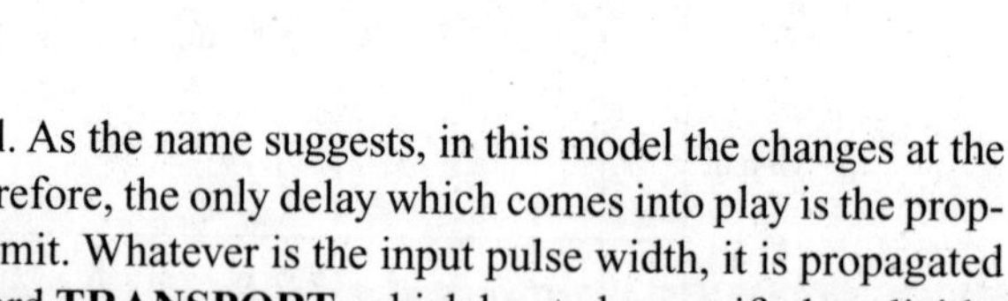

Figure 2.8 Timing Diagram for Waveform

2.6.2 Transport Delay Model

The other delay model is transport delay model. As the name suggests, in this model the changes at the input are simply transported to the output. Therefore, the only delay which comes into play is the propagation delay and there is no pulse rejection limit. Whatever is the input pulse width, it is propagated to the output. This delay model uses the keyword **TRANSPORT**, which has to be specified explicitly. Transport delay is purely the propagation delay based on routing delay of the circuit, irrespective of the pulse width of the input. So, all the spikes and glitches in the input are propagated to the output after the specified propagation delay.

SYNTAX:

```
signal <= TRANSPORT 'val' AFTER time_exp;
example:
a <= TRANSPORT input AFTER 10 ns;                    -- 'a' a BIT type signal
b <= TRANSPORT inp AFTER 5 ns;
```

In this example of transport delay, the keyword **TRANSPORT** is used to invoke the transport delay model along with the **AFTER** clause that is used to specify time for the delay.

For a <= **TRANSPORT** input **AFTER** 10 ns.

In this statement, the output signal 'a' is assigned the value of the 'input', after 10 ns from the time input is active. Similarly, in the statement a <= **TRANSPORT** input **AFTER** 5 ns, 'a' takes value of 'inp' after 5 ns.

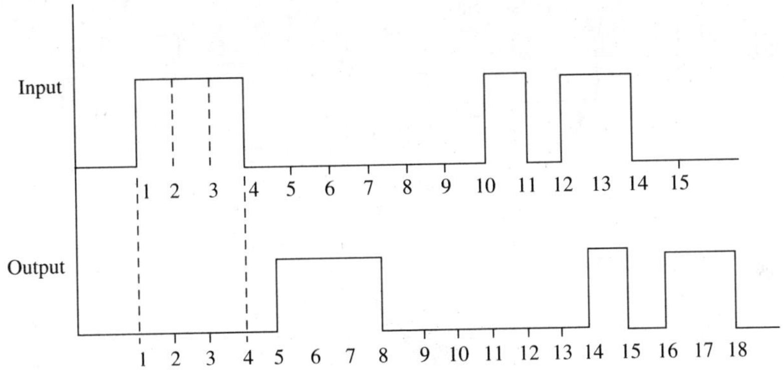

Figure 2.9 Transport Delay

Transport delay model is used to model circuit components with a very small inertia or propagation delays such as wires so that whatever signal is present at input is reflected at the output.

2.6.3 Delta Delay Model

If no delay is specified, then the delta delay model is used by default where infinitesimally small delay is automatically inserted to preserve the correct ordering of events.

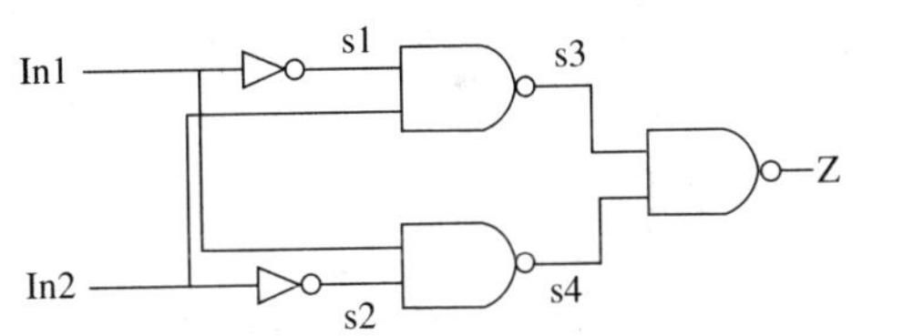

Figure 2.10 Circuit Diagram for a Combinational Circuit with Delta Delay

Example 2.7 Delta delay

```
LIBRARY IEEE;
USE IEEE.std_logic_1164.all;

ENTITY combinational IS
PORT (In1, In2: IN std_logic;
z : OUT std_logic);
END combinational;

ARCHITECTURE behaviour OF combinational
  SIGNAL s1, s2, s3, s4: std_logic:= '0';
  BEGIN
    s1 <= NOT In1;
    s2 <= NOT In2;
```

```
    s3 <= NOT (s1 AND In2);
    s4 <= NOT (s2 AND In1);
    z <= NOT (s3 AND s4);
END behaviour;
```

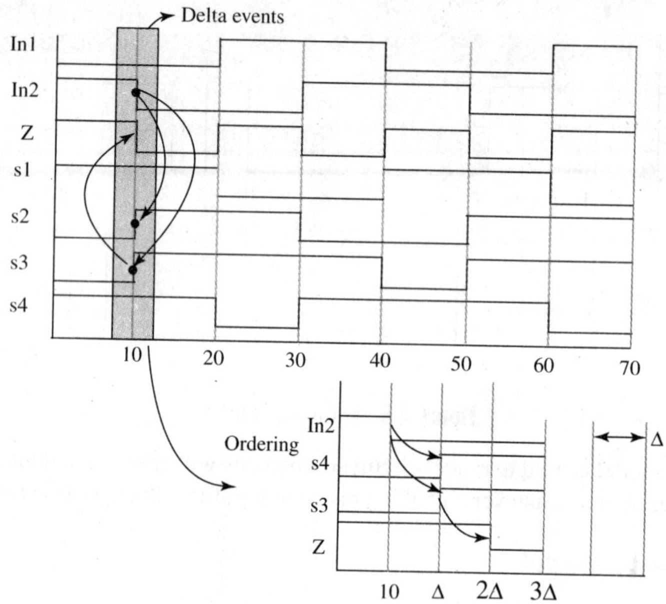

Figure 2.11 Timing Waveforms for Delta Delay Model of Combinational Circuit Shown in Figure 2.10

2.6.4 Differences Between Inertial Delay and Transport Delay

Both inertial delay as well as transport delay are used to model propagation delay in the digital circuits in VHDL, but there are differences in the considerations for propagation delay in the two models. They are as follows:

- The main difference is that the inertial delay models propagation delay due to components as well as pulse rejection width whereas transport delay models propagation delay due to wires or interconnections, that is, it accounts for routing delays.
- With inertial delay, the expression or signal behaves as a kind of filter where any pulse narrower than the specified pulse width are not allowed to reach the output. On the other hand, with transport delay expression or signal, it behaves like a transmission line, allowing every change to reach the output.
- Inertial model is the default model and most commonly used.

SUMMARY

- VHDL gives more flexibility to user when compared to Verilog.
- VHDL is platform-independent language.
- Entity and architecture are essential parts of any design modelled in VHDL.

- Entity is user interface of the design.
- Architecture defines the functionality of the design.
- Architecture can be modelled at three levels of abstraction—behaviour, the highest level; data-flow, the mid level, and structural, the lowest level.
- The levels of abstraction define three styles of modelling—behavioural, dataflow and structural.
- In addition, mixed style of modelling is also used which combines two or more styles in one design.
- Behavioural style just abstracts the functionality of design without going into component details of the design.
- Dataflow style concentrates on the flow of data from input towards output.
- Structural style models the design in terms of components that are comprising the design.
- VHDL is a concurrent as well as a sequential language; behavioural is sequential while dataflow and structural are concurrent styles of modelling.
- There are two delay models in VHDL—inertial and transport.
- Inertial is a default delay model which defines a pulse rejection limit.
- Transport is a simpler delay model which propagates an input to output allowing even glitches to pass.

QUESTIONS

Objective Questions

1. VHDL is a .... language than Verilog.
   a. grander b. bigger c. smaller d. faster
2. VHDL is a case-insensitive language. (True/False)
3. Architecture body describes the .... view of a circuit under design.
   a. external b. internal c. classic d. black box
4. .... style of modelling has the lowest level of abstraction.
   a. Behavioural b. Dataflow c. Structural d. Mixed
5. Lower the abstraction level, ... is the synthesis time.
   a. more b. lesser c. greater d. higher
6. Transport delay is a default delay in VHDL. (True/False)
7. Inertial delay model rejects the glitches and spikes. (True/False)
8. For modelling complex circuits, the level of abstraction is preferred to be ....
   a. higher b. any

Descriptive Questions

1. Write a note on the features of VHDL.
2. What are the basic design units in VHDL?
3. Discuss the different delay models in VHDL.
4. Compare and contrast the three modelling styles in VHDL.
5. What are configurations?
6. VHDL is a concurrent as well as a sequential language. Justify.
7. Discuss the different modes that are allowed for port specification in VHDL.

8. Write the entity specification for half subtractor.
9. Explain the significance of entity declaration and architecture with examples of design of a simple digital circuit. Write the architecture and entity specification for all gates given digital circuit as

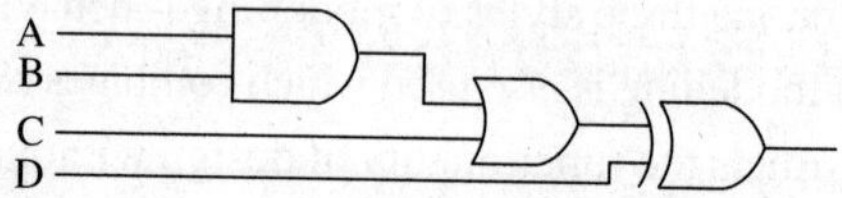

10. What is the difference between 'inout' and 'buffer' mode of ports in VHDL?
11. What is meant by component instantiation?
12. $Y = ABC + \overline{A}B\overline{C} + A\overline{B}\,\overline{C} + A\overline{B}C$

    Design a digital circuit in VHDL.
13. Compare the inertial and transport delay models.
14. What is meant by design unit?
15. What is meant by 'abstraction level'? Explain the different levels of abstraction.
16. For modelling of sequential circuits which delay model is preferred and why?

BASIC LANGUAGE ELEMENTS

3

CHAPTER OBJECTIVES

- To familiarize with the basic elements of VHDL
- Understand the rules governing user-defined names
- Learn data types and their use in VHDL
- Define and use user-defined data types as required
- Understand basic operators in VHDL

3.1 INTRODUCTION

This chapter introduces the basic language elements in VHDL. Before learning the circuit design of the circuits using VHDL, it is important to get familiar with the basic syntax, keywords and the allowed data types of VHDL. First of all, when writing a hardware model in VHDL, it is important to annotate the code with comments. The comment in VHDL is identified by two consecutive dashes (--) followed by a comment text.

For Example:

-- this is a comment

The comment extends from the two dashes to the end of the line and may include any text, as it is not formally a part of the VHDL model. The code of a model can include blank lines and the lines that contain comments alone, starting with two dashes. Long comments can be written on successive lines each starting with two dashes.

For Example:

-- This shows how to write long comments

-- Comment continued in next line

Some words are used by the language itself indicating specific built-in constructs. These words are called keywords or reserved words. Listed below are keywords of VHDL-93.

abs	**access**	**after**	**alias**
all	**and**	**architecture**	**array**
assert	**attribute**	**begin**	**block**
body	**buffer**	**bus**	**case**
component	**configuration**	**constant**	**disconnect**

downto	else	elsif	end
entity	exit	file	for
function	generate	generic	group
guarded	if	impure	in
inertial	inout	is	label
library	linkage	literal	loop
map	mod	nand	new
next	nor	not	null
of	on	open	or
others	out	package	port
postponed	procedure	process	protected
pure	range	record	register
reject	rem	report	return
rol	ror	select	severity
shared	signal	sla	sll
sra	srl	subtype	then
to	transport	type	unaffected
units	until	use	variable
wait	when	while	with
xnor	xor		

3.1.1 Identifiers

Identifiers are user-defined words that are used to name objects in VHDL models, for example, identifiers are used for input and output signals as well as the name of a design entity and architecture body. There are certain rules governing the identifiers in VHDL as follows:

- It should contain only alpha-numeric characters (A–Z, a–z, 0–9) and the underscore (_) character.
- The first character must be a letter and the last one cannot be an underscore.
- An identifier cannot include two consecutive underscores.
- An identifier is case-insensitive (e.g., And2 and AND2 or and 2 refer to the same object).
- An identifier can be of any length.

Examples of valid identifiers are: abc, ab_10, Gate_design.

Some invalid identifiers are: _a10, my_gate@input, gate-input.

The identifiers discussed here are known as basic identifiers. The rules for these basic identifiers are often too restrictive to indicate signals. For example, if one wants to indicate an active low signal such as an active low RESET, one cannot call it /RESET. In order to overcome these limitations, there is a set of extended identifier rules which allow identifiers with any sequence of characters.

- An extended identifier is enclosed by the backslash, "\", character.
- An extended identifier is case-sensitive.
- An extended identifier is different from reserved words (keywords) or any basic identifier (e.g., the identifier\identity\ is allowed)

Inside the two backslashes one can use any character in any order, except that a backslash as a part of an extended identifier must be indicated by an additional backslash only. As an example, to use the identifier BUS:\data, one writes: \BUS:\data\. Extended identifiers are allowed in the VHDL-93 version but not in VHDL-87.

Some examples of legal identifiers are Input, \Input\, and \input#1\.

3.2 DATA OBJECTS

VHDL is an object-based language and anything which is capable of holding some value in VHDL is known as an object. Every object has associated with it a unique name-identifier, a class and a data type. The class of object determines the nature of the data it holds and how the object is internally treated by the compiler. In VHDL, there are three types of data classes as follows:

- Constant
- Variable
- Signal

3.2.1 Constant

Constant is a class that is used to hold objects that have single specific value. As the name suggests the value once assigned remains static and cannot be changed throughout the simulation. It uses the keyword **CONSTANT** in its declaration.

SYNTAX:

```
CONSTANT const_name:TYPE [:= value];
Example:
CONSTANT a: INTEGER:= 2;
CONSTANT rise_time: TIME:= 5ns;
```

Every constant that is to be used has to be declared with a unique identifier, its data type, and its value.

For example, in statement **CONSTANT** a: **INTEGER**:= 2, **CONSTANT** is the keyword, 'a' is the identifier followed by :(colon), data type that is **INTEGER,** and finally its value 2. Naming the constant and specifying its data type is called declaration and assigning it a specific value is called initialization. In this statement both declaration as well as initialization are done in the same statement.

Example 3.1 Use of Constant

```
LIBRARY IEEE;
USE IEEE.
 ENTITY circle_area IS
     PORT (r: IN INTEGER
              area: OUT INTEGER);
 END circle_area;
ARCHITECTURE areaOF circle_area IS
```

```
CONSTANT pi: INTEGER:=3.14;
BEGIN
Area <=pi * r*r;
END simple;
```

It is possible that even if a constant is declared it may be initialized later. Such constants are called **deferred constants** and are initialized in package body.

A constant can be declared inside entity declaration, architecture declaration or inside a process-declarative part.

Inside an entity:

```
ENTITY abc IS
PORT (a : IN BIT);
CONSTANT x: INTEGER:= 2;
END ENTITY;
```

Inside an architecture:

```
ARCHITECTURE abc OF abc IS
CONSTANT x: INTEGER:= 2;
BEGIN
statements;
END abc;
```

Inside a process:

```
ARCHITECTURE abc OF abc IS
   BEGIN
P1:PROCESS (Sensitivity list)
    CONSTANT x: INTEGER:= 2;
     BEGIN
Sequential statements;
     END PROCESS P1;
END abc;
```

Constants help to improve the readability of the program and also make the design flexible by simply changing the values of design parameters as constants in the declaration itself, instead of modifying the actual values at each and every place in the design. Let us say, the rise time of input pulse for a design has to be set at 20 ns and then for same design results are to be simulated with a rise time of 50 ns, the same design can be used by using a constant parameter rise time with value 20 ns for the first case and then simply change its value to 50 ns without the need of changing the rise time value at each and every place in the design.

3.2.2 Variable

Variables are data objects that are used for local storage and can hold a single value. Usage of variables is allowed within sequential constructs, that is, inside a subprogram or inside a process.

This is attributed to the characteristic of the variables that their value is updated immediately without any delay which is important in the sequential programming. Variable declaration makes use of the keyword **VARIABLE** and unlike constants, its value can be changed any number of times.

SYNTAX:

```
VARIABLE var_name:TYPE [:=initial value];
Example:
VARIABLE a: INTEGER:= 2;
VARIABLE inp: BIT_VECTOR(3 DOWNTO 0);
```

Every variable that is to be used has to be declared with a unique identifier, its data type and its initial value (optional). For example, in statement **VARIABLE** a: **INTEGER**:= 2;

VARIABLE is the keyword, 'a' is the identifier followed by :(colon), data type that is **INTEGER** and finally it is initialized to value 2. Naming the variable and specifying its data type is called declaration and assigning it a specific value is called initialization. In the above statement both declaration as well as the initialization are done in the same statement. But it is possible for a variable to have initialization separately, though it will be automatically be initialized to garbage value by a compiler on declaration.

Example 3.2 Use of Variable

```
LIBRARY IEEE;
USE IEEE.STD_LOGIC_1164.all;

ENTITY use_var IS
PORT(a,b: IN INTEGER;
     c: OUT INTEGER);
END use_var;

ARCHITECTURE add OF use_var IS          -- behavioural architecture model of the entity
BEGIN
P1:PROCESS(a,b)
  VARIABLE x,y:INTEGER:=0;              -- declaration of variable
BEGIN
y:= a + b;
x:= 2*y;
c<= x ;
END PROCESS P1;
END add;
```

In Example 3.2, the two variables x and y are used which are assigned a value with an immediate effect.

Note: Different variable assignment operator (:=)

A variable can be declared inside a process (declarative region) or inside a subprogram.

Inside a process

```
ARCHITECTURE abc OF abc IS
  BEGIN
P1:PROCESS(Sensitivity list)
   VARIABLE a: INTEGER:= 2;              -- declaration + initialization
   VARIABLE x: INTEGER;                  -- declaration
  BEGIN
   x:= 4;                                -- variable assignment
   Sequential statements;
  END PROCESS
END abc;
```

Variables are the only data objects which can be updated immediately and have an assignment operator that is different from the other objects. A variable can be assigned a new value any number of times using a variable assignment statement as x:= 4;

3.2.3 Signals

Signals are data objects in VHDL that are used to model interconnections and are capable of holding a list of values. Every signal has associated with it a signal driver which determines the value and timing of changes of the signal, but a signal can have more than one drivers also. Each driver is a queue of events which indicate when and at what value a signal has to be changed. This requires a scheduling of future values that are to be assigned. It makes use of the keyword **SIGNAL.** A signal can be used anywhere inside the entity or architecture. By default all ports are data objects of class signal.

SYNTAX:

```
SIGNAL sig_name:TYPE [:=initial value];
Example:
SIGNAL a: INTEGER:= 2;
SIGNAL inp: BIT_VECTOR(3 DOWNTO 0);
```

Every signal that is to be used has to be declared with a unique identifier, its data type and its initial value (optional). For example, in statement **SIGNAL** a: **INTEGER**:= 2;

SIGNAL is the keyword, 'a' is the identifier followed by :(colon), data type that is **INTEGER,** and finally it is initialized to value 2. Naming the signal and specifying its data type is called declaration and assigning it a specific value is called initialization. In the above statement, both declaration as well as initialization are done in the same statement. Signals may not necessarily be initialized and can be assigned a value in the course of the program execution.

Example 3.3 Use of signal

```
LIBRARY IEEE;
USE IEEE.STD_LOGIC_1164.all;
ENTITY or_gateIS
PORT(a,b: IN BIT;
      c: OUT BIT);
END or_gate;
ARCHITECTURE or_gate OF or_gate IS
SIGNAL clk: BIT;                    -- behavioural architecture model of the entity
BEGIN
P1:PROCESS(clk)
BEGIN
c<= a OR b;
END PROCESS P1;
END or_gate;
```

In Example 3.3 a, b and c are all objects of the type signal. It should be noted that all ports are by default objects of class signal.

A signal can be declared inside an entity declaration, an architecture body (declarative part) or inside a package declaration.

Inside entity declaration

```
ENTITY abc IS
PORT (a : IN BIT);                  -- by default all ports are objects of class signal
SIGNAL x: INTEGER:= 2;
END ENTITY;
```

Inside architecture body

```
ARCHITECTURE abc OF abc IS
SIGNAL x,y: INTEGER;                -- signal declaration
BEGIN
x<= 2;                              -- signal assignment with delta delay
y<= 4 AFTER 10 ns;                  -- signal assignment with a delay of 10 ns
                                      statements;
END abc;
```

Signal values assigned are updated after delta delay or user-defined delay specified using **AFTER** clause.

3.3 DATA TYPES

Each data object has a type associated with it. The type defines the set of values that the object can have and the set of operations that are allowed on it. The concept of data type is very important in

VHDL as it is a strongly typed language that requires each object to be of a certain type. In general, one is not allowed to assign a value of one type to an object of another data type; so, there are no implicit conversions. Every type has associated with it a name and a range of values.

3.3.1 Predefined Types

VHDL has a series of predefined data types, specified in in-built libraries—IEEE 1076 standard and IEEE 1164 standard.

All these predefined types are discussed as follows:

- **Bit**: It is the simplest and most important data type for a digital system based on binary logic values of '1' and '0'. Any data object can be declared of the type bit before being used as:

```
SIGNAL x: BIT;             -- data object 'signal' of the type bit
VARIABLE s: BIT:= '1';     -- data object 'variable' of the type bit
CONSTANT a: BIT:= '1';     -- data object 'constant' of the type bit
```

All arithmetic, logical and relational operators are allowed on bit type.

- **Bit Vector:** It is an extension of two-level bit logic to present an array of values instead of a single value. The range of *n*-bit vector can be specified starting from '0' as index to *n*-1 as **BIT_VECTOR**(0 **TO** 7) or starting from *n*-1 to '0' as

BIT_VECTOR(7 **DOWNTO** 0).

BIT_VECTOR(0 **TO** 7) is of the form "$b_7b_6\ b_5b_4\ b_3b_2\ b_1b_0$"

LSB (at b_7) MSB (at b_0)

BIT_VECTOR(7 **DOWNTO** 0) is as " $b_7b_6\ b_5b_4\ b_3b_2\ b_1b_0$"

MSB (at b_7) LSB (at b_0)

Also note that bit _vector value is always included in double quotes ("").

For Example:

SIGNAL x: **BIT_VECTOR**(3 **DOWNTO** 0);

Note that arithmetic and relational operations are not allowed on bit vectors, for example, "1010" + "1100" is not allowed as it will give compiler error.

- **Std_ulogic:** It is a nine-value logic system defined in the std_logic_1164 package of the library of IEEE. It is an unresolved data type; that means if there are conflicting logical levels at the driver of any signal, the conflict is not automatically resolved. But it is a good alternative to bit data type for checking design errors though care should be taken that output wires are not connected directly. The nine-value system includes forced values, weak values, don't care and unresolved levels.

Table 3.1 Nine Logic Levels of STD_ULOGIC

Logic value	Interpretation
'X'	Forcing unknown
'0'	Forcing low
'1'	Forcing high
'Z'	High impedance (tri-state buffer)
'W'	Weak unknown
'L'	Weak low
'H'	Weak high
'-'	Don't care
'U'	Unresolved

FOR EXAMPLE:

SIGNAL x: **STD_ULOGIC**;

- **Std_logic:** Std_logic is a subtype of std_ulogic. It is an eight-value logic system which does not include 'U'(unresolved) logic level. It is a resolved data type which allows signals driven by more than one source to be used, that is, multiple driver signals.

The resolution of signal logic levels is in accordance with Table 3.2 given as follows:

Table 3.2 Resolution Table for STD_LOGIC

	X	**0**	**1**	**Z**	**W**	**L**	**H**	**–**
X	X	X	X	X	X	X	X	–
0	X	0	X	0	0	0	0	X
1	X	X	1	1	1	1	1	X
Z	X	0	1	Z	W	L	H	X
W	X	0	1	W	W	W	W	X
L	X	0	1	L	W	L	W	X
H	X	0	1	H	W	W	H	X
–	–	X	X	X	X	X	X	X

All arithmetic, logical and relational operations are allowed on std_logic as well as on std_ulogic types.

FOR EXAMPLE:

SIGNAL x: **STD_LOGIC**;

Example 3.4 To show the difference between std_logic and std_ulogic

```
LIBRARY IEEE;
USE IEEE.STD_LOGIC_1164.all;

ENTITY diff IS
PORT(a: IN STD_LOGIC;b:IN STD_ULOGIC; d: OUT STD_ULOGIC
      c: OUT STD_LOGIC);
END diff;
```

```
ARCHITECTURE diff OF diff
BEGIN
a<= '0';
a<= '0' OR '1';
c<=a;
b<= '1';
b<= d;
END diff;
```

The std_logic and std_ulogic types can be used instead of bit type, but vice versa may not always be true. A big advantage of using std_logic instead of bit type is resolution in case of multiple drivers. Apart from this, use of don't-care logic level is of particular importance in practical digital circuits which is not included in binary logic-based bit type. But note that, don't care cannot be used as "_ _ 11_ _" as simulation does not recognize don't-care logic. So instead, to use don't care std_match function has to be used.

Both std_logic as well as std_ulogic types allow vectors besides the single values.

Vectors are used to represent signal buses.

Examples:

SIGNAL x: **STD_ULOGIC_VECTOR(** 3 **DOWNTO** 0);

SIGNAL x: **STD_LOGIC_VECTOR(** 0 **TO** 7);

The rules governing std_logic_vector and std_ulogic_vector are the same as those for bit_vector, with the exception that arithmetic operations are allowed on std_logic_vector provided library package such as IEEE.std_logic_arith.all / IEEE.std_logic_unsigned.all / IEEE.std_logic_signed.all is included in the entity under design.

Note: Though std_logic type can be used instead of bit type, it is always good to use std_logic as it takes care of the unintialized values in the design and also in case of multiple drivers where std_logic alone works. These two types can be interchangeably used only for simple designs where no multiple drivers or high-impedance state exists and initialization to garbage values (in case of bit type) doesn't affect the design.

- **Integer:** This inbuilt data type with a range from −2,147,483,647 to +2,147,483,647 (32-bit integer) exists as part of the package 'standard' of STD library. All arithmetic and relational operators are allowed with integers. It uses the keyword **INTEGER**. There are also two predefined integer subtypes like **natural,** containing integers from 0 to the largest integer(0 to 2, 147,483,647) and **positive**, containing integers from 1 to the largest integer.(1 to 2,147,483,647).

For Example:

SIGNAL x: **INTEGER;**

VARIABLE a: **NATURAL;**

CONSTANT b: **POSITIVE**:= 3;

Note: NATURAL and **POSITIVE** are keywords.

- **Real:** Floating-point types in VHDL are used to represent real numbers with a mantissa part and an exponent part. The predefined floating point type real includes the greatest floating point range allowed by the system. The range of real numbers exists from –1.0e38 to +1.0e38. The keyword used is **REAL**. All arithmetic and relational operators are allowed with real numbers.

For Example:

SIGNAL x: **REAL**;

- **Character:** Single ASCII characters are represented in VHDL predefined-type **CHAR**. It represents single characters included in single quotes ('') or string of characters included in double quotes ("").

  For Example:

  SIGNAL x: **CHAR:= 'a'**;

- **Time:** Time is a predefined physical type. The range is implementation defined and is at least equal to the range of integers defined. The time-data type is used to represent simulation times and delays. A time value in VHDL can be written as a numeric literal followed by a time unit as 32 s, 5 ns, and so on. The type time includes both positive as well as negative values. VHDL also has a redefined subtype of time and delay_length, which only includes non-negative values. Many of the arithmetic operators can be applied to time values, but with some restrictions. The addition, subtraction, identity and negation operators can be applied to yield results of type time. A time value can be multiplied or divided by an integer or real value to yield a time value, and the two time values can be divided to yield an integer.

For Example:

SIGNAL clk: **TIME:=** 20ns;

All units of time ranging from smallest fs (femto second) to largest unit hr (hour) are predefined.

Units:

Fs;

ps = 1000fs;

ns = 1000ps;

µs = 1000ns;

ms = 1000µs;

sec – 1000ms;

min = 60sec;

hr = 60min;

Table 3.3 Summary of Predefined Data Types

Type	Range of values
Bit	'0', '1'
Bit_vector	an array with each element of type bit
Boolean	FALSE, TRUE
Character	any legal ASCII character, characters must be placed between single quotes

(Continued)

Type	Range of values
File_open_kind*	read_mode, write_mode, append_mode
File_open_status*	open_ok, status_error, name_error, mode_error
Integer	range is implementation dependent but includes at least $-(2^{31}-1)$ to +(231 – 1)
Natural	integer starting with 0 up to the maximum specified in the implementation
Positive	integer starting from 1 up to the maximum specified in the implementation
Real	Floating-point number in the range of –1.0 × 1038 to +1.0 × 1038 (implementation dependent)
Severity_level	note, warning, error, failure
String	array of which each element is of the type character
Time	an integer number of which the range is implementation defined; units can be expressed in sec, ms, us, ns, ps, fs, min and hr.
Std_logic	A resolved data type, 8-value system
Std_ulogic	Unresolved 9-valued logic system

Example 3.5 Use of predefined data types

```
LIBRARY IEEE;
USE IEEE.STD_LOGIC_1164.all;

ENTITY use_type IS
PORT(a,b: IN STD_LOGIC;
        C1: OUT REAL;
        C2:OUT INTEGER);
END use_type;
```

```
ARCHITECTURE add OF use_var IS            -- behavioural architecture model of the entity
BEGIN
P1:PROCESS(a,b)
SIGNAL b1:BIT;
VARIABLE x,y:INTEGER:=0;
VARIABLE r1:REAL:=0;

BEGIN
IF b1=='1' THEN
    y:= a + b;
    x:= 2*y;
    c2= x ;
ELSE
```

```
y:=a + b;
r1:= y/2;
c1= r1;
END IF;

END PROCESS P1;
END add;
```

Note: As VHDL is a strongly typed language no implicit data type conversions are allowed.

User-defined types: Apart from the predefined data types defined in the standard libraries of VHDL, the user has the flexibility to define a range of data types according to circuit requirements. Any new type can be introduced by using the type declaration, which names the type and specifies its value range. The syntax is

TYPE *identifier* **IS** type_definition;

and any data object of the defined type can be created as:

SIGNAL x: *identifie*r;

It should be taken care that before using the data object of a particular user-defined type, the type is defined in the same design unit or inside a package which is included inside the design unit using that type.

User-defined types are classified as follows:

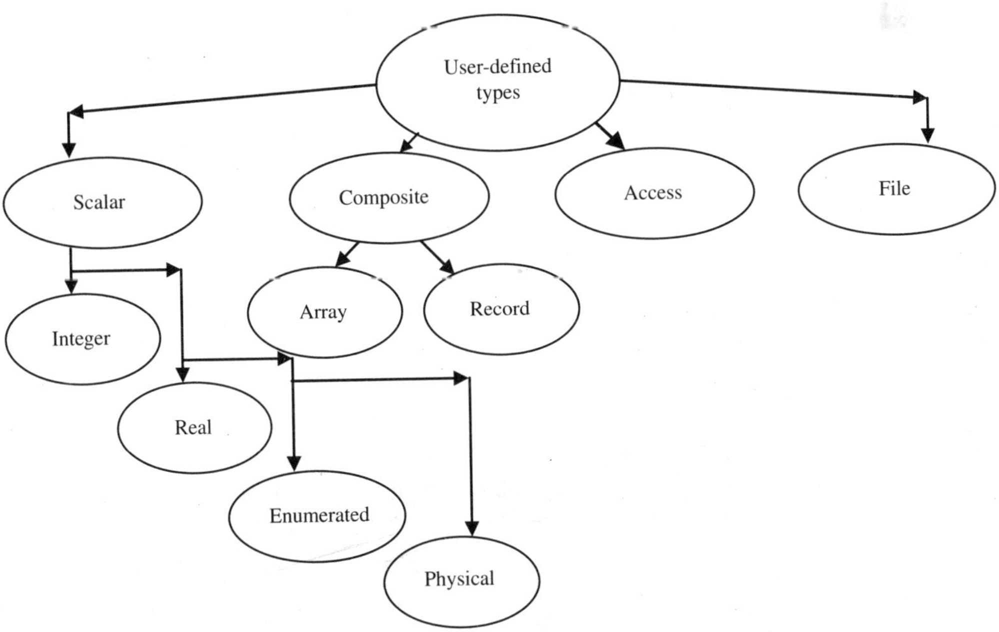

Figure 3.1 Classification of VHDL Data Types

Also for each user-defined type created, a subtype can also be created. A subtype is a subset of a defined type.

For Example:

TYPE my_type **IS RANGE** 31 **DOWNTO** 0;

SUBTYPE data_word **IS** my_type **RANGE** 7 **DOWNTO** 0;

3.3.2 Scalar Type

Scalar data types can hold only one value at a time. Scalar type is further classified into four types, namely,

- Integer
- Real
- Enumerated
- Physical

Integer Type: Though Integer is a predefined type ranging from -2^{31} to $+2^{31}$, VHDL gives the user the flexibility to create its own integer type over a required range of values. This allows the user to have integer values over a small range as needed.

For Example:

TYPE my_int_type **IS RANGE** –32 **TO** 32;

In this example, a type identified by my_int_type is created which includes integers from –32 to 32. Once created/declared it is just treated as a predefined range and any data object of the type can be created as

SIGNAL a: my_int_type;

Integer subtypes can also be created from the integer types, which is nothing but type with constraints.

SUBTYPE small_int **IS** my_int_type **RANGE** from 0 to 32;

This creates a subtype small_int of the type my_int_type which includes only positive values of the type my_int_type.

Real Type: User defined is similar in concept as integers; it is created as a subset of predefined real type to limit the range.

For Example:

TYPE real_range **IS RANGE** –5.1 **TO** 10.3;

SUBTYPE sub_real **IS** real_range **RANGE** 0.0 to +1.8;

Real numbers can also be expressed in exponential notation as

Mantissa E *exponential_part* example:31.3 E -2.

Further, any numeric value can be represented in different base systems, that is, VHDL allows a number to be represented in different base systems as hexadecimal number, binary number, decimal number, octal number, and so on.

A number can be represented as

Base#number#

Or

Base#number#**E** exponent.

FOR EXAMPLE:

2#10010# represents number 10010 in base 2 system or binary value 10010.

16#F# represents number F as a hexadecimal number.

16#A# E 1 represents A * $(16)^1$

Enumerated Type: Enumerated data type represents a type which can take value from a defined list only. These values may be integers, identifiers or single character literals. Internally, all are treated as integers.

DECLARATION SYNTAX:

```
TYPE name IS (list);

Example:
TYPE color IS (Red, blue, green);
TYPE binary IS (0,1);
```

Example 3.6 Use of enumerated data types

```
LIBRARY IEEE;
USE IEEE.STD_LOGIC_1164.all;

ENTITY enum_type IS
PORT(a,b: IN STD_LOGIC;
      c: OUT STD_LOGIC);
END enum_type;

ARCHITECTURE enum_type OF enum_type IS
TYPE opcode IS (and-op,or-op,not-op);
SIGNAL x: opcode;
BEGIN
PROCESS(a,b)
BEGIN
  IF x== and-op THEN
    c<= a and b;
  ELSIF x== or-op THEN
      c<= a or b;
  ELSE
      c<= not(a);
  END IF;
END PROCESS;
END enum_type;
```

In Example 3.6, depending on the opcode, the operation is selected and required output is given on output port c.

The use of enumerated data type in VHDL is of particular importance in state machine design as new enumerated type consisting of allowed states can be created as follows:

TYPE state **IS** (A,B,C,D);
VARIABLE x: state;

Now variable 'x' can take any value from the list of values defined for the type 'state', that is, (A,B,C,D). Enumerated types have to be defined in the architecture body or inside a package.

Physical Type: Physical types are used to represent physical quantities like current, voltage, and so on, and defining the physical quantity provides the base unit for expressing the quantity. Successive higher units may be defined depending on the range specified.

For Example:

```
TYPE voltage IS RANGE 0 TO 440
Units
      V;
      mV= (1/1000)V;
      μV= (1/1000)mV;
END units;
```

Here, voltage is physical type which can represent values in range 0 to 440V and recognizes three units of type voltage: Volt, Millivolt(mV) and Microvolt(μV).

Example 3.7 Using physical types

```
LIBRARY IEEE;
USE IEEE.STD_LOGIC_1164.all;

ENTITY phy_type IS
PORT( c: OUT STD_LOGIC);
END phy_type;

ARCHITECTURE phy_type OF phy_type IS
TYPE current IS RANGE 0 to 10
Units
    A;
    mA= (1/1000)A;
    μA= (1/1000)mA;
END Units;

SIGNAL x: current;
SIGNAL trig:STD_LOGIC;
BEGIN
PROCESS(trig)
BEGIN
  IF x = =5mA THEN
    c<= '1';
```

```
ELSE
  c<= '0';
 END IF;
END PROCESS;
END phy_type;
```

3.3.3 Composite Type

Composite type is used for abstraction to represent group of values of similar or different type. Composite type in VHDL is composed of two types: Array and Record. Array represents a collection of similar type of values whereas a record is used to represent a group of different type of values. Composite type finds its significance in object-based languages.

Array: Array is used to represent an object containing more than the values of similar type. For example, a storage register of eight bits can be represented as an object of the type array consisting of eight-bit values. So, eight values of type bit are grouped into a single object of the type array.

TYPE reg **IS ARRAY** (0 **TO** 7) **OF BIT**;

A subtype of existing predefined types can also be created as

SUBTYPE positive **IS INTEGER RANGE** 1 **TO INTEGER'HIGH;**

In Example 3.7, a subtype named positive is created as an array of integers with positive value ranging from 1 to highest positive value available in the system.

Note the use of High which gives highest value of INTEGER supported by the system. So array is nothing but a collection of similar type of objects. An array can be single- or multidimensional. Both the above subtype declarations are one-dimensional arrays. Multidimensional arrays can also be created in VHDL.

Example 3.8 To design a 32-bit memory with word size 8 and show, read and write operations

```
LIBRARY IEEE;
USE IEEE.STD_LOGIC_1164.all;

ENTITY arr_type IS
PORT( rd,wr: IN STD_LOGIC;
         Outp: OUT STD_LOGIC_VECTOR (0 TO 7));
END arr_type;

ARCHITECTURE mem OF arr_type IS
TYPE reg_width IS ARRAY (0 TO 7) OF STD_LOGIC;
TYPE mem IS ARRAY (0 TO 31) OF reg_width;

PROCESS (rd,wr)
BEGIN
  IF rd = = '1' THEN
   out <= mem(28);
```

```
ELSE IF wr == '1' THEN
    mem(4) <= "11100110";
    out <= mem(4)
  END IF;
END PROCESS;
END mem;
```

Arrays can be constrained or unconstrained. The 2-d array, mem in this example is a constrained array because its range is specified from 0 to 31 at the time of its declaration. In some cases, the array range is not completely specified in their declaration. Such arrays are called unconstrained arrays. Unconstrained arrays are useful in situations where a common base type has to be provided for different subtypes with different ranges.

Example: **TYPE** pos_val **IS ARRAY(INTEGER RANGE<>) OF INTEGER**
SUBTYPE ten_val **IS** pos_val(1 **TO** 10);

Record Type: Record type is very similar to structures in C. It is used to group objects of different type together which can be referenced as a single type.

Syntax:

```
TYPE rec_name IS RECORD
    N1: TYPE;
    N2 : TYPE;
..
..
N5: TYPE;
END RECORD;
```

Here, rec_name is a user-defined name of the record type to be created, N1,N2……..N5 are the objects of different types included in the record rec_name. Once the record type is created any object of the type can be created as

```
VARIABLE r1: rec_name ;
To access the members of record r1, '.' Operator is used as
r1.N1, r1.N2 and so on.
```

3.3.4 Access Type

The concept of access type is similar to that of pointers in the C language. Access type points to address a particular type of object and is useful for accessing dynamically allocated objects. It is used to facilitate large memory designs. Only variables can be of access type, and they must point to a value allocated dynamically using **new** (not to another variable).

An access variable is initialised to the value **null**. Just like pointers in C, a deallocation of allocated memory is done using DEALLOCATE(Ptr), which is implicitly defined and can be called to release the storage allocated using **new**.

Use of only variable type of objects limits the use of access types to sequential modelling alone.

SYNTAX:

```
TYPE ptr1 IS ACCESS : INTEGER;
VARIABLE r1: ptr1 ;
r1:= NEW INTEGER(3)
or
-- dynamic allocation
```

An object of type access can refer to three types of values such as:

- Scalar object

  For a scalar object of access type ,

  ```
  TYPE ptr1 IS ACCESS : INTEGER;
  VARIABLE r1: ptr1 ;
  r1:= NEW INTEGER(3)
  ```

- Array object

  ```
  TYPE arr_ex IS ARRAY(0 TO 9) OF INTEGER;
  TYPE ptr1 IS ACCESS: arr_ex;

  VARIABLE r1: ptr1;
  r1.all := NEW arr_ex( 0,1,2,3,4,5,6,7,8);
  ```

  Or

  To access individual array members

  ```
  r1.all(3) := NEW arr_ex(2);
  ```

- Record object

  ```
  TYPE rec1 IS RECORD
     ROLL_No: INTEGER;
     MARKS: REAL;..
     GRADE: CHAR;
  END RECORD;

  TYPE ptr1 IS ACCESS rec;
  VARIABLE r1 :ptr1;
  r1:= NEW rec1(2260 ,67.5, 'C');
  ```

  for accessing individual members

  ```
  r1.ROLL_No:= 2213;
  ```

By default the access type variable is initialized to NULL.

Note: Access types are not synthesizable.

Example 3.9 Use of access and record types

```
LIBRARY IEEE;
USE IEEE.STD_LOGIC_1164.all;

ENTITY rec_access IS
PORT( den: IN STD_LOGIC;
        Outm: OUT REAL);
END rec_access;

ARCHITECTURE rec_access OF rec_access IS
TYPE stu_rec IS RECORD
     ROLL_No: INTEGER;
     MARKS : REAL;..
     GRADE: CHAR;
END RECORD;
TYPE ptr1 IS ACCESS stu_rec;

BEGIN
PROCESS (den)
VARIABLE r1 :ptr1;
BEGIN
  IF den = = '1' THEN
    r1.all:= NEW stu_rec(1,71.3, 'A');
  ELSE
    Outm<= r1.MARKS;
  END IF;
END PROCESS;
END rec_access;
```

Incomplete Types: Incomplete types are the types which are not specified completely at the time of declaration. Incomplete types are needed when recursive reference or self-reference to a type is required.

FOR EXAMPLE:

```
TYPE arr ;
TYPE rec1 ;
```

These are generally needed to model memory types.

Using incomplete types

```
TYPE stac;                          -- incomplete type
TYPE stack_ptr IS ACCESS stac;      -- using incomplete type
```

```
TYPE stac IS RECORD
    input: Integer;
    ptr: stack_ptr;
END RECORD;
```

3.3.5 File Type

A file is a stream of values of a specified type which can be read or written during simulation. Files belong to a special kind of data type called a file type.

SYNTAX:

TYPE *identifier* **IS FILE OF** *data_type.,*

Here, identifier is a user-defined name of the file and data-type is the type of data contained in the file which can be of any predefined type or user-defined composite type, such as an array or record.

Using file type object of type file can be declared which can be read from, written or be modified. To declare an object of type file the following syntax is used:

FILE file_name: file-type **IS** MODE "physical path";

For file type object declaration three things are required to be specified:

- User-defined name of the file along with its type.
- MODE of file: Two modes **IN** and **OUT** are allowed with file objects. While with 'IN' mode file can be read from with 'OUT' mode file can be written to.
- Physical path of the file on the disk, where it is saved.

On the file type objects three operations are allowed to be performed with the help of inbuilt procedures:

- READ: allows dat to be read from a file. It is a two-argument procedure
  READ(file_name,data).
- WRITE: allows to write data to a file. It is a two-argument procedure
  WRITE(file_name,data).
- ENDFILE: returns true if end of the file is reached. It is single argument procedure ENDFILE(file_name).

Note: File type is not synthesizable and generally used in testbenches for simulation purposes. Package TEXTIO is used for file types.

Example 3.10 Use of file type

```
LIBRARY IEEE;
USE IEEE.STD_LOGIC_1164.all;

ENTITY file_type IS
PORT( den: IN STD_LOGIC);
END file_type;
```

```
ARCHITECTURE ex OF file_type IS

BEGIN
PROCESS (den)
TYPE roll_no_file IS FILE OF INTEGER;
VARIABLE i : INTEGER:=0;
TYPE data IS ARRAY(0 TO 49) OF INTEGER;
VARIABLE roll_no: data;
FILE r_list : roll_no_file IS OUT "c:\ roll_list.dat";
BEGIN
  IF den = = '1' THEN
   FOR i=0 TO 49 LOOP
      Write(r_list, roll_no(i));
   END LOOP;
  END IF;
END PROCESS;
END ex;
```

VHDL does not support implicit data conversions which makes it a strongly typed language. To convert from one data type to another explicit conversions are required. Some of the basic conversions between built-in types are defined in 1164 package while conversions between user-defined types or composite types are needed to be defined by users in the form of functions if required.

Table 3.4 Predefined Conversion Functions in std_logic_1164 Package

Conversion	Function
std_ulogic to bit	to_bit(*expression*)
std_logic_vector to bit_vector	to_bitvector(*expression*)
std_ulogic_vector to bit_vector	to_bitvector(*expression*)
bit to std_ulogic	to_stduLogic(*expression*)
bit_vector to std_logic_vector	To_StdLogicVector(*expression*)
bit_vector to std_ulogic_vector	To_StdUlogicVector(*expression*)
std_ulogic to std_logic_vector	To_StdLogicVector(*expression*)
std_logic to std_ulogic_vector	To_StdUlogicVector(*expression*)

The IEEE std_logic_unsigned and the IEEE std_logic_arith packages allow additional conversions such as from an integer to std_logic_vector and vice versa.

The syntax of a type conversion is as follows:

type_name (*expression*);

In order for the conversion to be legal, the *expression* must return a type that can be converted into the type *type_name*. Following are the conditions that must be fulfilled for the conversion to be possible:

- Type conversions between integer types or between similar array types are possible.
- Conversion between array types is possible if they have the same length and if they have identical element types or convertible element types.
- Enumerated types cannot be converted.

3.4 OPERATORS

An operator is a logical or mathematical function which takes one or two values and produces a single result. Operators are built into the syntax of the language and cannot be user-defined. They can only be overloaded.

Operators in VHDL are of five types as follows:

- Arithmetic operators
- Logical operators
- Relational operators
- Shift-and-rotate operators
- Miscellaneous operators

3.4.1 Arithmetic Operators

Arithmetic operators support basic operations of addition and multiplication.

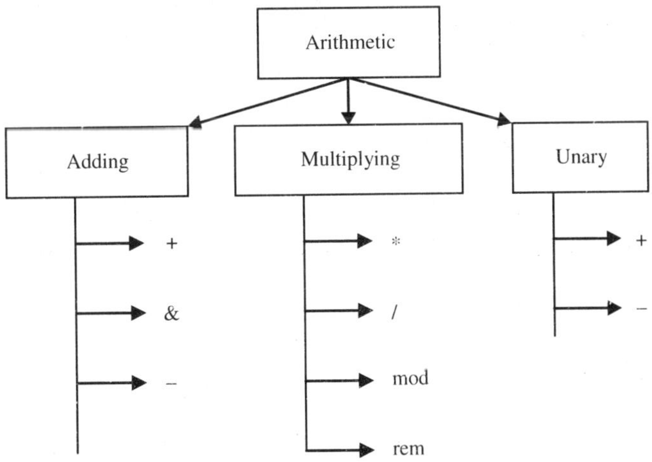

SYNTAX:

```
op<= first_operand OPERATOR Second operator;        -- binary operators
op<= OPERATOR value;                                 -- unary operator
```

All adding operators are two-operand operators.

There are three addition operators:

- Addition(+) and Subtraction(−): It is basic arithmetic addition. It works on integer type and floating-point types. Both operands are of same type. Arithmetic addition is defined for std_logic type in std_logic_arith_unsigned. Else, for other types concept of operator overloading is used.

For Example:

c<= a + b, where c, a and b are integer/real values

c<= 8–5

- Concatenation (&): Concatenation operator is used to join together two or more 1-d arrays or single elements.

  "0101" & '1'

  "ABC" & "DEF"

  'A' & 'B'

Multiplying operators work on numeric types.

- Multiplication (*): Multiplication is a binary operator which requires both operands to be of same type—either integers or real type with exception of multiplication of physical type with a scalar value with a result of type physical.

  For Example:

  c<= a *b;

  til<= 10 * 1 ns;

- Division (/): Division is a binary operator which requires both operands to be of the same type—either integers or real type with an exception of the division of physical type with a scalar value with result of type physical or division of physical type with another physical type resulting in a scalar value.

  c<= a/b;

  res<= 10ns/1ns;

- Remainder (rem): Remainder is a binary operator which returns the remainder of the division of two integer values. It is defined as

  A **REM** B = A – (A/B)*B (A/B in an integer)

  The result is the same as operand type but with the sign of first operator.

  For Example:

  17 **REM** 4 =17 – (17/4) *4 = 17 – 4*4 = 1

  –17 **REM** 4 = –17 – (–17/4) *4 = –17 + 4*4 = –1.

- Modulus (mod): Modulus operator is also a binary operator which works on integer values. It is defined as

  A **MOD** B = A – B * N (in which N is an integer). N is the smallest value for which the sign of the result is that of the second operand, that is, 'B'.

  For Example:

  17 **MOD** 4 =17 – (17/4) * 1= 17 – 4*1=13

  –17 **MOD** 4 = –17 – (–17/4) *5= – 17 + 4*5= 3

  17 **MOD** – 4 =17 – (17/–4) *–5= 17 – 4*5= –3

Unary Operators: Unary operators '+' and '–' are sign operators where + indicates positive value and – indicates a negative value.

3.4.2 Logical Operators

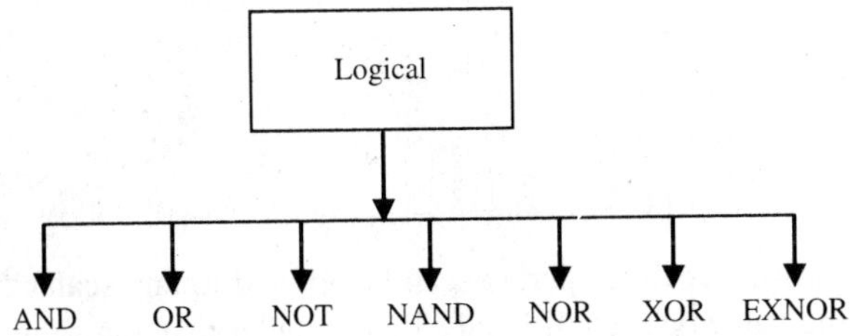

Logical operators are defined for the "bit", "boolean", "std_logic" and "std_ulogic" types and their 1-d vectors. They are used to define Boolean logic expression or to perform bit-per-bit operations on arrays of bits. They give a result of the same type as the operand (Bit or Boolean). These operators can be applied to signals, variables and constants.

Note: NAND and NOR operators are not associative. Parentheses should be used in a sequence of nand or nor operators to prevent a syntax error.

For Example:

X **NAND** Y **NAND** Z will give a syntax error and should be written as (X **NAND** Y) **NAND** Z.

3.4.3 Relational Operators

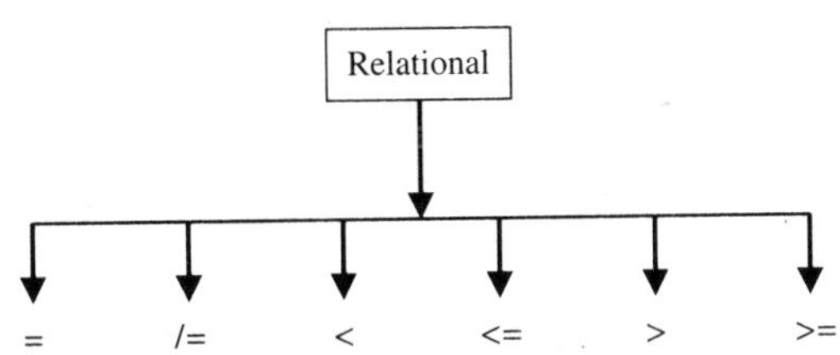

All relational operators return Boolean true or false after assessing the relative relation between the operands.

- Equality (=): Equality operator can be applied to any type. Returns true if two values on the left- and right-hand side of equality operator are equal, else false.

```
IF a=b THEN
  c<='0';
ELSE
  c<= '1';
END IF;
```

- Inequality (/=): Inequality operator can be applied to any type. Returns true if two values on the left- and right-hand side of inequality operator are inequal, else false.

```
IF a /=b THEN
  c<= '1';
ELSE
  c<= '0';
END IF;
```

- Smaller than (<): Smaller than operator can be applied to any scalar type or discrete arrays. Returns true if the value on the left-hand side is smaller than the value on the right hand side of the operator, else false.

```
IF a < b THEN
  c<= '1';
END IF;
```

- Smaller than or equal to (<=): Smaller than or equal to operator can be applied to any scalar type or discrete arrays. Returns true if the value on the left-hand side is smaller or equal to the value on the right-hand side of the operator, else false.

```
IF a <= b THEN
  c<= '1';
END IF;
```

- Greater than (>): Greater than operator can be applied to any scalar type or discrete arrays. Returns true if the value on the left-hand side is greater than the value on the right-hand side of the operator, else false.

```
IF a > b THEN
  c<= '1';
END IF;
```

- Greater than or equal to (>=): Greater than or equal to operator can be applied to any scalar type or discrete arrays. Returns true if the value on the left-hand side is greater or equal to the value on the right-hand side of the operator, else false.

```
IF a >= b THEN
  c<= '1';
END IF;
```

Note: For discrete array types, the comparison is done on an element-per-element basis, starting from the left towards the right.

Example 3.11 To show the use of relational operators: Find the largest of the three integers

```
LIBRARY IEEE;
USE IEEE.STD_LOGIC_1164.all;
ENTITY rel_op IS
PORT(en: IN STD_LOGIC; a,b,c: IN INTEGER;
        d: OUT INTEGER);
END rel_op;
ARCHITECTURE ex OF rel_op IS
BEGIN
PROCESS (a,b,c)
BEGIN
 IF en= = '1' THEN
  IF a >b THEN
   IF a>c THEN
      d<= a;
   ELSE
      d<= c;
    END IF;
   ELSE
    IF b>c THEN
      d<= b;
   ELSE
      d<= c;
   END IF;
END IF
END IF;
END PROCESS;
END ex;
```

3.4.4 Shift-and-Rotate Operators

Shift-and-rotate operators perform a bit-wise shift or rotate operation on a 1-d array of bit , Boolean or std_logic.

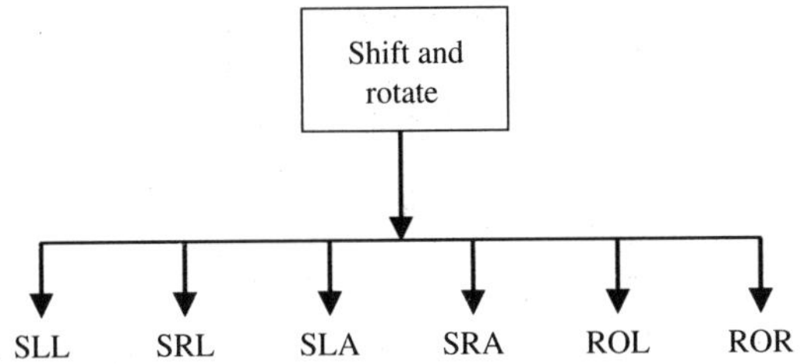

SYNTAX:

Op <= data **OPERATOR** integer;

- Shift left logical (SLL): It is a logical shift operation which shifts the bits to left and thus, the vacated bits on the right are filled with zeros.

FOR EXAMPLE:

a<= 101001

b<= a **SLL** 2 = 100100.

- Shift right logical (SRL): It is a logical shift operation which shifts the bits to right and thus, the vacated bits on the left are filled with zeros.

FOR EXAMPLE:

a<= 101001

b<= a **SRL** 2 = 001010.

- Shift left arithmetic (SLA): It is an arithmetic shift operation which shifts the bits to left and thus, the vacated bits on the right are replicated with the rightmost bit.

FOR EXAMPLE:

a<= 101001

b<= a **SLA** 2 = 100111.

- Shift right arithmetic (SRA): It is an arithmetic shift operation which shifts the bits to right and thus, the vacated bits on the left are replicated with the leftmost bit.

FOR EXAMPLE:

a<= 101001

b<= a **SLL** 2 =111010.

- Rotate Left (ROL): It is a circular rotation operation which rotates the bits to left.

FOR EXAMPLE:

a<= 101001

b<= a **ROL** 2 = 100110.

- Rotate Right (ROR): It is a circular rotation operation which rotates the bits to right.

FOR EXAMPLE:

a<= 101001

b<= a **ROR** 2 = 011010.

3.4.5 Miscellaneous Operators

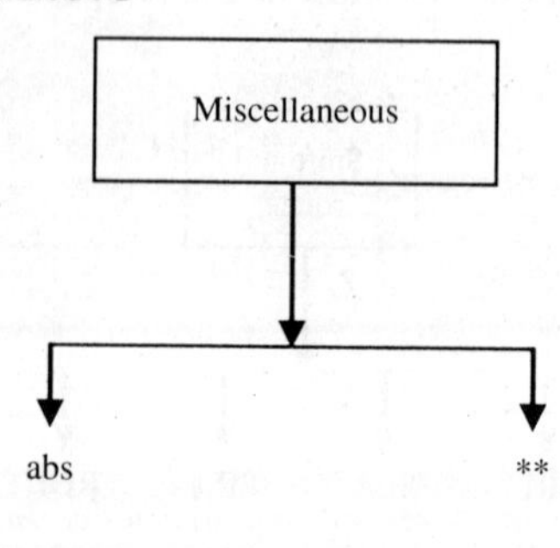

SYNTAX:

```
Op <= ABS (numeric_data)          -- Syntax for operator 'abs'
Op <= operand ** integer          -- Syntax for exponentiation '**'
```

- Absolute (abs): It works with any numeric data type and returns the absolute value of the input numeric value.

FOR EXAMPLE:

abs(−2) = 2

- Exponentiation (**): It is a binary operator, where the left operand is a numeric value, either an integer or a floating type; and the right operand is an integer, which indicates the exponent value.

FOR EXAMPLE:

b<= 3 **2 = 9.

> **Note:** The logical negation operator '**NOT**' is considered as a miscellaneous operator as it shares the same precedence as the miscellaneous operators 'abs' and '**'.

Operator Precedence

Table 3.5 Precedence Order for Operators in VHDL

Class	Operators						
Logical operators	AND	OR	NAND	NOR	XOR	XNOR	Lowest precedence
Relational operators	=	/=	<	<=	>	>=	↓
Shift operators	sll	srl	sla	sra	rol	ror	↓
Addition operators	+	=	&				↓
Unary operators	+	−					↓
Multiplying operators	*	/	mod	rem			↓
Miscellaneous operators	**	abs	not				Highest precedence

Unless the parentheses are used the operator precedence is the highest for miscellaneous operators and the lowest for logical operators. Within a class, all operators have the same precedence.

Summarizing Operators

Table 3.6 Types of Operands Compatible with Operators

		Logical	
Operator	**Description**	**Operand types**	**Result type**
And	Logical and	Bit, Std_logic, Boolean	Same as operand type
Or	Logical or	Bit, Std_logic, Boolean	Same as operand type
not	Logical negation	Bit, Std_logic, Boolean	Same as operand type
Nand	Logical negation of and	Bit, Std_logic, Boolean	Same as operand type
Nor	Logical negation of or	Bit, Std_logic, Boolean	Same as operand type
Exor	Exclusive OR	Bit, Std_logic, Boolean	Same as operand type

(Continued)

Table 3.6 (Continued)

		Relational		
=	Equality		any type	Boolean
/=	Inequality		any type	Boolean
<	Smaller than		scalar or discrete array types	Boolean
<=	Smaller than or equal		scalar or discrete array types	Boolean
>	Greater than		scalar or discrete array types	Boolean
>=	Greater than or equal		scalar or discrete array types	Boolean
		Shift and Rotate		
sll	Shift left logical (fill right vacated bits with the 0)		Left: Any one-dimensional array type with elements of type bit or Boolean; Right: integer	Same as left type
srl	Shift right logical (fill left vacated bits with 0)		same as above	Same as left type
sla	Shift left arithmetic (fill right vacated bits with rightmost bit)		same as above	Same as left type
sra	Shift right arithmetic (fill left vacated bits with leftmost bit)		same as above	Same as left type
rol	Rotate left (circular)		same as above	Same as left type
ror	Rotate right (circular)		same as above	Same as left type
		Adding		
+	Addition	Numeric type	Same as left operand	Same type
–	Subtraction	Numeric type	Same as left operand	Same type
&	Concatenation	Array or element type	Same as left operand	Same array type
		Unary		
+	Identity		Any numeric type	Same type
–	Negation		Any numeric type	Same type
		Multiplying		
*	Multiplication	Any integer or floating point	Same type	Same type
		Any physical type	Integer or real type	Same as left
		Any integer or real type	Any physical type	Same as right
/	Division	Any integer or floating point	Any integer or floating point	Same type
		Any physical type	Any integer or real type	Same as left
		Any physical type	Same type	Integer
Mod	Modulus	Any integer type		Same type
Rem	Remainder	Any integer type		Same type

		Miscellaneous		
**	Exponentiation	Integer type	Integer type	Same as left
		Floating point	Integer type	Same as left
Abs	Absolute value	Any numeric type		Same type

SUMMARY

- Identifiers are user-defined names, which are different from keywords.
- Identifiers can contain alphabets, numbers as well as special characters but should always start with an alphabet.
- Extended identifiers are case-sensitive.
- Objects in VHDL can be: SIGNAL, VARIABLE or CONSTANT.
- Signals are the main mode of communication and interconnection.
- Variables are allowed only with sequential modelling and are immediately updated without any delay.
- Data types in VHDL can be predefined or user-defined.
- From predefined types std_logic is the only resolved type and can be used in case of multiple drivers for a signal.
- Std_ulogic is the unresolved form of std_logic.
- User-defined type can be scalar, composite, access or file type.
- Incomplete types are allowed in VHDL to facilitate self-reference or recursive reference to a type.
- Operators in VHDL are broadly of five types.

QUESTIONS

Objective Questions

1. …………….. in VHDL are case-sensitive.

   a. Identifiers b. Reserved words c. Extended identifiers d. Objects
2. 4t1mux is legal identifier. (True/False)
3. Which operator has the highest precedence?

   a. NAND b. '+' c. '**' d. rem
4. If a = 1110110, then a**SLL**3 is ……………..
5. Which logical operator does not obey associativity rule?

   a. NOT b. AND c. NOR d. EXOR
6. By default, the ports of VHDL entity belong to which class?

   a. signal b. constant c. variable d. none of these
7. Initialization of objects is allowed along with declaration for all VHDL objects. (True/False)

8. Modulus and remainder operators give the same value, if for same values of operands, if ....
   a. both operands have the same sign
   b. both have different signs
   c. operand with greater value is positive
   d. operand with greater value is negative
9. 10#4#E2 represents ....
   a. $(4)^2$ b. 4 *10 *2 c. $4*(10)^2$ d. (E2)* 10/ 4
10. 7 mod −4 evaluates to ....
    a. 3 b. −3 c. 1 d. −1
11. 9 rem 2 evaluates to ....
    a. 1 b. −1 c. 0 d.2

Descriptive Questions

1. VHDL is a strongly typed language. Justify with an example.
2. Discuss the different classes of VHDL objects.
3. What is the difference between std_logic and std_ulogic?
4. Discuss the eight levels allowed in std_logic.
5. Both data types bit and std_logic represent values '1' and '0'. How to decide which data type is to be used and why?
6. Discuss all the predefined data types in VHDL.
7. Write a note on the composite types in VHDL.
8. What are incomplete types? What is their use?
9. What are subtypes? Give an example of constrained array subtype.
10. What is the advantage of using subtypes?
11. What is difference between remainder (rem) and modulus (mod) operator?
12. Discuss the different operators in VHDL. Give their order of precedence.
13. Write the entity declaration for a 8 :1 Mux.
14. Define a physical type to represent resistance values.
15. Write a code to implement ROM as an eight-word array of four bits each.
16. Write a note on type casting in VHDL.
17. Declare an unconstrained array of integers.

4

BEHAVIOURAL MODELLING

CHAPTER OBJECTIVES

- To understand the sequential modelling and its difference from the concurrent
- Use of process statement
- Use of EVENT attribute
- Differences between signals and variables
- Learn conditional statements
- Learn looping constructs
- To know about postponed processes
- Differences between an active process and a passive process
- To be able to design sequential circuits

4.1 INTRODUCTION

Behavioural modelling describes the highest level of abstraction in the circuit design using VHDL, where the circuit to be designed is described in terms of behaviour of output in response to the changes in the input. It describes a circuit at a functional level. It is a sequential style of modelling wherein a sequence of statements in design is very important. This style of modelling is generally suitable for complex, synchronous/asynchronous sequential circuits. It helps in faster simulation when compared with other styles, but requires a complete understanding of the circuit behaviour for the circuit to be designed. This style is always preferred for circuit design where functional and simulation analysis of circuit is required. It is also referred to as algorithmic style of modelling.

4.2 SEQUENTIAL Vs CONCURRENT STATEMENTS

VHDL is sequential as well as concurrent language as it supports both sequential and concurrent statements. In general, it is considered to be concurrent, event-triggered language, but it supports sequential statements in which the order or sequence of writing statements is important unlike in concurrent. for example, process and subprograms are examples of sequential statements in VHDL. In fact, behavioural style of modelling is based on these sequential statements whereas dataflow and structural style of modelling are concurrent.

Consider the following pieces of coding in VHDL:

CODE 1	CODE 2
Entity.....	**Entity**.....
....	....
END;	**END**;
ARCHITECTURE	**ARCHITECTURE** ....
BEGIN	**BEGIN**
Statement 1;	Statement 1;
Statement 2;	Statement 3;
Statement 3;	Statement 2;
END .....;	**END** .....;

In sequential programming, Code 1 and Code 2 are expected to give different results because of the difference in order of statements 2 and 3 in two pieces of code, because in sequential programming is just like procedural languages as C, where order or sequence of writing the statements is very important and defines a step by step execution, followed one after the other. On the other hand, in concurrent programming, Code 1 and Code 2 are equivalent and are expected to give same results, as the order of statements is not important.

Example 4.1 Example of behavioural modelling/sequential modelling

```
LIBRARY IEEE;
USE IEEE.STD_LOGIC_1164.all;

ENTITY and_gate IS
PORT(a,b: IN  STD_LOGIC;
        c: OUT STD_LOGIC);
END and_gate;

ARCHITECTURE and_gate OF and_gate IS    -- behavioural architecture model of the entity
BEGIN
P1:PROCESS(a,b)
                                        -- declarative part of process
BEGIN
c <= a and b;                           -- executable part of process
END PROCESS P1;
END and_gate;
```

In the Example 4.1, simple and gate have been modelled using behavioural style. Note the use of process statement. Everything written inside a process is treated as sequential.

4.3 PROCESS STATEMENT

Process statement is the primary construct used for behavioural modelling. All VHDL statements written inside a process are sequential though the process statement by itself is a concurrent statement.

SYNTAX:

```
[Process label]: PROCESS(Sensitivity List)
Variable, signal declarations; (local to the process)          -- declarative region of process

BEGIN

Sequential statements1;
Sequential statement 2;
..............
Sequential statement n;

END PROCESS [Process label];
```

A process is declared inside the architecture body of the entity. Every process may be identified with an optional process label, followed by colon and then by the keyword **PROCESS**. In example 4.1 P1 is the process label. Even though process label is optional it is good for programming practice to use process labels to identify processes, especially in case of multiple processes; and it also helps to improve the readability of the code. Next to keyword **PROCESS** is sensitivity list of the process. Every process has associated with it a sensitivity list.

4.3.1 Sensitivity List

Sensitivity list is set of signals to which process is sensitive (responsive), that is, whenever an event (change in value) occurs on any of the signals in sensitivity list, process comes into execution. The process is suspended only after executing all the statements inside the process in sequence and remains suspended till the next event occurs in any of the signals in the sensitivity list.

So a process can either be in an executed or in a suspended state. The transition from suspended to executed state occurs on event change in any of the sensitivity list signals while transition from execution to suspension state occurs on completion of the process.

In the Example 4.1 of *and gate* behavioural model, both signals ‘a’ and ‘b’ are included in the sensitivity list. So whenever the value of ‘a’ or ‘b’ or both changes from ‘0’ to ‘1’ or ‘1’ to ‘0’, the process will start its execution; and output ‘c’ of and gate will be updated according to expression c<= a **AND** b;

Next in process are two parts—a declarative section where all the declarations of signals, variables, and constants with scope local to the process are declared; and the other section is the executable part marked with the keyword **BEGIN** inside the process, where all the executable statements are written in a sequence in which they are expected to be executed. Finally, every process ends with keyword **END PROCESS** followed by optional process label, if any.

4.4 VARIABLES

Variables are the class of VHDL objects allowed only with the sequential style of modelling. Variables are objects that are used for local storage within a process and subprograms alone. Inside a process local variables can be declared in the declarative part of process before the keyword **BEGIN** to represent its local, temporary values.

SYNTAX:

```
[Process label]: PROCESS(Sensitivity List)
VARIABLE var1[,var2]:TYPE:=initial_value                -- declarative region of process
BEGIN
Statements;
END PROCESS [Process label];
```

Example 4.2 To swap two integer values

```
LIBRARY IEEE;
USE IEEE.STD_LOGIC_1164.all;

ENTITY swap IS
PORT(a,b: INOUT  INTEGER);
END swap;

ARCHITECTURE behave OF swap IS          -- behavioural architecture model of the entity
 SIGNAL E:STD_LOGIC;
BEGIN
P1:PROCESS(E)
VARIABLE temp:INTEGER:=0;
 BEGIN
temp:= a;
a<= b;
b<=temp;
END PROCESS P1;
END behave;
```

Variables once declared in the process can be used inside the executable part of process as part of an expression or assignment statements just as signals. The scope of the variable declared in a process is limited to the process alone and cannot be passed or used outside the process directly, even in the architecture that is containing the process. In the example to swap two integer values, the variable temp of the type integer is declared and assigned the value of signal 'a'. **Note:** Variable assignment uses ':=' unlike signal assignment which uses '<=' operator.

This difference between the variable and the signal assignment follows from the inherent differences between a signal and a variable.

4.4.1 Differences Between a Signal and a Variable

Although non-static values inside a process can be used either in terms of a signal or a process there is a difference in which these non-static values affect the output due to the difference between a signal and a variable.

Signals are used to model interconnection wires in the entity and require event scheduling and handling, that is, whenever a signal assignment statement is executed, the value is scheduled to be assigned after a default delta delay or a user-specified delay, whereas in case of variable assignments, immediate assignment takes place without any delay. Therefore, signal values are updated at the end of the process while for the process it is immediate.

The event scheduling and handling of signals costs in terms of processing overheads such as memory required and speed of execution. In case of multiple drivers for a signal in a process, resolving of signals has extra overheads.

Let us see examples to illustrate the differences between variables and signals.

Example 4.3 Use of variables

```
LIBRARY IEEE;
USE IEEE.STD_LOGIC_1164.all;

ENTITY ex_var IS
PORT(a: IN  INTEGER:=4; trig: IN STD_LOGIC;
       b: OUT INTEGER);
END ex_var;

ARCHITECTURE ex_var OF ex_var IS            -- behavioural architecture model of the entity
SIGNAL res :INTEGER:=0;
BEGIN
    P1: PROCESS(trig)                       -- declarative part of process
          VARIABLE var1: INTEGER:=1;
          VARIABLE var2: INTEGER:=2;
          VARIABLE var3: INTEGER:=3;
    BEGIN
       var1:= var2 + var3;
       var3:= 2*var1;
       res<=var3;
       b<=a + res;                          -- executable part of process
    END PROCESS P1;
END ex_var;
```

In the Example 4.3, whenever an event occurs on input ‘a’ which acts as a trigger, process P1 is executed. Inside the process, three variables var1, var2 and var3 are declared. When the process executes, first var1 is updated to given by sum var2 + var3. In NEXT statement, var3 is assigned value d*var1 when uses new updated value of var1 obtained from previous statement.

res <= var3 are a signal assignment statement and b <= a + res

which give res = 0 and b = unintialized in first simulation and get their new value in subsequent simulation cycle.

At t=0, a=4, b=−2147…..(unintialized), res =0;

At t=1, a=4, b=4, res=10(var1:=2+3=5, var3:=2*5=10)

At t=2, a=4, b=14 (10 + 4), res=24 (var1:=2+3=5, var3:=2*5=10) and so on in each progressive simulation cycle.

Let us see some examples using signals.

Example 4.4 Use of signals

```
LIBRARY IEEE;
USE IEEE.STD_LOGIC_1164.all;

ENTITY ex_sig IS
PORT(a: IN  INTEGER:=4; trig: IN STD_LOGIC;
        b: OUT INTEGER);
END ex_sig;

ARCHITECTURE ex_sig OF ex_sig IS              -- behavioural architecture model of the entity
SIGNAL res :INTEGER:=0;
BEGIN
   P1: PROCESS(trig)                          -- declarative part of process
        SIGNAL sig1: INTEGER:=1;
        SIGNAL sig2: INTEGER:=2;
        SIGNAL sig3: INTEGER:=3;
BEGIN
   sig1<= sig2 + sig3;
   sig3<= 2* sig1;
   res<= sig3;
   b<=a + res;                                -- executable part of process
  END PROCESS P1;
END ex_sig;
```

Output is:

At t=0, a=4, b=−2147…..(unintialized), res=0;

At t=1, a=4, b=4, res=3(sig3=3)

At t=2, a=4, b=7(3+ 4), res=3(var1:=2+3=5,var3:=2*5=10) and so on in each progressive simulation cycle. In the first simulation cycle only the initialized values are taken and in the next cycle, the previous values are used though sig1 should be having value 2 + 3 =5 and sig3 should have been 2*sig1=2*5=10 as expected, but sig3 = 3 (initialized value) and so on.

So, as seen, signals get updated at the end of the process and after a default delta simulation delay. Therefore, the value of signal gets its updated value in the next cycle unlike the variables where the new values are immediately taken and reflected in the results. In case of variables in the same simulation cycle, the new value from the previous statement is used in the NEXT statement. Therefore, value updation is immediate.

Differences Between Signal and Variable

Signal	Variable
1. Signals are VHDL data objects used to represent wires and interconnections.	Variables represent temporary storage in VHDL.
2. Assignment operator for signals is '<='.	Assignment operator for variables is ':='.
3. Values of signals are updated only after default delta delay or specified user delay.	Values of variables are updated immediately on execution of variable assignment statement.
4. Signals require event scheduling and synchronizing of signal drivers.	No event scheduling and synchronizing is required in case of variables.
5. Use of signals has greater processing overheads and consumes more memory.	Use of variables consumes less memory due to no event scheduling.
6. Use of signal is allowed with all styles of modelling.	Use of variables is allowed only with sequential style (inside a process) or subprograms.

4.5 SEQUENTIAL STATEMENTS

Sequential, conditional and loop statements are allowed to be used inside a process. Unlike concurrent statements, these statements follow the sequence of execution.

4.5.1 IF Statement

'IF' is the simplest statement used for evaluation of a condition. IF condition holds true a set of statements is executed, else no or different statements are executed.

IF statement exists in three forms:

```
1. IF condition THEN                                   -- SIMPLE IF
       Statements;.
   END IF;
2. IF condition THEN                                   -- IF-ELSE
       Statements;.
   ELSE
       Statements;.
   END IF;
3. IF condition THEN                                   -- IF-ELSIF LADDER
       Statements;.
   ELSIF condition THEN
       Statements;
   ELSE
       Statements;
   END IF;
```

Note: In all the three forms every IF has to be ended with an END IF.

Example 4.5 Design of D flip-flop with asynchronous reset

```
LIBRARY IEEE;
USE IEEE.STD_LOGIC_1164.all;

ENTITY dff IS
PORT(d, clk,rst: IN  STD_LOGIC;
       q: OUT STD_LOGIC);
END dff;

ARCHITECTURE behave OF dff IS              -- behavioural architecture model of the entity

BEGIN
   P1: PROCESS(rst,clk)                    -- declarative part of process
       BEGIN
         IF rst= '0' THEN
           q<='0';
         ELSIF clk'EVENT AND clk='1' THEN
           q<=d;
         END IF;
     END PROCESS P1;
END behave;
```

Example 4.6 Design of a 4-bit up counter

```
LIBRARY IEEE;
USE IEEE.STD_LOGIC_1164.all;
USE IEEE.STD_LOGIC_UNSIGNED.all;

ENTITY counter IS
PORT(E, clk,rst: IN STD_LOGIC;
       count: INOUT STD_LOGIC_VECTOR(3 DOWNTO 0));
END counter;

ARCHITECTURE behave OF counter IS          -- behavioural architecture model of the entity

BEGIN
P1: PROCESS(rst,clk)                       -- declarative part of process
  BEGIN
    IF rst='0' THEN
       count<= "0000";
    ELSIF clk'EVENT AND clk='1' THEN
       count<= count + 1;
```

```
      ELSE
        count <= count;
      END IF;                                    -- executable part of process
    END PROCESS P1;
END behave;
```

Note: In Example 4.5, **USE IEEE.STD_LOGIC_UNSIGNED.all;** is included to allow arithmetic addition of **std_logic_vector (count).**

4.5.2 CASE Statement

CASE statement is a multiple condition statement which can be an effective replacement for IF-ELSIF ladder for multiple conditions, to check a single parameter for against multiple values.

```
CASE expression IS
   WHEN choice1 => Statements;.
   WHEN choice2 => Statements;

    WHEN OTHERS => Statements;                   -- default value
END CASE;
```

Example 4.7 Design of 2:4 decoder using CASE statement

```
LIBRARY IEEE;
USE IEEE.STD_LOGIC_1164.all;
ENTITY d2t4 IS
PORT(E: IN STD_LOGIC;W: IN STD_LOGIC_VECTOR(1 DOWNTO 0)
       q: OUT STD_LOGIC_VECTOR(0 TO 3);
END d2t4;
ARCHITECTURE d2t4 OF d2t4 IS          -- behavioural architecture model of the entity
BEGIN
   P1: PROCESS(W,E)
      BEGIN
        IF E='1' THEN
           CASE W IS
              WHEN "00" =>y<= "1000";
              WHEN "01" =>y<= "0100";
              WHEN "00" =>y<= "0010";
              WHEN OTHERS =>y<= "0001";
```

```
            END CASE;
        END IF;
    END PROCESS P1;
END d2t4;
```

4.5.3 LOOP Statement

LOOP statement is used for iterations of a set of sequential statements. LOOP statement exists in three forms as follows:

```
1. FOR identifier IN RANGE LOOP                                -- FOR LOOP
        Statements;
   END LOOP;
2. WHILE condition LOOP                                        -- WHILE LOOP
        Statements;
   END LOOP;
3. LOOP                                                        -- SIMPLE LOOP
        Statements;
   END LOOP;
```

FOR LOOP:

```
  FOR i IN 1 TO 10 LOOP
      i=2*i;
END LOOP;
```

Example 4.8 To compute the factorial of a number

```
LIBRARY IEEE;
USE IEEE.STD_LOGIC_1164.all;

ENTITY fact IS
PORT(clk: IN STD_LOGIC; n: IN INTEGER
      op: OUT INTEGER);
END fact;
ARCHITECTURE fact OF fact IS          -- behavioural architecture model of the entity
BEGIN
  P1: PROCESS(clk)
        VARIABLE x:INTEGER:='1';
      BEGIN
        IF clk'EVENT AND clk='1' THEN
           FOR x IN 3 DOWNTO 0 LOOP
```

```
            x:=x*x;
            END LOOP;
        END IF;
        op<= x;
    END PROCESS P1;
END fact;
```

WHILE LOOP:

```
WHILE i<9 LOOP
    i=2*i;
  END LOOP;
```

While loop differs from FOR-loop as it repeats the sequential statements until a particular condition is met and is useful where it is not known how many times a loop is to be repeated. But FOR-loop iterates for a fixed number of times which should be known. Also in while loop the loop identifier has to be incremented inside the loop explicitly unlike in FOR-loop where the loop identifier is automatically incremented/decremented.

Example 4.9 To compute the sum of even integers that are less than 10

```
LIBRARY IEEE;
USE IEEE.STD_LOGIC_1164.all;

ENTITYsumeven IS
PORT(trig :IN STD_LOGIC;
       op: OUT INTEGER);
END sumeven;

ARCHITECTURE sumeven OF sumeven IS          -- behavioural architecture model of the entity

BEGIN
   P1: PROCESS(trig)
        VARIABLE i:INTEGER:=1;
        VARIABLE sum_partial:INTEGER:=0;
     BEGIN
       IF clk'EVENT AND clk='1' THEN
         WHILE i <10 LOOP
            IF I mod2=0 THEN
              sum_partial:=sum_partial + i;
            ELSE
              sum_partial:=sum_partial;
            END IF;
                i:=i+1;
```

```
            END LOOP;
          END IF;
          op<=sum_partial;
      END PROCESS P1;
END sumeven;
```

4.5.4 WAIT Statement

WAIT statement is a sequential statement, used to specify timing inputs for simulation or synthesis, or explicitly suspend a process without using the sensitivity list of the process.

WAIT statement exists in three forms as follows:

SYNTAX:

```
1. WAIT ON signal1[,signal2];
2. WAIT UNTIL Boolean Expression;
3. WAIT FOR time expression;
     or combined form
4. WAIT ON signal UNTIL Boolean Expression FOR time expression;
Examples:
   WAIT ON A,B;
   WAIT UNTIL enable='1';
   WAIT FOR 10 ns;
   WAIT ON A UNTIL enable='1' FOR 10 ns;
```

WAIT ON: WAIT ON is a mimic of the sensitivity list. Any number of signals can be specified using a WAIT ON statement. If WAIT ON statement is included in a process, whenever an event occurs on any of the signals included in WAIT ON statement, the process gets into execution mode. To exactly mimic the sensitivity list it is generally included as the first or the last statement inside the process body.

Example 4.10 D flip-flop with asynchronous clock and reset

```
LIBRARY IEEE;
USE IEEE.STD_LOGIC_1164.all;
ENTITY dff IS
PORT(d, clk,rst: IN STD_LOGIC;
       q: OUT STD_LOGIC);
END dff;
ARCHITECTURE behave OF dff IS              -- behavioural architecture model of the entity
BEGIN
   P1: PROCESS                             -- no sensitivity list
       BEGIN
          IF rst='0' THEN
             q<='0';
```

```
    ELSIF clk'EVENT AND clk='1' THEN
        q<=d;
    END IF;
    WAIT ON clk, rst;                    -- use of WAIT ON instead of sensitivity list
END PROCESS P1;
END behave;
```

WAIT UNTIL: This form of WAIT allows the process to be suspended until a particular condition specified is satisfied. The condition expression should return a Boolean value. This creates an implicit sensitivity list.

Example 4.11 Design of 4-bit shift register

```
LIBRARY IEEE;
USE IEEE.STD_LOGIC_1164.all;

ENTITY shiftreg IS
PORT(d, clk,rst: IN STD_LOGIC;
        q: OUT STD_LOGIC);
END shiftreg;

ARCHITECTURE behave OF shiftreg IS          -- behavioural architecture model of the entity
SIGNAL intl :STD_LOGIC_VECTOR(3 DOWNTO 0);
BEGIN
   P1: PROCESS                              -- declarative part of process
      BEGIN
        IF rst='1' THEN
           intl<= "0000";
        ELSE
           intl<= d & intl'LEFT DOWNTO 1;
        END IF;
      WAIT UNTIL clk'EVENT AND clk='1';    -- use of WAIT UNTIL
    END PROCESS P1;
    q<= intl(0);
 END shiftreg;
```

WAIT FOR: WAIT FOR is used to delay the execution of the process for a specified time. It takes value which is in units of time or as a time expression.

Example 4.12 Design of D-latch

```
LIBRARY IEEE;
USE IEEE.STD_LOGIC_1164.all;
```

```
ENTITY latch IS
PORT(d, clk: IN STD_LOGIC;
      q: OUT STD_LOGIC);
END latch;
ARCHITECTURE behave OF latch IS          -- behavioural architecture model of the entity
BEGIN
  P1: PROCESS                            -- declarative part of process
      BEGIN
        IF clk'EVENT AND clk='1' THEN
          q<=d;
        END IF;
        WAIT FOR 15 ns;                  -- use of WAIT FOR
      END PROCESS P1;
END behave;
```

Sensitivity List Vs WAIT Statement: Both sensitivity list as well as WAIT statement can be used to control the execution of a process. Both are mutually exclusive, that is, a process can either have a sensitivity list or any form of a WAIT statement. Both if used together in a process compiler flags an error. If a process is using WAIT statements for process execution control, it can support multiple WAIT statements in the same process.

To mimic the behaviour of sensitivity list, WAIT statement is generally included at the end of the process, as all the processes are executed at least once during simulator initialization. But sensitivity list being statically allocated has less overhead during simulation against WAIT statement.

WAIT Vs Transport Delay: WAIT statements can also be used for specifying delays for scheduling events and assertions. Transport delay is also used for scheduling events but with a difference that WAIT statements advance simulation in time for all signals whereas transport delay controls timing of events on a single signal.

4.5.5 ASSERT Statement

ASSERT statement is used for exception handling based on self-checking model that reports violations in the design immediately. ASSERT statement checks the value of a Boolean expression. If true, no action is taken; but if false, a user-specified message is reported to the user, indicating level of error. ASSERT statement is a non-synthesizable statement.

SYNTAX:

```
ASSERT condition
[REPORT "Message"] (no semicolon)
[SEVERITY expression];
```

In ASSERT statement, **ASSERT** is the keyword followed by the Boolean expression condition, whose violation is to be checked. Next is an optional REPORT statement which is used to give the text message to the user when violation occurs. Although the text string is given by the designer, it has default text string message "Assertion Violation".

Note report statement is not ended in semicolon. Next is an optional SEVERITY expression which is a predefined clause with enumerated values indicating the level of the error due to assertion violation, that is, it indicates how severe the violation is.

It takes any of the four predefined values as follows:

- NOTE: the lowest level of severity used to convey information to user on assertion violation.
- WARNING: Warns the user of unexpected results due to assertion violation.
- ERROR: Indicates serious errors and improper results can be expected due to assertion violation. It is the default value for severity.
- FAILURE: Highest level of error, indicating disastrous results due to assertion violation.

So, as the severity level moves from NOTE to FAILURE the level of error caused too increases. ASSERT statement is used for checking timing violations in designs or checking undefined operating conditions, such as division by zero.

Example 4.13 Design of S-R flip-flop

```
LIBRARY IEEE;
USE IEEE.STD_LOGIC_1164.all;

ENTITY rsff IS
PORT(r,s,clk: IN STD_LOGIC;
      q,qbar: OUT STD_LOGIC);
END rsff;

ARCHITECTURE behave OF rsff IS          -- behavioural architecture model of the entity
  BEGIN
    P1: PROCESS (clk)
        BEGIN

          ASSERT (s='1' NAND r='1')     -- use of ASSERT statement
            REPORT "Undefined state"
            SEVERITY ERROR;

        IF s='1' THEN
            q<='1';
        ELSIF r='1' THEN
            q<='0';
        END IF;
            qbar<=NOT (q);
      END PROCESS P1;
  END behave;
```

In S-R flip-flop, the output is not defined for a condition when both r='1' and s='1'. So, ASSERT statement is used to check this condition whenever this condition occurs—user is reported an error "Undefines state".

4.5.6 EXIT Statement

EXIT statement is used to jump out of loop conditionally or unconditionally and terminate the loop.

SYNTAX:

```
LOOP
EXIT or EXIT [LOOP LABEL] or EXIT WHEN condition;
END LOOP;
```

Example 4.14 To find the sum of first 10 integers

```
LIBRARY IEEE;
USE IEEE.STD_LOGIC_1164.all;
ENTITYsum10 IS
PORT(clk :IN STD_LOGIC;
      op: OUT INTEGER);
END sum10;
ARCHITECTURE sum10 OF sum10 IS          -- behavioural architecture model of the entity
BEGIN
  P1: PROCESS(clk)
      VARIABLE i,sumint:INTEGER:=0;

BEGIN
          LOOP
            sumint:=sumint + i;
             i=i+1;
            EXIT WHEN i=11;
          END LOOP;
        op<=sumint;
      END PROCESS P1;
END sumeven;
```

4.5.7 NEXT Statement

The NEXT statement is used to skip a set of sequential statements or a loop iteration. Inside a loop when NEXT statement is encountered, all the statements after NEXT inside the loop are skipped and processing jumps to next iteration, if any.

SYNTAX:

```
LOOP
Statement1;
NEXT or NEXT WHEN condition;
Statement2;
.......
END LOOP;
```

4.5.8 NULL Statement

NULL statement is no-action statement used when after a certain condition is met and no action is required to be performed, say, in a CASE statement when for a particular choice no action is to be taken, then NULL statement is used.

Example 4.15 Use of NULL statement

```
LIBRARY IEEE;
USE IEEE.STD_LOGIC_1164.all;
ENTITY ex_null IS
PORT(clk: IN STD_LOGIC;inp: IN INTEGER RANGE 0 TO 10;)
      op: OUT STD_LOGIC_VECTOR(3 DOWNTO 0);
END ex_null;
ARCHITECTURE ex_null OF ex_null IS          -- behavioural architecture model of the entity
SIGNAL A,B:STD_LOGIC_VECTOR(3 DOWNTO 0);

BEGIN
P1: PROCESS(clk)
    BEGIN
        CASE inp IS
          WHEN 1 => op<= A OR B;
          WHEN 2 => op<= A AND B;
          WHEN 3 => op<= A NAND B;
          WHEN 4 => op<= A NOR B;
          WHEN 5 => op<= A XOR B;
          WHEN 6 => op<= A EXNOR B;
          WHEN 7 => op<= A EXOR B;
          WHEN OTHERS => NULL;
        END CASE;
      END IF;
   END PROCESS P1;
END ex_null;
```

4.6 POSTPONED PROCESSES

The concept of postponed process was introduced in VHDl'93. It is a process which does not get executed till the last possible moment in the simulation cycle. A typical use of such process is to perform timing check. It allows all the input transients to die out and let the inputs be stabilized before the execution of the process. The postponed process is the last one to get executed in an entity design. It waits till all the other non-postponed processes get executed.

To make a process postponed the keyword **POSTPONED** is added before the keyword **PROCESS** in the beginning of the process itself.

SYNTAX:

```
POSTPONED PROCESS (Sensitvity List)
BEGIN
Statement1;
Statement2;
.......
END PROCESS;
```

4.7 PASSIVE PROCESSES

Passive processes are the processes with no-signal assignment statement either inside the process or in any of the subprograms called inside the process. Passive processes are generally defined inside the entity declaration. They are used for checking purposes such as time check or condition violation check, as done by ASSERT statements. Although their function is very similar to ASSERT statement but use of passive processes has an advantage over ASSERT statement that they are included in all the architectures of an entity and need not be written for each architecture or process individually.

SUMMARY

- Behavioural modelling is sequential style of modelling.
- Process statement is the main construct which contains the sequential code in behavioural modelling.
- Every process has a sensitivity list.
- The process is written inside the architecture of the entity.
- Process uses data objects VHDL—variables which are used for temporary storage inside the process.
- Variables allow an immediate updation of values whereas signals update values after delta delay.
- A number of sequential statements are allowed in VHDL like IF, CASE, WAIT, ASSERT, LOOP, NEXT, NULL and EXIT.
- WAIT statements exist in three forms and can be used as an alternative to sensitivity list.
- WAIT statement and sensitivity list are mutually exclusive.
- Multiple WAIT statements are allowed inside a process.
- Postponed process is executed in the last delta of simulation cycle and allowed for timing check.
- Passive process is a process which contains no-signal assignment statement.

QUESTIONS

Objective Questions

1. Behavioural modelling is a .... style of modelling in VHDL.

   a. concurrent b. sequential c. default d. none

2. Process statement is ....

   a. concurrent b. sequential

3. Multiple WAIT statements are allowed inside a process statement. (True/False)
4. Every process has associated with it: ....

   a. sensitivity list, declarative section, executable section and process end.

   b. sensitivity list, WAIT statement, declarative section, executable section and process end.

   c. sensitivity list, declarative section, executable section and process label.

   d. WAIT statement, declarative section, executable section and process end.

5. Variables consume .... memory than signals.

   a. less b. more c. equal d. none

6. Passive process can be a postponed process. (True/False)
7. If no value is assigned to a default case in a CASE statement, it is ....

   a. null b. unaffected c. others d. of particular value

8. Level of abstraction for behavioural modelling is .... to actual hardware

   a. closest b. intermediate c. farthest d. none

9. NEXT statement is a/an .... statement.

   a. conditional jump b. unconditional jump c. termination d. no operation

10. Consider the following code:

```
L1: For i IN RANGE 0 to 4 LOOP
     L2: FOR j IN RANGE 0 To 3 LOOP
     Statements;
       IF j= 2 THEN
         EXIT;
       END IF
   END LOOP L2;
END LOOP L1
```

    In this code, the EXIT statement will terminate which iteration and exit which loop?

11. The default delay model is ....

    a. delta b. transport c. inertial d. none of these

Descriptive Questions

1. Explain the differences between the following:

   a. combinational and sequential circuits b. sensitivity list and WAIT statement

   c. signal and variable d. IF and CASE

2. Explain the various loops in VHDL with examples.
3. Differentiate between a process and a WAIT statement. Can they be used simultaneously in a program?

4. How are sequential statements different from the concurrent statements?
5. Write a short note on transport delay.
6. Design all the flip flops: D, J-K, S-R and T in VHDL.
7. Design a decade counter which counts from 0 to 9 and then back from 9 to 0.
8. Design a 4-bit ripple-carry adder. What is the hardware generated for this circuit?
9. Design a finite state machine with input ‘x’ and output ‘z’. It has the following state transition table:

Present state	**x=0**	**x=1**
a	f,0 (next state, output)	b,0
b	d,0	c,0
c	f,0	e,0
d	g,1	a,0
e	d,0	c,0
f	f,1	b,1
g	g,0	h,1
h	g,1	a,0

10. Design a 4-bit shift register with parallel load. Draw its truth table, circuit and waveforms.
11. Design a bus structure with registers.
12. Write the VHDL code for the design of an even parity generator for a 4-bit data.
13. Discuss various control statements.
14. What is a passive process. What is its use? Give one example of design which uses passive process.
15. Design a 4-bit binary to gray-code converter with minimum NAND gates. Write the behavioural code for the same.
16. Compare and contrast FOR loop and while loop.

DATAFLOW MODELLING

5

CHAPTER OBJECTIVES

- Introduction to dataflow modelling
- Understand the differences between concurrent and sequential signal assignment statements
- Explore the concurrency of VHDL
- Learn about conditional assignment statements
- Use of blocks in VHDL
- Use of generate statements to allow iterations and looping in concurrent modelling
- Concurrent assertion statements

5.1 INTRODUCTION

Dataflow modelling describes the architecture of the entity under design without describing its components in terms of flow of data from input towards output. This style is nearest to RTL description of the circuit. Dataflow modelling is concurrent style of modelling in VHDL, that is, unlike behavioural modelling the order of statements is not important in the coding of dataflow modelling. All statements are executed in parallel. This makes the concurrent signal assignment statements event-triggered which implies that any concurrent assignment statement is executed only when any event occurs on the signals that are used in the expression of concurrent signal assignment statement.

For Example:

```
ENTITY dataflow IS
PORT(a,b,e,f : IN BIT; c,d: OUT BIT);
END dataflow;
ARCHITECTURE ex OF dataflow IS          -- declarative section of architecture
BEGIN
      c<= a OR b;                       -- statement 1
      d<= e AND f;                      -- statement 2
END ex;
```

This shows the dataflow architecture. It uses two concurrent statements—one for the expression of 'c' and another for 'd', respectively. Whenever an event occurs on signals 'a' or 'b', statement 1 will be executed assigning a new value to the output signal 'c' after a delay of Δ; and if any of the signals 'e' or 'f' changes it value, statement 2 is executed assigning a new value to signal 'd' after a delay of Δ(delta). Thus, every concurrent statement is event-triggered. Whenever an event occurs on its input signals it is executed.

As seen in this example, no delay is specified. This is because concurrent statements by default works on delta delay, which defines an infinitesimally small delay of Δt between event change and its effect on the output. Otherwise, in order to schedule a correct order of signals, 'AFTER' clause can be used to specify a user-defined delay with a concurrent statement.

For Example:

```
ARCHITECTURE datafl OF half_adder IS            -- declarative section of architecture
BEGIN
    sum<= a XOR b after 15 ns ;                 -- statement 1
    carry<= a AND b after 20 ns;                -- statement2
END dataflow;
```

Now, in this example of half-adder, both concurrent statements depend on the same input signals 'a' and 'b'. Whenever an event occurs on signal 'a' or signal 'b', both the statements will be executed. To ensure that sum is computed first and then only carry, the after clause can be used to insert the user-defined time delay between even change at input and its effect on output.

The difference between concurrent signal assignment and sequential signal assignment statement is that a concurrent signal assignment statement is an event-triggered statement whereas a sequential assignment statement is executed in sequence only when the process in which they are contained is active. Consider an example as follows:

```
ENTITY dataflow IS
PORT(a,b : IN INTEGER; c,d: OUT INTEGER);
END dataflow;
ARCHITECTURE ex OF dataflow IS                  -- declarative section of architecture
BEGIN
    c<= a + b;                                  -- statement 1
    d<= 2 *c;                                   -- statement2
END ex;

ENTITY behavioural IS
PORT(a,b : IN INTEGER; c,d: OUT INTEGER);
END behavioural;
ARCHITECTURE ex OF behavioural IS               -- declarative section of architecture
BEGIN
PROCESS(a,b)
BEGIN
    c<= a + b;                                  -- statement 1
    d<= 2 *c;                                   -- statement2
END PROCESS;
END ex;
```

In the dataflow style, for this example, whenever any change in value of signals 'a' and 'b' occurs, 'c' will be get its new value after a delta delay. Due to the change in the value of 'c', the next statement will be invoked assigning a new value to signal 'd' using the new value of signal 'c'.

On the other hand, in the sequential style of modelling , even if the value of 'c' changes, signal 'd' will not get its new value till the process is activated, which is only when values on signals 'a' and 'b' change. Once the process is active, 'c' will get its new value from 'a' and 'b' after the delta delay and 'd' will get its value from the previous value of 'c' after the delta delay.

Table 5.1 Comparison Between Sequential and Concurrent Signal Assignment Statements

Concurrent signal assignment	Sequential signal assignment
1. It is used with the dataflow style of modelling.	1. It is used with the sequential style of modelling.
2. It is event-triggered.	2. It is not event-triggered but executed in sequence.
3. The order of statements is not important.	3. The order of statements is important as all the statements are executed in sequence.
4. It works on the delta delay model by default; 'after' clause can be used to specify the user-defined delay.	4. The use of 'after' clause is not common with sequential signal assignment statements.
5. The problem of multiple drivers arises in this style.	5. Being sequential in nature, the problem of multiple drivers is not there.

5.2 CONDITIONAL ASSIGNMENT STATEMENTS

VHDL allows conditional constructs in dataflow modelling analogous to if-else/case in the sequential modelling. Two types of conditional assignment statements are used to check conditions in the data-flow style as follows:

- WHEN-ELSE conditional statement
- WITH-WHEN-ELSE conditional assignment

5.2.1 When-Else

It is a concurrent conditional selection statement which checks the condition and assigns a value to the output if condition is true; else, the next condition is checked and so on. It is very similar to case statement.

SYNTAX:

```
Output <=Val1 WHEN [ after time expression] condition1 ELSE
          Val2 WHEN [ after time expression] condition2 ELSE
                .
                .
                .
          Valn WHEN [ after time expression] condition ELSE
          Val [ after time expression] [WHEN condition];
```

As seen the output signal is assigned the value in accordance with the condition it satisfies—val1 if condition1 is satisfied, val2 if condition2 is satisfied and so on. In a default case when none of the conditions from 'n' set of possible conditions is satisfied, 'val' is assigned to the output which is the last value in WHEN-ELSE statement.

This statement can be used inside the architecture body, a block or a generate Loop. In special cases, 'unaffected' may be used instead of a value for the default case. Use of 'Unaffected' is equivalent to NULL statement.

For Example:

```
Output <=  Val1 WHEN [ after time expression] condition 1 ELSE
           Val2 WHEN[ after time expression] condition 2 ELSE
           .
           .
           .
           Valn WHEN [ after time expression] condition n ELSE
           UNAFFECTED;
```

Conditional and selected signal assignments are a concise way to describe combinational logic in register transfer level descriptions.

Example 5.1 Use of conditional assignment statement

```
LIBRARY IEEE;
USE IEEE.STD_LOGIC_1164.all;

ENTITY mux IS
PORT(a,b,c,d: IN BIT;
     sel:IN BIT_VECTOR(1 DOWNTO 0);
     y: OUT BIT);
END mux;

ARCHITECTURE mux1 OF mux IS          -- bhavioural architecture model of the entity
BEGIN

     y <=a  WHEN sel= "00" ELSE
          b WHEN sel= "01" ELSE
          c WHEN sel= "10" ELSE
          d;

END mux1;
```

5.2.2 With-When-Select

It is a conditional selection statement.

Syntax:

```
WITH parameter SELECT
          Output<= val1 WHEN condition 1,
                   Val2 WHEN condition 2,
                   .
                   .
                   .
                   UNAFFECTED WHEN OTHERS;
```

Example 5.2 Use of with-when-select statement

```
LIBRARY IEEE;
USE IEEE.STD_LOGIC_1164.all;

ENTITY mux IS
PORT(a,b,c,d: IN BIT;
      sel:IN BIT_VECTOR(1 DOWNTO 0);
      y: OUT BIT);
END mux;

ARCHITECTURE mux2 OF mux IS             -- behavioural architecture model of the entity
BEGIN

WITH sel SELECT
        y <= a WHEN "00" ,
             b WHEN "01" ,
             c WHEN "10" ,
             d WHEN OTHERS;
END mux2;
```

5.3 BLOCK STATEMENT

Block statement is used to group self-contained logical areas. Block statements are used to perform the following functions:

(i) to disable some signals using guard expression
(ii) to limit the scope of declarations
(iii) to group logical areas in the program to improve readability

It helps to partition the program into separate sections. Each section identified by a separate block can have its own local signals, types, and so on. This helps reduce complexity of program incase number of signals local to the architecture body required is more. With respect to declaration of objects and types, the block behaves similar to architecture declarative part. Anything which can be declared inside an architecture body can be declared inside a block also.

SYNTAX:

```
BLOCK LABEL: BLOCK [guard expression] IS
            Block Header
            Block declarations
     BEGIN
            Concurrent statements
                      .
                      .
     END BLOCK [BLOCK LABEL];
```

Every block has a block label which is optional if only an 'e' block exists in a program. In case of multiple blocks it is necessary to include block headers for each block. This is followed by the keyword **BLOCK**. A block can have an optional guard expression. Every block has a declarative section in which all its local signals, types and constants can be declared and a block body in which all the concurrent statements (guarded as well as simple) occur. It is marked with the keyword **BEGIN** and closed with the keyword **END BLOCK** followed by an optional block label. One block may be contained in another block.

For Example:

```
blk1 : BLOCK
      BEGIN
              blk2: BLOCK
              BEGIN
                      Concurrent statements;
              END BLOCK blk2;
         Concurrent statements;
      END BLOCK blk1;
```

In this example, blk2 is contained in blk1. blk1 is the parent block while blk2 is the nested block. The upper-level block blk1 cannot reference the lower-level block blk2 signals and its local objects, but a lower-level block can reference the upper-level block signals/objects.

Same signal/object names can be used indifferent blocks as each object/signal declared inside a block is local to that block. But in case of nested blocks, if the same name for a signal/object is used in both the blocks, then inside the nested block, its own local signal/object overrides the signal/object of the parent block. To access the parent block signal/object block qualifier is used as:

Blk1.signal/object _name

Parent block name (qualifier)

Guarded Blocks: If the guard expression is included in a block, then it is said to be a guarded block. Guard expression is a Boolean expression, which if true, enables the drivers of statements in the block which are guarded and if false, guarded statements are not activated. Else, for any block containing guarded statements, an implicit guard signal is declared which serves as a trigger signal for the guarded statements in the block.

Example 5.3 Use of guarded block: Implicit guard signal

```
LIBRARY IEEE;
USE IEEE.STD_LOGIC_1164.all;

ENTITY latch IS
PORT(clk,d: IN STD_LOGIC;
        q: OUT STD_LOGIC);
END latch;

ARCHITECTURE latch OF latch IS              -- behavioural architecture model of the entity
BEGIN
B1: BLOCK (clk= '1')
```

Example 5.4 Use of guarded block: Explicit guard signal

```
LIBRARY IEEE;
USE IEEE.STD_LOGIC_1164.all;

ENTITY latch IS
PORT(clk,d: IN STD_LOGIC;
        q: OUT STD_LOGIC);
END latch;

ARCHITECTURE latch OF latch IS              -- behavioural architecture model of the entity
BEGIN
B1: BLOCK (clk= '1')
    SIGNAL GUARD: BOOLEAN;
    BEGIN
     q<= '1' WHEN NOT GUARD ELSE '0';
   END BLOCK b1;
END latch;
```

Block Header: Block header contains port and generic declarations and their mappings, which associate objects (signals/ports) outside the block to objects and generic parameters inside the block. It helps the code to be reused.

Let us say an entity 'blk' that exists is defined as

```
LIBRARY IEEE;
USE IEEE.STD_LOGIC_1164.all;

ENTITY blk IS
PORT(a,b: IN BIT;
       c,d: OUT BIT);
END blk;

ARCHITECTURE ex OF blk IS              -- behavioural architecture model of the entity
BEGIN
B1: BLOCK (clk= '1')
      BEGIN
         c<= a AND b;
         d <= a XOR b;
      END BLOCK ;
END ex;
```

```
LIBRARY IEEE;
USE IEEE.STD_LOGIC_1164.all;

ENTITY blk_exp IS
PORT(a,b: IN BIT;
      c: OUT BIT);
END blk_exp;

ARCHITECTURE ex1 OF blk_exp IS            -- behavioural architecture model of the entity
BEGIN
B1: BLOCK (clk= '1')
    PORT (a1,b1:IN BIT; c1: OUT BIT);
    PORT MAP(a1=>a,b1=>b,c1=>c);
    BEGIN
      c1<= a1 AND b1;
    END BLOCK ;
END ex1;
```

In the above example, Block 'B1' has block header containing port declarations for ports a1, b1 and c1 which are local to block b1. Also the block header contains mapping of local ports of 'B1' to ports of the entity 'blk_exp' using PORT MAP statement.

5.4 CONCURRENT ASSERT STATEMENT

An assert statement that is used for exception handling can be used with both sequential as well as concurrent modelling of VHDL entities. Its use with behavioural modelling has been discussed in the previous chapter. Here its use with dataflow modelling is discussed. It has the same syntax as in the behavioural modelling.

SYNTAX:

```
ASSERT condition
[ REPORT "Message"] (no semicolon)
[ SEVERITY expression];
```

Only difference is its usage in style of modelling.

5.5 CONCEPT OF DELAY IN CONCURRENT LOGIC

As discussed in the beginning of this chapter, dataflow modelling is based on delta delay model. Delta (Δ) is an infinitesimal unit of time which is required for the correct execution of concurrent statements that are executed at the same time though the actual simulation time remains unaffected.

For Example:

Consider the following circuit:

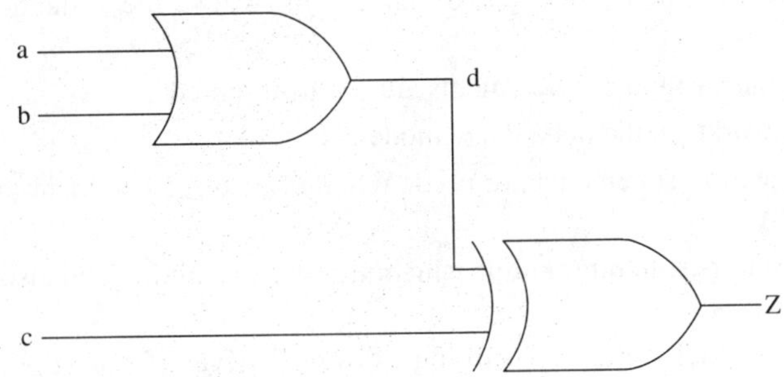

Figure 5.1 A Combinational Circuit

The dataflow model for the circuit given in Figure 5.1 is

```
ENTITY dataflow IS
PORT(a,b,c : IN STD_LOGIC; Z: OUT STD_LOGIC);
END dataflow;
ARCHITECTURE ex OF dataflow IS
SIGNAL d:STD_LOGIC                      -- declarative section of architecture
BEGIN
        d<= a OR b;                     -- statement1
        z<= c XORx d;                   -- statement2
END ex;
```

Now, when the value of either ‘a’ or ‘b’ changes ‘d’ gets its new value, which in turn triggers statement2 giving a new value to ‘z’. These changes take place in the same simulation cycle but infinitesimal delay Δ is inserted between the execution of two statements. Let us say initially when all inputs are ‘0’, then signal ‘b’ changes from ‘0’ to ‘1’ at time t=5 ns and ‘d’ will assume its new value at time t= 5 + Δ ns. Due to a change in the value of ‘d’ the second statement is triggered,by giving a new value to output ‘z’ at time t= 5 + 2Δ ns.

But all this happens in the same simulation cycle so that the event sequence for one simulation cycle will be

Table 5.2 Simulation Details for a Dataflow Design

Time	0 ns	5 ns	5 ns(5 + Δ) ns	5(5 + 2Δ) ns
a	0	0	0	0
b	0	1	1	1
c	0	0	0	0
d	0	0	1	1
z	0	0	0	1

So, ‘z’ assumes its new value in the same simulation cycle in which b changes.

SUMMARY

- Dataflow style is concurrent style of modelling where the order of statements is not important.
- Concurrent signal assignment statements are event-triggered.
- Dataflow style works on the delta delay model.
- Conditional selection statements are used: When-else-select statement and with when-select statement.
- Blocks are used to partition programs which improve readability and also allow local signals/ objects.
- Use of guarded blocks allows disabling of signal driver if a particular condition is not satisfied.
- Nesting of blocks is allowed at any level.
- In the nested block, the upper-level block cannot influence the signals/objects of its nested block, but a nested block can influence its parent block signals/objects.
- For same signals, the names in the nested block qualifiers are used.
- Block headers are used to declare port mappings and generic mappings.
- Concurrent assertion statements are used to check violations for conditions which are yet to be satisfied.
- Delta delay is infinitesimal delay which is required for the correct scheduling of statements.

QUESTIONS

Objective Questions

1. Dataflow style is .... style of modelling in VHDL.
   a. sequential b. procedural c. concurrent d. graphical
2. Concurrent statements are ....
   a. sequential b. event-triggered c. transaction-triggered d. none of these
3. Dataflow style is closest to ....
   a. RTL simulation b. gate-level simulation
   c. post-layout simulation d. functional simulation
4. Dataflow style is suitable for combinational design rather than sequential circuit design. (True/False)
5. Delta delay model is based on infinitesimal delay, which is a/an ....
   a. delay between input change at its effect on output b. propagation delay of circuit
   c. interconnection delay d. none
6. Assertion statement is used for ....
   a. exception handling b. semantics errors c. syntax errors d. delay

7. Find the error:

```
ARCHITECTURE ex OF ex IS     -- behavioural architecture model of the entity

BEGIN
B1: BLOCK
IF (clk= '1') THEN
  BEGIN
    q<= GUARDED d;
  END BLOCK b1;
END ex;
```

Descriptive Questions

1. Discuss the differences between concurrent signal assignment statement and sequential signal assignment statement.
2. Compare and contrast the dataflow style of modelling with sequential style of modelling.
3. Write a program to design 4:2 priority encoder using dataflow style. Write all possible descriptions for 4:2 encoder using dataflow modelling.
4. Design a four-operation basic arithmetic calculator using dataflow style.
5. Discuss the conditional selection statements in VHDL.
6. What is the difference between WHEN-ELSE and WITH-WHEN-SELECT statement?
7. Explain the term 'unaffected'. What is its use?
8. Discuss delta delay model for concurrent assignment statements.
9. Where and when is 'after' clause used?
10. Write the VHDL code for a design of 3: 8 decoder using dataflow style of modelling.
11. Write the VHDL program for a design of comparator in the dataflow style.
12. Design the decimal to BCD encoder in the dataflow style.
13. For question numbers 10, 11 and 12 design using sequential style. Compare and contrast the two styles. Can it be concluded that dataflow style is more suitable for combinational logic.
14. Write an entity design in dataflow style to show the use of concurrent assertion statement.
15. Justify with example "dataflow style is based on the delta delay model".
16. Write the VHDL coding for demultiplexer.
17. What are guarded blocks?
18. Discuss the block statement. What are the advantages of using a block statement?

STRUCTURAL MODELLING 6

CHAPTER OBJECTIVES

- Understand hierarchical modelling in VHDL
- Advantages of using components in design
- To define and instantiate a component
- Using generics

6.1 INTRODUCTION

Structural modelling is the simplest style of modelling and the lowest level of abstraction. It describes the circuit design in terms of its components. Structural style is nearest to gate-level simulation of the circuit. In this, the circuit is represented as a set of interconnected components. It facilitates hierarchical modelling of circuit's particularly large complex designs. For complex designs, top-level entity is usually modelled in the structural form, whereas lower-level entities depending on the level of complexity, are modelled at different levels of abstraction. With increasing number of components on a single chip, hierarchical modelling is being preferred over other flat models.

In the structural form, the top-level entity is partitioned into smaller lower-level entities, each of which is known as a component.

Each lower-level entity is referred as a component which may be further partitioned into next lower-level entities as incase of lower-level entity 1 in the Figure 6.1, which is partitioned into components entity11 and entity12.

Every component in VHDL is specified with its component declaration and mapped to top-level entity using component instantiation.

6.2 COMPONENT DECLARATION

An entity before using the components is required to declare the components in its architecture-declarative section, or alternatively in a package which needs to be included in the top-level entity with 'USE' clause. The declaration of a component includes the specifications for its input–output ports on similar lines as an entity declaration with the type of data accepted at ports and mode of port.

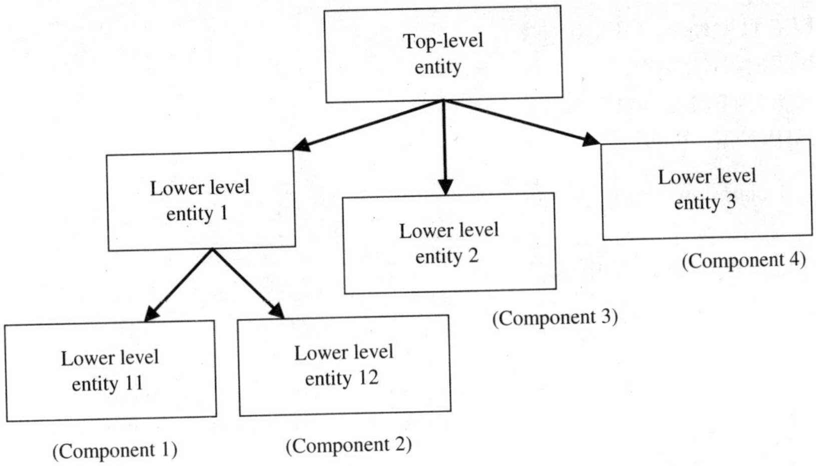

Figure 6.1 Structural Modelling Hierarchies

SYNTAX:

```
COMPONENT name
PORT ( port_name1: MODE TYPE;[ port_name2: MODE TYPE;] port_name:
MODE TYPE );                                          -- Local PORT declarations;
  END COMPONENT;
```

COMPONENT is the keyword followed by a legal identifier which is a user-defined name of the component. This is followed by port declaration for the local ports of the component and finally ends the component using keywords **END COMPONENT**.

Example 6.1
Let us say an entity Or_g is defined as

```
ENTITY Or_g IS
  PORT (a,b:IN BIT; c: OUT BIT);
END Or_g;

ARCHITECTURE ex OF Or_g IS
  BEGIN
        c<= a OR b;
END ex;
```

Now this entity is used as a component in the top-level entity Nor_g
Assuming Or_g and Nor_g to be in the same library:

```
ENTITY Nor_g IS
  PORT (x,y:IN BIT; z: OUT BIT);
END Nor_g;
```

```
ARCHITECTURE ex1 OF Nor_g IS
  COMPONENT Or_g IS
  PORT (a,b:IN BIT; c: OUT BIT);
END COMPONENT;

        -- declarations if any
BEGIN
        Component instantiations
END ex1;
```

The component declared should exist as an entity by the same name in another or same design file but in same/visible libraries, where its functionality in the form of architecture body is defined. The top-level entity which is using the component contains only the declaration of the component in its architecture declaration as in the Example 6.1. If the entity used as a component is included in a different library that is not visible to design, then the configuration statement is used in the architecture declaration of the top-level entity after the declaration of the component, which binds the component with the specific entity in the specified library.

In the latter case, when the entity used as a component is contained in a different library, it is always a good practice to place all configuration statements in a separate module and include the complete module in top-level entity. It saves a lot of compilation time, particularly for large designs for modifications made to any of the component entity. An example of component binding using configuration is covered in Chapter 8 describing configurations.

Using structural style of modelling with components also helps in code reusability by using entities once defined as components for different top-level entities as many number of times as required.

6.3 COMPONENT INSTANTIATION

Component instantiation is creating an instance of the specific component by associating its local ports to signals of top-level entity. It puts the components used in executable form for the top-level entity. PORT MAP statement is used for component instantiation. Component instantiation is included in the executable part of the architecture body marked by the keyword **BEGIN**.

SYNTAX:

```
ENTITY name IS
  PORT declarations
END name;

ARCHITECTURE name OF name IS
-- declarations;
  BEGIN
        Label : Component name PORT MAP ( list of actual signals ) statements;
-- Component instantiation
END name;
```

Component instantiation is always identified by a *Label*, which is a legal identifier in VHDL followed by colon and then name of the component '*Component name*' whose instance is being created. The keywords **PORT MAP** is used to associate the local ports of the component with the actual signals of the entity using the component. This association between the component locals and the entity actuals can be specified using two ways:

- Positional association
- Named association

6.3.1.1 Positional Association

In the positional association only the list of actuals is specified, that is, the list of actual entity signals is specified in such an order that its position in the list is the same as the position of its corresponding local port of the component in the component declaration.

The positional association is of the form:

Label: Component name **PORT MAP** (*actual1, actual2,... actual n*) where, actual1, actual2 ... actual n are signals/ports of the top-level entity.

The example 6.2 given below deals with entity Nor_g using component Or_g.

Example 6.2

```
ENTITY Or_g IS
  PORT (a,b:IN BIT; c: OUT BIT);
END Or_g;

ARCHITECTURE ex OF Or_g IS
  BEGIN
        c<= a OR b;
END ex;
```

Now this entity is used as a component in the top-level entity Nor_g, assuming Or_g and Nor_g to be in the same library as follows:

```
ENTITY Nor_g IS
  PORT (x,y:IN BIT; z: OUT BIT);
END Nor_g;

ARCHITECTURE ex1 OF Nor_g IS
  COMPONENT Or_g IS
  PORT (a,b:IN BIT; c: OUT BIT);
END COMPONENT;

BEGIN
        O1: Or_g PORT MAP( x, y,z); -- positional association of port signals x,y and z for entity
                                       Nor_g with local ports of component Or_g
END ex1;
```

In this example,

The statement identified by label 'O1' is the component instantiation of component Or_g.

Now, using positional association,

'x' will be associated or mapped to component port 'a'

'y' will be associated or mapped to component port 'b'

'z' will be associated or mapped to component port 'c'

First actual with first port of component, second actual with second port and so on.

Care should always be taken that associating ports from entity to component should have the same mode and type, else an error will be flagged by the compiler. Therefore, the order of actual list is very important in the positional association.

6.3.1.2 Named Association

Named association is mapping local to actual in any order with their names specified. It has lesser chances of error in mapping between the local ports of component and the actual signals of entity due to wrong mapping order as mapping now is between ports identified by their names.

Named association is of the form

Label: Component name **PORT MAP** *(local_port2=>actual_sig2, local_port 1=>actual_sig1,...... local_port n=>actual _signal n);*

where, actual1 actual2 .... actual n are signal/ports of top-level entity and local_port1, local_port2.... are the local ports of the component.

Consider again the example of Nor_g using components Or_g and component Not_g as:

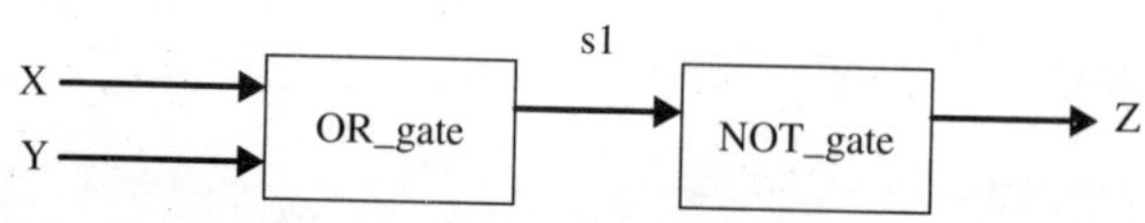

Figure 6.2 NOR Gate Using Components of OR Gate and NOT Gate

```
ENTITY Or_g IS
  PORT (a,b:IN BIT; c: OUT BIT);
END Or_g;

ARCHITECTURE ex OF Or_g IS
  BEGIN
        c<= a OR b;
END ex;

ENTITY Not_g IS
  PORT (l:IN BIT; n: OUT BIT);
END Not_g;
ARCHITECTURE Not_g OF Not_g IS
  BEGIN
        n <= NOT (l);
END Not_g
```

Now, this entity is used as a component in top-level entity Nor_g, assuming Or_g and Nor_g to be in the same library, we have

Example 6.3 Structural model for Nor gate using both forms of association

```
LIBRARY IEEE;
USE IEEE.STD_LOGIC_1164.all;
ENTITY Nor_g IS
  PORT (x,y:IN BIT; z: OUT BIT);
END Nor_g;

ARCHITECTURE ex1 OF Nor_g IS
  COMPONENT Or_g IS                                   -- component Or gate
  PORT (a,b:IN BIT; c: OUT BIT);
END COMPONENT;

COMPONENT Not_g IS
  PORT (l:IN BIT; n: OUT BIT);
END COMPONENT;

SIGNAL s1:BIT;

BEGIN
      O1: Or_g PORT MAP( x, y,s1); -- positional association of port signals x,y and z for
                                      entity Nor_g with local ports of component Or_g
      N1: Not_g PORT MAP(l => s1, n=>z); -- named association of signals s1and z entity
                                      Nor_g with local ports of component Not_g
END ex1;
```

Note the second statement in the architecture body. It is an instance of component Not_g created using label N1 and having named association between local ports 'l' and 'n' of component Not_g with actual signals 's1' and port 'z' of entity Nor_g;

Note: 's1' is an intermediate signal local to the architecture of Nor_g.

While mapping the local ports of the component to the actual ports of entity care should be taken, that mapped ports have the same mode; if the actual mode is 'INOUT' then it can be mapped to the formal ports having modes 'IN', 'INOUT' or 'OUT'.

At times, a component port may not be connected to any signals; so when mapping to actuals it is left open using the keyword **OPEN**.

For Example:

Consider the component and_gate as

```
  COMPONENT and_g IS
  PORT (a:IN BIT; b: IN BIT := '0'; c: OUT BIT);
END COMPONENT;
```

Note that the second input port is initialized to '0' so that its component instantiation may be written as:

A1: and_g **PORT MAP**(a,**OPEN**,c);

6.4 GENERICS

6.4.1 Introduction

The concept of 'generics' in VHDL is used to parameterize a design. It is useful as the same design can be used by creating different instances—each having different parameter values. These parameter values which differ in various instances of the same design are called generic parameters. So, generics helps to generalize a design.

SYNTAX:

GENERIC (generic_name: **TYPE** := value)

Generic parameters in a design can be declared using the keyword **GENERIC**. A user-defined name of the generic parameter is specified followed by the type of parameter and finally, a constant value that is assigned to the generic parameter which is optional, that is, initialization of the constant generic parameter at the time of declaration is optional. But it is necessary to initialize the generic parameter value before any of its instances is used. A generic parameter can be declared inside an entity declaration, inside a component declaration or inside a block declarative section, and can be used in associated architecture bodies, configurations or component instantiations.

FOR EXAMPLE:

```
ENTITY Nor_g IS
  GENERIC( prop_delay: TIME := 10 ns);
  PORT (x,y:IN BIT; z: OUT BIT);
END Nor_g;
```

6.4.2 Generic Map

If a generic parameter is declared inside a component, then a GENERIC MAP is used for mapping the generic parameters of the component with actual parameters for top-level entity.

SYNTAX:

Label: Component_name **GENERIC MAP** (mapping of local parameters to actual);

GENERIC MAP statement is similar to **PORT MAP** statement. Mapping of local generic parameters to parameters of entity can be done by positional association or named association. Generic map statement can be used in component instantiation, in entity-component binding configuration and inside block declarative section.

Note: Most of the synthesis tools support only integer values for generics.
Example: Consider the example of Nor_g.

Note: Generic map statement does not end with a semicolon but is followed by a PORT MAP statement of the component used in case of component instantiations.

```
ENTITY Or_g IS
  GENERIC (delay1, delay2: TIME := 0ns);
PORT (a,b:IN BIT; c: OUT BIT);
END Or_g;

ARCHITECTURE ex OF Or_g IS
  BEGIN
        c<= a OR b;
END ex;
ENTITY Nor_g IS

  PORT (x,y:IN BIT; z: OUT BIT);
END Nor_g;

ARCHITECTURE ex1 OF Nor_g IS
  COMPONENT Or_g IS                          -- component Or gate
    GENERIC (prop_delay, rise_time);
PORT (a,b:IN BIT; c: OUT BIT);
END COMPONENT;

COMPONENT Not_g IS
  PORT (l:IN BIT; n: OUT BIT);
END COMPONENT;

SIGNAL s1:BIT;

BEGIN
        G1:Nor_g GENERIC MAP(15 ns, rise_time=> 3 ns)
        O1: Or_g PORT MAP( x, y,s1); -- positional association of port signals x, y and z for entity
                                        Nor_g with local ports of component Or_g
        N1: Not_g PORT MAP(l => s1, n=> z); -- named association of signals s1and z entity
                                        Nor_g with local ports of component Not_g
END ex1;
```

Example 6.4 Use of generics with components

```
  LIBRARY IEEE;
USE IEEE.STD_LOGIC_1164.all;

ENTITY half_adder IS
      GENERIC rise_t, fall_t: TIME:= 5ns;          -- entity declaration
    PORT (a,b: IN STD_LOGIC;
            sum, carry: OUT STD_LOGIC);
END half_adder;
```

```
ARCHITECTURE struct OF half_adder IS                    -- declarative section of architecture

COMPONENT and_gate IS
GENERIC rise_t, fall_t : TIME;
        PORT (x,y: IN STD_LOGIC;
                z: OUT STD_LOGIC);
END and_gate;

COMPONENT  xor_gate IS
   GENERIC rise_t, fall_t: TIME;
     PORT (l,m: IN STD_LOGIC;
      n: OUT STD_LOGIC);
END xor_gate;

BEGIN
X1:xor_gate  GENERIC MAP(10 ns,10 ns) PORT MAP(a,b,sum);
A1:and_gate GENERIC MAP(rise_t=> 10 ns, fall_t => 10 ns) PORT MAP(a,b,carry);

END struct;
```

6.4.3 Generics with Behavioural Modelling

Generics can be used with behavioural modelling to create general purpose designs. With behavioural modelling it requires no generic map statement but only generic declaration.

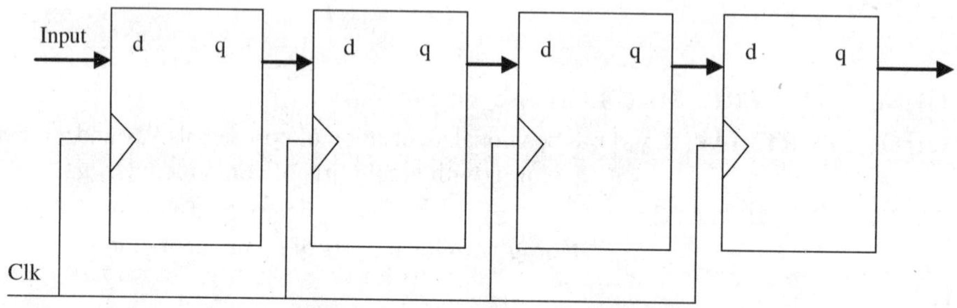

Figure 6.3 4-Bit Shift Register

Example 6.5 Generic shift register

```
LIBRARY IEEE;
USE IEEE.STD_LOGIC_1164.all;

ENTITY shift_reg IS
  GENERIC n: INTEGER:= 4;
PORT(d, clk,rst: IN STD_LOGIC;
        q: OUT STD_LOGIC);
END shift_reg;
```

```
ARCHITECTURE behave OF shift_reg IS        -- behavioural architecture model of the entity
  SIGNAL intmed: STD_LOGIC_VECTOR(n-1 DOWNTO 0);
BEGIN
  P1: PROCESS(rst,clk)                     -- declarative part of process
    BEGIN
      IF rst= '1' THEN
        intmed<= (OTHERS =>0);
      ELSIF clk'EVENT AND clk='1' THEN
        intmed<=d & intmed'LEFT DOWNTO 1;
      END IF; –
    END PROCESS P1;
END behave;
```

SUMMARY

- Structural modelling is the simplest modelling style which represents circuit as set of interconnected components.
- It is the lowest level of abstraction.
- It is a hierarchical modelling style.
- All lower-level entities are used as component in upper-level entity.
- Every component has a component declaration and a component instantiation.
- Component declaration declares the interface of the component while the component instantiation maps its local ports to signals of top-level entity.
- Component instantiation is identified by a label statement.
- Both positional association and named association can be used in the component instantiation.
- Generics is used to generalize a design.
- Generics refers to parameters which are likely to differ in instances of same design.
- Generic parameters are declared in the entity declaration and may or may not be initialized along with its declaration.
- Initialization of generic parameter should be done before using it.
- Generic map statement is required in case of generics of components to map local parameters of the component to actual parameters of the top-level entity.
- Generics can be used with behavioural modelling but no generic map is required for behavioural modelling.

QUESTIONS

Objective Questions

1. Structural style is .... level of abstraction.
   a. lowest b. highest c. intermediate d. upper
2. Structural modelling supports.............
   a. hierarchy b. flat models c. inheritance d. none of these

3. Component .... uses labels.
   a. declaration   b. instantiation   c. interface   d. definition
4. Structural modelling supports modularity. (True/False)
5. Port not connected to any signal is left as ....
   a. open   b. close   c. short   d. floating
6. .... are used for design parameterization.
   a. Generics   b. Components   c. Configurations   d. Entities
7. Generics are .... objects.
   a. signal   b. variable   c. constant   d. file
8. Default generic values are .... by values specified in a generic map.
   a. overwritten   b. overridden   c. unaffected   d. none of these

Descriptive Questions

1. Compare a structural style with other styles of modelling.
2. Explain positional and named association for binding actuals with component locals.
3. Write the VHDL structural code for 4:1 multiplexer. Draw the waveforms.
4. Design a decade counter using J-K flip-flop as the component.
5. What are generics? Explain how a component can be generalized using generics.
6. Write the VHDL code for generic parity generator and detector circuit.
7. Design the binary code to gray-code converter. Draw its truth table. Design it using structural style.
8. What is meant by component? Explain component instantiation.
9. Design a byte adder circuit using structural style alone.
10. Design a decoder circuit (generic) in structural style as well as in dataflow style. Simulate and synthesize both the designs and compare the results.
11. Write the VHDL code for 2t4 decoder. Draw the hardware generated and the waveforms of inputs and output.
12. Compare and contrast component declaration and component instantiation.
13. What are the advantages of generics?
14. Generics help in creating timing models of VHDL entities. Justify.
15. Design a n-bit UP counter inVHDL.
16. Design a guarded latch and use generic parameter to specify its set_up and hold _time.
17. In how many ways an entity can be bound with a component?
18. Write a VHDL code to design all gates using NAND gate as a component.

7

SUBPROGRAMS AND OVERLOADING

CHAPTER OBJECTIVES

- Introduction to modular programming
- Subprograms—functions and procedures; defining and using them
- Differences between usage of functions and procedures
- Pure and impure functions
- Function overloading
- Operator overloading

7.1 INTRODUCTION

Function and procedures are collectively known as subprograms. Subprograms are sequential in nature just like a process statement. They support the modular programming approach, where a piece of code is intended of performing a particular function, or tasks are written as a general purpose code; and these modules are stored as procedure and procedures which can be called whenever and whereever needed. Although procedure and procedures are collectively known as subprograms sharing same sequential, modular nature there exist some major differences between them. Subprograms allow reusability of code by allowing the storage in packages from where they can be called across different design units. Alternatively, subprograms can also be written in the main code.

Functions are user-defined modules for returning a single value whereas procedures may return zero or more than one values.

7.2 FUNCTIONS

Functions are sequential modules of code which perform a particular task and return a single output value. Every function has associated with it the following:

(i) **Function declaration:** Describes the interface of function with a unique function name, list of input values and their types and type of output expected.

(ii) **Function definition:** Defines the logic behind function performed by the function.

(iii) **Function call:** Calls the function for execution with actual parameters as inputs from its stored location.

(iv) **Actual parameters:** The run-time values which need to be processed by function and given at the time of function call.

(v) **Formal parameters:** The values of input as identified in the function declaration with a unique name and their data type.

(vi) **Return statement:** The last statement in the function body which returns the single expected output from the function.

SYNTAX:

```
FUNCTION Function_name(CLASS param1:TYPE; CLASS param2
[,param3 :TYPE) RETURN TYPE;                    -- FUNCTION DECLARATION

FUNCTION Function_name(CLASS param1: TYPE; CLASS param2
[,param3]:TYPE) RETURN TYPE                     -- FUNCTION DEFINITION
Declarations local to function body
BEGIN
Sequential statements;
RETURN output_val;
END Function_name;
```

As seen every function is identified by a unique user-defined name: *Function_name* after the keyword **FUNCTION**. Function name is followed by parenthesis which includes a list of input parameters, their class and their types. The object class of input parameters allowed are either SIGNAL or CONSTANT, default being CONSTANT (in case the class of object is not mentioned). By default the mode for all input parameters in the function is IN so that no explicit mention is required for the mode specification. More than one input parameters of different types can be separated by semicolons, but if there are more than one inputs required to function of the same data type then they can be separated by commas.

FOR EXAMPLE:

```
FUNCTION func_name (SIGNAL  a,b  :INTEGER; SIGNAL c :STD_LOGIC)
RETURN INTEGER;
```

Two inputs of the same data type separated by a comma.

Inputs of different data type separated by semicolons.

This is followed by the **RETURN** keyword and type-of-return output argument. In the function body after the declarator and before the beginning of the executable part of the the function body, marked by the keyword **BEGIN** is the declarative section of the function in which all the signals, variables and constants local to the function can be declared. Then the keyword **BEGIN** marks the beginning of the actual executable function body in which different logic statements are written in sequence, each statement ended by a semicolon. The last statement in the function body is the RETURN statement which returns a single value as the output of the function. Finally, the END statement is used to end the function.

Note: Variables are allowed with subprograms due to sequential nature.

Example 7.1 To show a simple function: To find the square of a number

```
FUNCTION sq_num( SIGNAL a: INTEGER) RETURN INTEGER
VARIABLE res: INTEGER;                  -------------
BEGIN
res:= a * a;
RETURN res;
END sq_num;

Function Call
......
x=5; where x, y are signals of type integer;
Y<= sq_num(x);
```

In the Example 7.1, the function named *sq_num* is created which takes only one input value of the type Integer a is formal input parameter and the function returns one value of type integer.

In the declarative section, a local variable 'res' of type integer is declared to be used inside the function itself. Inside the executable part of function body there are two sequential statements. First, res:=a*a which assigns the value of square of input parameter a to local variable res. In the next statement, Return is used to give the output when the type integer is returned.

A function is always called in the main code and the function call is generally an expression statement or assignment statement in which the value returned by the function is assigned to the signal or variable, provided the data types of assigned value (returned value of function) and assigning object (signal/variable) are the same. Function call statement can be a part of sequential code or concurrent code.

Note: All the input parameters have positional association between formal and actual parameters, that is, the order of passing formal parameters in function call should be the same as the order of actual parameters in the function declaration.

7.3 USING FUNCTIONS

A function can be written either as a part of a main code where it is required to be used or it can be stored in the form of a package and called in the main code by including the package containing the function in the entity under design. The function stored in the package allow reusability, code sharing and code partitioning whereas inside the main code the function has a limited use for the entity containing it alone.

7.3.1 Function in the Main Code

Function can be written as a part of main code either inside the entity declaration or inside the architecture body or inside the process itself.

7.3.1.1 Function Definition Inside the Entity Declaration

Inside the entity the whole function definition is written after a declaration of ports. No function declaration is required.The declarator of function definition is sufficient and the function call is made in architecture body executable part.

Example 7.2 Showing a function defined in entity for D flip-flop with asynchronous reset

```
LIBRARY IEEE;
USE IEEE.STD_LOGIC_1164.all;

ENTITY fnc_main IS
PORT(d,clk,rst: IN STD_LOGIC;
        q: OUT STD_LOGIC;
FUNCTION edge_rst( SIGNAL a,b: STD_LOGIC) RETURN STD_LOGIC
SIGNAL res: STD_LOGIC;                          -- FUNCTION DEFINITION
BEGIN
IF a = '1' THEN
  IF b'EVENT AND b='1' THEN
   res<='1';
  ELSE
   res<='0';
   END IF;
  END IF;
RETURN res;
END edge_rst;);
END fnc_main;

ARCHITECTURE fnc_main OF fnc_main IS
BEGIN
      IF edge_rst(rst,clk) THEN                  -- FUNCTION CALL
          q<=d;
     ELSE
          q<='0';
    END IF;
END fnc_main;
```

In the Example 7.2, function *edge_rst* is defined inside the entity declaration to check the aynchronous reset signal and clock signal. When reset signal is set and positive clock edge occurs, value of input 'd' is assigned to output 'q' of D flip-flop.

7.3.1.2 Function Definition Inside Architecture Body

Another alternative of writing the function as a part of the main code is to write it inside the architecture body inside its declarative region before the keyword **BEGIN**. Function call is then given inside the executable part of the architecture body. The following example (Example 7.3) shows function definition inside the architecture body.

Example 7.3 Function written inside the main code as part of architecture

```
LIBRARY IEEE;
USE IEEE.STD_LOGIC_1164.all;

ENTITY fnc_main IS
PORT(d,clk,rst: IN STD_LOGIC;
        q: OUT STD_LOGIC );
END fnc_main;

ARCHITECTURE fnc_main OF fnc_main IS

FUNCTION edge_rst( SIGNAL a,b: STD_LOGIC) RETURN STD_LOGIC
SIGNAL res: STD_LOGIC;                                  -- FUNCTION DEFINITION
BEGIN
IF a = '1' THEN
IF b'EVENT AND b='1' THEN
    res<='1';
  ELSE
    res<='0';
    END IF;
  END IF;
RETURN res;
END edge_rst;

BEGIN                                   -- beginning of executable part of architecture
     IF edge_rst(rst,clk) THEN          -- FUNCTION CALL
         q<=d;
     ELSE
         q<='0';
     END IF;
END fnc_main;
```

Similarly, a function can also be defined inside the process declarative region as a part of the main code.

7.3.2 Function Inside Package

Function can be written and stored inside a package which allows modular programming, code sharing across different design units and reusability of code. The entity under design which needs to use the function has to include the package containing the function with USE clause, and the function call with actual values is made inside the main code. Package is nothing but a storehouse of user-defined codes which need to be reused or shared across different design units. Every package has a package declaration where all declarations are made and package body where subprograms: function and function definitions are given. Generally, functions are written inside packages instead of main code allowing it to be reused.

Example 7.4 To show function written inside a package

```
LIBRARY IEEE;
USE IEEE.STD_LOGIC_1164.all;

PACKAGE fnc_pack IS
FUNCTION vect_int (SIGNAL x:BIT_VECTOR) RETURN INTEGER;
END fnc_pack;

PACKAGE BODY fnc_pack IS

FUNCTION vect_int (SIGNAL x:BIT_VECTOR) RETURN INTEGER IS
VARIABLE res:INTEGER:=0;
BEGIN
  FOR i IN 0 TO x'HIGH – 1 LOOP
    res:= res +( 2 ** i) * x(i);
  END FOR;
  RETURN res;
END vect_int;
END fnc_pack;
```

7.3.2.1 To Call the Function in Main Code in Entity Under Design

In the Example 7.4, a package named *fnc_pack* contains the function vect_int for converting a bit_vector into its integer equivalent. Function is first declared specifying its number and type of inputs and type of output, in the package declaration. The input to function is a bit_vector whose range is not specified making the function's general purpose, that is, it can be used to convert the bit_vector of any size into its integer equivalent. The function is then defined inside the package body. A local variable 'res' of the type integer is declared inside the function to store the results and is initialized to 0. In the FOR loop, the iterations are specified from 0 to s'high-1, where 0 is the lower index of bit_vector and high is the attribute which returns upper index of signal 's'. Finally, the logic for conversion to integer equivalent is given and the result is returned to variable 'res'.

To use this function an entity bit_int is created which inputs an 8-bit vector to be converted to its integer equivalent. For the conversion function *vect_int* is called inside the architecture body of entity. But note that before call is made to function in the beginning USE clause **USE WORK.fnc_pack.all;** is written which tells the design to include package *fnc_pack* from the design library work (default-user library).

7.3.3 Scope of Function

The scope of function depends on the location of the function definition—either as a part of the main code or inside a package. Function defined in the main code scope of function is limited to its containing design unit alone, that is, if inside an architecture body alone that architecture of the entity can access the function, if inside entity declaration all the architectures of that entity can access the function and if inside a process alone that particular process alone can access that function. On the other hand, the functions defined inside a package are accessible across different design units.

7.4 TYPES OF FUNCTION

In VHDL, any function can be classified either as

(i) PURE
(ii) IMPURE

A function is said to be pure if it returns the information through a return statement and for the same values of input parameters, the value returned by the function is always the same, irrespective of number of times the function call is made; for example, a function to compute the square of a number is a pure function because any number say 4, will have its square 16 always. So, according to the logic implemented in the function, pure procedure gives the same fixed output for a fixed set of inputs. Though keyword **PURE** is used for pure procedure before keyword **FUNCTION** in the function declaration, there is no need for explicitly making a function pure as by default all functions are considered to be pure. Pure functions have a restriction imposed on them that they cannot access any variables and signals declared in its parent unit (architecture/subprogram/process) in which the function is nested, else variables/signals may change values between calls.

On the other hand, impure procedures are procedures which generally give different, random output for the same set of inputs each time a function call is made. Unlike the pure procedure, impure procedure can access the signals, variables in the parent unit (architecture, subprogram/process) in which the function is nested having the scope outside the function. For a function to be impure keyword **IMPURE** has to be included before the keyword **FUNCTION** in the declaration. For example, of IMPURE function is random number generator function which produces a random number in the given range every time the function call is made.

```
PURE/IMPURE FUNCTION Function_name( CLASS param1:TYPE; CLASS
param2 [,param3 :TYPE ) RETURN TYPE;              -- FUNCTION DECLARATION

PURE/IMPURE FUNCTION Function_name(CLASS param1: TYPE; CLASS
param2 [,param3]:TYPE) RETURN TYPE                -- FUNCTION DEFINITION
Declarations local to function body
BEGIN
Sequential statements;
RETURN output_val;
END Function_name;
```

As all procedures by default are pure all the procedures used previously are also pure procedures. Now let us see Example 7.5 for an impure function.

Example 7.5 For impure function

```
LIBRARY IEEE;
USE IEEE.STD_LOGIC_1164.all;

ENTITY fnc_impure IS
PORT(clk,rst: IN STD_LOGIC;
        op: OUT NATURAL );
END fnc_impure;

ARCHITECTURE fnc_impure OF fnc_impure IS
BEGIN
PROCESS
CONSTANT seq_mod:NATURAL:=2**5;
SUBTYPE seq_no IS NATURAL RANGE 0 TO seq_mod-1;
VARIABLE num: seq_no ;

IMPURE FUNCTION rand_gen RETURN seq_no IS
VARIABLE n: seq_no ;              -- FUNCTION DEFINITION
BEGIN
   num:=(num + 1) MOD seq_mod;
   n:= num;
RETURN n;
END rand_gen;

BEGIN
IF rst = '0' THEN
   op<= 0;
ELSIF clk'EVENT AND clk='1' THEN
   op <= rand_gen( );
END IF;
END PROCESS;
END fnc_impure;
```

7.5 PROCEDURES

Procedures are subprograms with difference from a procedure that procedures can return zero or more values. Rules for procedures are similar to that of a procedure with minor differences governing the mode of input parameters and output values returned. Every procedure is associated with following:

(i) **Procedure declaration:** Describes the interface of a function with a unique function name, list of values, their mode and their types.

(ii) **Procedure definition:** Defines the logic behind a function performed by the procedure.

(iii) **Procedure call:** Calls the procedure for execution with actual parameters as inputs from its stored location.

(iv) **Actual parameters:** The run-time values which need to be processed by procedure and given at the time of procedure call.

(v) **Formal parameters:** The values passed as arguments to procedure declaration with their mode specified, identified with a unique name and their data type. Even the output expected is passed as an argument to the procedure.

SYNTAX:

```
PROCEDURE proc_name( CLASS param1:MODE TYPE; CLASS param2
[,param3 : MODE TYPE );                              -- PROCEDURE DECLARATION

PROCEDURE proc_name( CLASS param1:MODE TYPE; CLASS param2
[,param3 : MODE TYPE ) IS                            -- PROCEDURE DEFINITION
Declarations local to procedure body
BEGIN
Sequential statements;
END proc_name;
```

As seen, every procedure is identified by a unique user-defined name: *proc_name* after the keyword **PROCEDURE**. Procedure name is followed by parenthesis which includes list of input parameters, their class, their mode and their data types. All four object classes are allowed, namely, SIGNAL, VARAIBLE, CONSTANT or FILE with input parameters, default being CONSTANT (in case the class of object is not mentioned). Unlike in procedure MODE of input parameters to procedure has to be specified. Three modes are allowed—IN , INOUT and OUT, default mode for all input parameters in procedure is IN for objects of class SIGNAL, CONSTANT or FILE. For object of class VARIABLE only modes INOUT or OUT are allowed. More than one input parameters of different types can be separated by semicolons, but if there are more than one inputs required to procedure of same data type they can be separated by commas.

FOR EXAMPLE:

```
PROCEDURE func_name (SIGNAL a,b :IN INTEGER; SIGNAL c : OUT
STD_LOGIC);
```

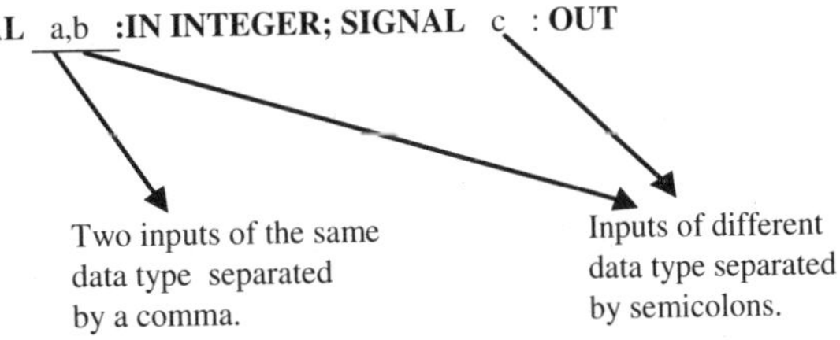

Note: There is RETURN statement.The output value to be returned is passed as an input parameter itself. In the procedure body after the declarator before the beginning of an executable part of the procedure body, marked by keyword **BEGIN** is the declarative section of the procedure in which all the signals, variables and constants local to the procedure can be declared. Then keyword **BEGIN** marks the beginning of actual executable procedure body in which different logic statements are written in sequence, each statement ended by a semicolon. END statement is used to end the procedure.

Note: Variables allowed with subprograms due to sequential nature.

Example 7.6 Example of a simple procedure to find the square of a number

```
PROCEDURE sq_num ( SIGNAL a: IN INTEGER;VARIABLE b: OUT
INTEGER) IS

        -------------

BEGIN
b:= a * a;
END sq_num;

Procedure Call
....
x=5; where x are signals of type integer;
VARIABLE y : INTEGER:=0;

sq_num(x,y);            -- procedure call
```

In the Example 7.6, the procedure named *sq_num* is created which takes two arguments— 'a' is the formal parameter of type integer that serves as input due its mode IN and 'b' is the formal parameter of type Integer with mode as OUT used to give output. Inside the executable part of procedure body there are two sequential statements. First, b:=a*a, assigns the value of square of input parameter 'a' to variable 'b'. END is used to end procedure.

Procedure is always called in the main code and note that procedure call is not part of the expression or assignment statement as in the case of function; instead, procedure call is an independent sequential statement.

7.6 USING PROCEDURES

Like functions, procedure can be written either as part of the main code where it is required to be used or it can be stored in the form of a package and called in the main code by including the package containing the procedure in the entity under design. Procedure stored in package allow reusability, code sharing and code para titioning whereas inside the main code function has limited use for the entity containing it alone.

Example 7.7 Using procedure for comparing two integer values

```
LIBRARY IEEE;
USE IEEE.STD_LOGIC_1164.all;

ENTITY proc_main IS
PORT(inp1,inp2: IN INTEGER;
      High,low: OUT INTEGER);
END proc_main;
```

```
ARCHITECTURE proc_main OF proc_main IS

PROCEDURE compare( SIGNAL n1,n2: IN INTEGER;SIGNAL h_op,l_op: OUT
INTEGER) IS

BEGIN
 IF (n1>n2) THEN
   h_op<=n1;
   l_op <= n2;
 ELSE
   h_op<=n2;
   l_op <= n1;
 END IF;
END  compare;

BEGIN

 compare (inp1,inp2,high,low);                    -- procedure call

END proc_main;
```

In the Example 7.7, the procedure compare is defined inside the main code inside the architecture body. An entity is declared with ports to input two integer values and output ports to give higher of the two values at port 'high' and lower of two values at port 'low'. Inside the declarative part of architecture body procedure compare is defined which takes four arguments – two input n1,n2 of type integer and two arguments in OUT mode h_op and l_op again of type integer. In the procedure body IF –ELSE is used to check the higher of the two values. The greater of the two values is assigned to h_op and lower to l_op.

Once the procedure is defined it is to be called inside the executable part of architecture body .The procedure call is made using sequential statement—compare (inp1,inp2,high,low);

> **Note:** It is an independent sequential statement and not part of any expression. Also the entity ports are passed as actual arguments to the procedure in the call in the same order as arguments are specified in the procedure definition, that is n1=>inp1,n2=>inp2,h_op=>high,l_op=>low.

Example 7.8 To show procedure located in the package body

```
LIBRARY IEEE;
USE IEEE.STD_LOGIC_1164.all;

PACKAGE proc_pack IS
PROCEDURE seven _seg_ctrl (SIGNAL x: IN STD_LOGIC_VECTOR
(3 DOWNTO 0); SIGNAL y: OUT STD_LOGIC_VECTOR(7 DOWNTO 0));
END proc_pack;
```

```
PACKAGE BODY  proc_pack IS
PROCEDURE seven _seg_ctrl (SIGNAL x: IN STD_LOGIC_VECTOR
(3 DOWNTO 0) ; SIGNAL y: OUT STD_LOGIC_VECTOR(7 DOWNTO 0)  IS
BEGIN
  CASE x WHEN "0000"=>  y<= "00000001";  -- 0( all outputs are active low)
          WHEN "0001"=>  y<="11001111"; -- 1
          WHEN "0010"=>  y<= "10010010"; -- 2
          WHEN "0011"=>  y<= "10000110"; -- 3
          WHEN "0100"=>  y<= "11001100"; -- 4
          WHEN "0101"=>  y<= "10100100"; -- 5
          WHEN "0110"=>  y<= "10100000"; -- 6
          WHEN "0111"=>  y<= "10001111"; -- 7
          WHEN "1111"=>  y<= "10000000"; -- 8
END CASE;
END seven _seg_ctrl;
END proc_pack;
```

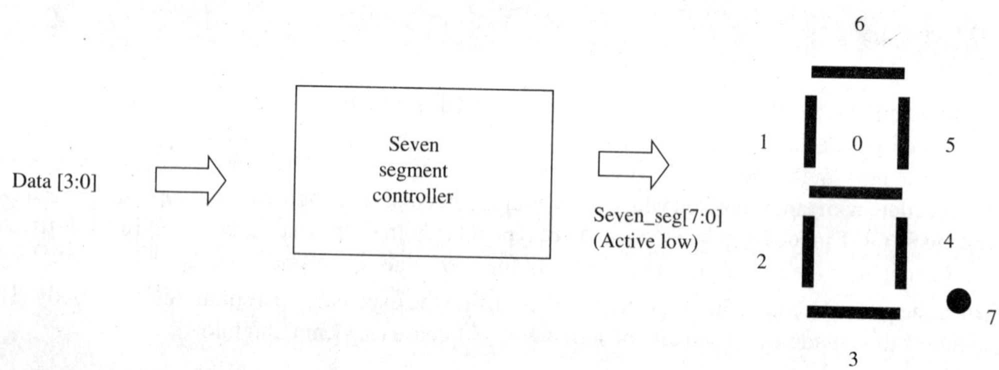

To call the procedure in the main code

```
LIBRARY IEEE;
USE IEEE.STD_LOGIC_1164.all;
USE WORK.proc_pack.all;          -- including the package containing required function
ENTITY sev_seg IS
PORT(a: IN STD_LOGIC_VECTOR(3 DOWNTO 0);
      b: OUT STD_LOGIC_VECTOR(7 DOWNTO 0));
END sev_seg;
ARCHITECTURE sev_seg OF sev_seg IS
BEGIN
  seven _seg_ctrl(a,b);                          -- procedure call
END sev_seg;
```

In the Example 7.8, a procedure named *seven_seg_ctrl* is created and stored inside a package *proc_pack*. The procedure is created to control seven-segment display using a 4-bit vector. Since seven segments are to be controlled, 4-bit controlling vector is sufficient assuming all inputs of seven-segment display to be active low. CASE statement is used in procedure to control the display. To call the procedure in the main code, USE clause is used to include package *proc_pack* from the design library work (default-user library). Since package contains only one procedure and no other items the whole of the package needs to be included; so, *.all* is used after the package name. Procedure is called inside the architecture body with its inputs mapped to actual arguments drawn from the entity with 'a' serving as input 'x' and 'b' serving as output 'y' of procedure.

A disadvantage of using procedures over functions is that procedures can access values of objects from their parent unit. Say a procedure is embedded inside architecture or process. It can use all the objects declared inside its parent design unit, even if the object is not passed as an input argument to the procedure. So the procedure may access signals from the architecture even if those signals are not present in the argument list of procedure. This may lead to undesirable, unintentional changes or accidental changes in such accessed objects.

7.7 DIFFERENCES BETWEEN FUNCTIONS AND PROCEDURES

Though procedures and functions are collectively referred to as subprograms and share a common sequential nature, there are subtle differences between functions and procedures which help the user to decide when and where to use. Both functions and procedures support modular programming, code sharing and code reusability. Both have same locations for definition.

The differences are listed in a tabular form as follows:

Functions	Procedures
1. Function returns a single value, through RETURN statement.	Procedures return zero or more values which are returned through parameters passed to procedure and no RETURN statement is used.
2. Functions may accept zero or more input arguments with default mode IN, only class and type of parameters need to be specified in function declaration.	Since procedures return values through its passed arguments only IN, INOUT and OUT mode of arguments is allowed. For every argument passed mode, class and type need to be specified in the declaration.
3. Input parameters passed to functions are essentially a signal or a constant (default) object.	Parameters passed to procedure can be a signal, constant, variable or file object. Default for IN mode: constant and for OUT/INOUT: Variable.
4. A function call is always a part of an expression or assignment statement.	A procedure call is an independent sequential statement.
5. Executes in zero simulation time as wait statements are not to be used in functions.	May or may not simulate in zero simulation time as procedures can have a wait statement.

Note: Wait and components are both not synthesizable in functions as well as procedures.

7.8 OVERLOADING

Overloading is the advanced concept of VHDL which helps to improve code readability and extends the capabilities of VHDL to help the designer overcome the cumbersome process of using and remembering different names for same operation being performed on different types of data or extending the basic operation to different number or user-defined operands. So it allows the same design name to be used in multiple forms. It is the concept of polymorphism in procedural languages. Overloading can be implanted either as

(i) Subprogram overloading
(ii) Operator overloading

7.8.1 Subprogram Overloading

In subprogram overloading same function/procedure can be used with different arguments. Arguments in overloaded form can differ either in

(i) Number of arguments
(ii) Data type of arguments
(iii) Both number and data type of arguments
(iv) Different return value, if any

Say a function to compare two integer values and higher of two values is made. The same function can also be written with the same name for arguments of type Bit_vector.

```
FUNCTION compare(a,b:INTEGER) RETURN INTEGER;
FUNCTION compare(x,y:BIT_VECTOR) RETURN BIT_VECTOR;
```

The form of function used is decided by the compiler at the compile time by matching the type and number of arguments in the function call with existing forms. The form that matches with the forma arguments is invoked at run-time.

Let us see Example 7.9 to understand the concept.

Example 7.9 Overloading a function with different types of arguments

```
LIBRARY IEEE;
USE IEEE.STD_LOGIC_1164.all;

ENTITY fnc_overload IS
PORT(n1,n2: IN INTEGER;
     x1,y1:IN BIT_VECTOR;
     Z_high:OUT BIT_VECTOR;
     high: OUT INTEGER );
END fnc_overload;
```

```
ARCHITECTURE fnc_overload OF fnc_overload IS

FUNCTION compare ( a,b: INTEGER) RETURN INTEGER IS
VARAIBLE res:INTEGER;
BEGIN                                              -- FUNCTION DEFINITION
   IF a>b THEN
      res:= a;
   ELSE
      res:=b;
   END IF;
RETURN res;
END compare;

FUNCTION compare ( a,b: BIT_VECTOR) RETURN BIT_VECTOR IS
VARAIBLE res: BIT_VECTOR;
BEGIN

FOR i IN RANGE 0 TO a'HIGH-1 LOOP
                                                   -- FUNCTION DEFINITION
   IF a(i)>b(i) THEN
      res:= a;
      EXIT;
   ELSIF b(i) >a(i) THEN
      res:=b;
      EXIT;
   ELSE
      NEXT;
END IF;
END FOR;

RETURN res;
END compare;

BEGIN
   high <= compare(n1,n2);                  -- call function to compare two integer values
   z_high <= compare(x1,y1);                -- calls function to compare two bit vectors

END fnc_overload;
```

In Example 7.9, two functions with same name 'compare' but with different type of input and return parameters are defined inside the same architecture body. The first function 'compare' compares two integers and returns the greater of two as output whereas second function 'compare' compares two bit_vectors and returns greater of two as output.

Example 7.10 To overload a function with different number of arguments

```
LIBRARY IEEE;
USE IEEE.STD_LOGIC_1164.all;

PACKAGE pack_over IS
FUNCTION compare ( a,b: INTEGER) RETURN INTEGER;
FUNCTION compare ( a,b,c: INTEGER) RETURN INTEGER;
FUNCTION compare ( a,b,c,d: INTEGER) RETURN INTEGER;
END pack_over;

PACKAGE BODY pack_over IS

FUNCTION compare ( a,b: INTEGER) RETURN INTEGER IS
VARAIBLE res:INTEGER;
BEGIN                                         -- FUNCTION DEFINITION
   IF a>b THEN
     res:= a;
   ELSE
     res:=b;
   END IF;
RETURN res;
END compare;

FUNCTION compare ( a,b,c: INTEGER)v RETURN INTEGER IS
VARAIBLE res1,res:INTEGER;
BEGIN                                         -- FUNCTION DEFINITION
   res1:= compare(a,b);
   res:=compare(res1,c);
RETURN res;
END compare;

FUNCTION compare ( a,b,c,d: INTEGER) RETURN INTEGER IS
VARAIBLE res1,res2,res:INTEGER;
BEGIN                                         -- FUNCTION DEFINITION
   res1:= compare(a,b);
   res2:=compare(res1,c);
   res:=compare(res2,d);
RETURN res;
END compare;
END pack_over;
```

To call the functions in the main code

```
LIBRARY IEEE;
USE IEEE.STD_LOGIC_1164.all;
USE WORK.pack_over.all;            -- including the package containing the required function

ENTITY over_load IS
PORT (a0,a1,a2,a3: IN INTEGER;
        o1,o2, o3: OUT INTEGER);
END over_load;
ARCHITECTURE over_load OF over_load IS

BEGIN
   o1<= compare (a0, a1);
   o2<= compare (a0, a1, a2);
   o3<= compare (a0, a1, a2, a3);                              -- function call
END over_load;
```

In the Example 7.10, three forms of function compare are created and stored in the package pack_over. The three forms of function have the same type of arguments, that is, integer and same return-type integer, but the number of arguments is different. The first form accepts two inputs, second accepts three arguments and third form accepts four input arguments.

To call the function in the main code all the three forms are invoked by the compiler by matching the number of arguments with the available function definitions.

7.8.2 Overloading Operators

Another form of overloading is to overload the existing built-in operators to be used with user-defined data types or extending the operation to operands meeting user requirements. A very common example is bit_vector addition which is not supported in VHDL, so addition operator can be overloaded to add two bit_vectors with the help of operator overloading.

SYNTAX:

```
FUNCTION "operator symbol" ( CLASS param1:TYPE; CLASS
param2 :TYPE ) RETURN TYPE;                       -- DECLARATION

FUNCTION "operator symbol" ( CLASS param1:TYPE; CLASS
param2 :TYPE ) RETURN TYPE IS                     -- DEFINITION
Declarations local to function body
BEGIN
Sequential statements;
RETURN output_val;
END "operator symbol";
```

Syntax for operator overloading is the same as creating a function with a difference that instead of function the name symbol of operator to be overloaded is used. They are included in double quotes. "*operator symbol*" can be any operator to be overloaded say, "+", "and", "*", and so on, and parameters passed are the type and number of operands for the operator to be overloaded. The return type is the type of result expected from the overloaded operator (generally the same as the type of inputs used), say for the function to overload to add 2-bit_vectors the result will be of type bit-vector only. Example 7.11 shows overloading of addition operator '+' to add two-bit vectors.

Example 7.11 Overloading an operator

```
LIBRARY IEEE;
USE IEEE.STD_LOGIC_1164.all;
ENTITY op_overload IS
PORT (x1,y1:IN BIT_VECTOR;
      Z:OUT BIT_VECTOR);
END op_overload;

ARCHITECTURE op_overload OF op_overload IS

   FUNCTION "+" ( a,b: BIT_VECTOR) RETURN BIT_VECTOR IS
    VARAIBLE carry: BIT:='0';
    VARAIBLE sum: BIT_VECTOR;
   BEGIN
    FOR i IN RANGE 0 TO a'HIGH-1 LOOP
        sum(i):=a(i) XOR b(i) XOR carry;
        carry:= (a(i) AND b(i) ) OR (a(i) AND carry ) OR (carry AND b(i) );
    END FOR;
   RETURN sum;
END "+";
BEGIN
   z<= x1 + y1;                                    -- function call
END op_overload;
```

SUMMARY

- Procedures and functions are sequential codes collectively known as subprograms.
- Functions return single value while procedures return zero or more values.
- Rules governing the use and location of procedures and functions are the same.
- Subprograms can be defined inside a main code or inside a package.
- Functions call is always an expression or an assignment statement while procedure call is an independent statement.

- Overloading of subprograms allows same names to be used for performing same operation on different type of arguments.
- Overloaded programs differ in arguments in data type or in number or in the type of values returned.
- Predefined operators can also be overloaded providing language an extensibility to be used with different data types including user-defined types.

QUESTIONS

Objective Questions

1. Subprograms are .... constructs.

   a. concurrent b. sequential c. hierarchical d. none of these
2. Subprograms refer to ....

   a. functions b. procedures c. both d. none
3. Functions and procedure differ in ....

   a. input arguments b. output arguments c. subprogram call d. all of the above
4. Function call is always a/an .... statement.

   a. concurrent b. assignment c. sequential d. none
5. Wait statement can be used inside ....

   a. function b. procedure c. both d. none
6. .... functions have a restriction imposed on them that they cannot access any variables and signals declared in its parent unit.

   a. Recursive b. Pure c. Impure d. Nested
7. Default class of objects for input arguments of procedure is ....

   a. variable b. constant c. signal d. file
8. .... overloading features allow the reusability of operators with different types.

   a. Operator b. Function c. Procedure d. Subprogram
9. .... parameters are used at the time of function call, while .... parameters are used in a function definition.

   a. Formal, actual b. Actual, formal c. Formal, formal d. Input, output

Descriptive Questions

1. What is the difference between functions and procedures? Explain with example.
2. What is meant by operator overloading. Give example.
3. Discuss the use of function for type conversion.
4. What is overloading? Explain overloading in VHDL.
5. Write down the syntax for creating a function. Show how to create function and call it with an example.
6. What are the advantages of using subprograms?
7. Write short notes on:

   a. Procedures b. Subprogram overloading

8. Design an entity to convert an integer to bit_vector such that the architecture of an entity is modelled using behavioural and dataflow modelling and both architectures make use of the same function for vector to integer conversion.
9. Write down a VHDL code for generic n-bit multiplier creating a function for multiplication.
10. Write a procedure to convert J-K flip-flop to D flip-flop.
11. Write a VHDL code to overload ">"l operator to find the smaller of two std_logic_vector numbers. Call this overloaded operator to arrange an array of STD_LOGIC_VECTOR (0 to 7) of 10 values.
12. Compare and contrast the modes of input parameters that are allowed in functions and procedures. Justify with examples.
13. What is the difference between variables declared inside a process and inside a function?
14. Write a function to perform arithmetic and logical operations on 8-bit std_logic_vector. Define it inside a process as a ALU module.
15. Write a VHDL module for memory (ROM). What is used: function or aprocedure and why?
16. Write a function to convert std_logic_vector to its integer equivalent. Generalize it so that any length vector can use it.

8 CONFIGURATIONS AND PACKAGES

CHAPTER OBJECTIVES

- Concept of configurations and their significance
- Using configuration for entity–architecture binding
- Using configuration for entity–component binding
- Learning about packages and their use
- Concept of deferred constants

8.1 INTRODUCTION TO CONFIGURATIONS

Configurations are advanced concepts introduced in VHDL which extends the flexibility and usability of VHDL language. Though it is not a simulation concept it provides the late binding capability to VHDL by associating an entity with multiple architecture views and binding an entity–architecture pair at run time as specified by the user. As the name suggests it configures at the execution or simulation time the binding between entity and one of its many existing architecture models.

8.1.1 Need for Configurations

In HDL any entity to be designed has two essential parts—an entity declaration specifying its external interface and an architecture body which forms the basis of its internal structure or functionality. Depending on the level of abstraction required for an entity to be designed, the user has a choice to use any of the allowed styles of modelling for architecture and at times, different architecture implementations of the same entity may be needed. In such a case, the configuration provides the mechanism for binding the entity with its particular architecture for execution. Therefore, configuration is a collective description of bindings between an entity and an architecture which can act as a single design unit at the time of simulation; in case of absence of default, entity–architecture multiply the architecture bodies of the same entity. Suppose a simple full adder is to be designed in VHDL.

As shown in the Figure 8.1, the entity has a declaration common to its all four styles of modelling, namely, structural, dataflow, mixed and behavioural. But at the time of execution, the entity needs to be binded with only one architecture and this function is done by configuration, as Configuration A binds the entity with its structural model.

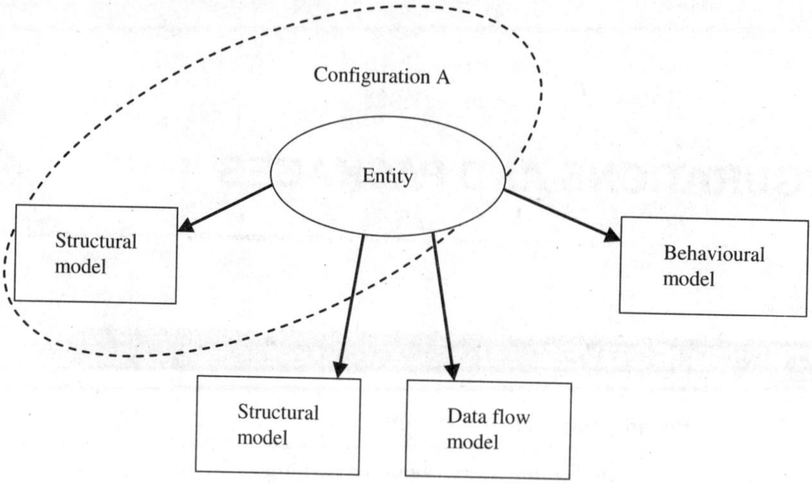

Figure 8.1 Configuration Binding

Secondly, configurations have a special functional with regard to structural style, as they are also used for binding a component with an entity. For example, let us say a basic XOR gate entity is designed as a component to be used in half adder but there is another entity design of a comparator in the same project which also requires XOR gate as its component. So, instead of declaring the XOR gate separately for both the entities, namely, half adder and comparator, the single component of XOR gate can be defined and in the entity under design be binded with the entity using configuration.

8.2 WRITING CONFIGURATION

Every configuration is comprised of a configuration specification which defines a unique design description from various design units by specifying an entity–architecture pair association.

Configuration specification is of two types:

(i) Specification for entity–architecture pair (default configurations)
(ii) Specification for component–entity binding (component configurations)

Syntax for Configuration Specification

CONFIGURATION *name* **OF** *Entity_name* **IS**
FOR *name of architecture associated*
  [**FOR** *label of component in port map: name of the component to be binded]* to be used only for structural designs
  [**USE ENTITY**.*name of design Library. Entity-name2(Architecture name)];*
[**END FOR**;]
END FOR;
END *name;*

Every configuration binding is identified by a unique user-defined name: *name* followed by the name of the entity: *entity_name*. Body of configuration specification consists of FOR clause which tells the name of the architecture that is to be binded with the entity. Further, in case of structural style components are to be binded with entity under design. A 'FOR' clause is included for every instance of the component to be binded with the entity. This FOR clause includes label of the component instantiation, followed by the name of the component. This is followed by USE clause specifying the location of that component definition as an entity in terms of *design library.entity name2(architecture_name)*. Every FOR clause has to be ended with an END FOR statement. And finally configuration specification is ended with an END configuration *name*.

8.2.1 Entity–Architecture Pair Configuration

Entity–architecture pair configuration is also known as default configuration specification wherein an entity is associated with one of its many existing different architecture models through configuration specification. It is written in the entity design file itself after the architecture body.

Example 8.1 Configuration specification for entity–architecture pairing

```
LIBRARY IEEE;
USE IEEE.STD_LOGIC_1164.all;
ENTITY config IS
PORT(a: IN STD_LOGIC;
      b: OUT STD_LOGIC);
END config;
ARCHITECTURE config1 OF config IS          -- dataflow architecture model of the entity
BEGIN
b<=not(a) ;
END config1;
ARCHITECTURE config2 OF config IS          -- behavioural architecture model of the entity
SIGNAL clk: STD_LOGIC;
BEGIN
PROCESS(clk)
IF clk'EVENT AND clk='1' THEN
b<=not(a) ;
END IF;
END PROCESS;
END config2;
CONFIGURATION dataflow OF config IS        -- configuration specification for dataflow
                                              architecture and entity binding

FOR Config1
  END FOR;
  END dataflow;
CONFIGURATION behave OF config IS          -- configuration specification for behavioural
                                              architecture and entity binding

FOR Config2
END FOR;
END behave;
```

SYNTAX:

In this example, the entity *config* has two architecture models—dataflow and behavioural. To bind the entity with architecture, two separate configuration specifications have been written—one for each architecture style, namely, dataflow and behave. At the time of simulation, the user will be prompted to choose one of the two available configuration specifications and one selected will be put to execution. Since no components are to be binded in both the specifications only one FOR clause is used which tells the name of the architecture to is to be binded, for example, in the configuration *dataflow* entity config is binded with architecture *Config1* and in configuration *behave* the entity config is binded with architecture *Config2*.

8.2.2 Component–Entity Binding Configuration Specification

The binding between a component and an entity is another function of configuration specification that is particular to the structural style of architecture modelling. Component–entity binding can be specified in two ways as follows:

(i) Binding for entity–architecture pair
(ii) Binding for lower-level configurations

Binding for Entity–Architecture Pair: If a configuration specification binds the structural description to an entity then the information about the instantiated components of the entity is required to be given in the configuration specification.

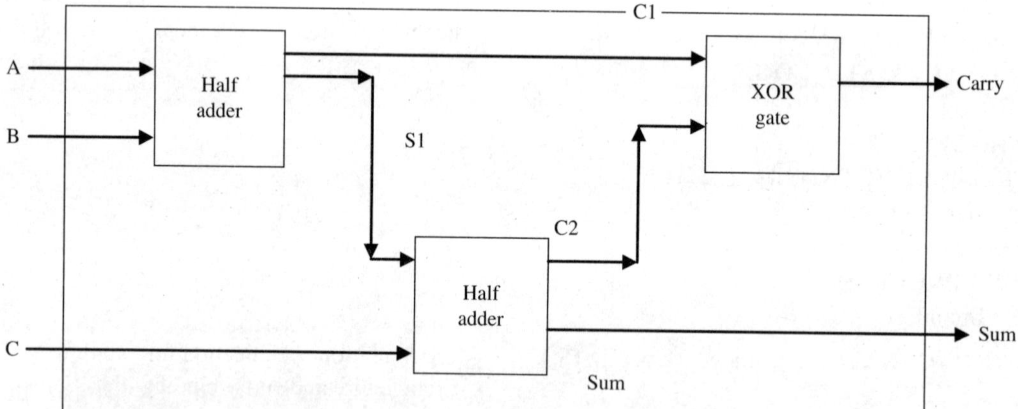

Figure 8.2 Full-adder Circuit Made of Components: Two Half Adders and a XOR Gate

Consider the structure of a full adder circuit as shown in Figure 8.2, comprising two half adders and a XOR gate. In this example, the half adder exists as an separate entity and is used in full adder as a component twice. Similarly, XOR gate exists as an entity and is used as a component in the full adder.

Using configuration specification for binding components half adder and XOR gate to full adder, following VHDL coding, is written assuming we have the following code for half adder and XOR gate as an entity in the same design file.

Example 8.2
Code for Half Adder

```
LIBRARY IEEE;
USE IEEE.STD_LOGIC_1164.all;
ENTITY Half_adder IS
PORT(I1,I2: IN STD_LOGIC;
        S,Ca: OUT STD_LOGIC);
END Half_adder;
ARCHITECTURE dataflow OF Half_adder IS
BEGIN
S <= I1 XOR I2;
Ca<= I1 AND I2;
END dataflow;
```

Code for XOR gate

```
LIBRARY IEEE;
USE IEEE.STD_LOGIC_1164.all;
ENTITY Xor_gate IS
PORT(x,y: IN STD_LOGIC;
        z: OUT STD_LOGIC);
END Xor_gate;
ARCHITECTURE dataflow OF Xor_gate IS
BEGIN
z <= x XOR y;
END dataflow;
```

Based on these coding for half adder and XOR gate, a configuration specification is written to bind these entities, namely, half adder and XOR_gate as components in Full adder design.

Example 8.3 Full adder design based on component binding for structural modelling

```
LIBRARY IEEE;
USE IEEE.STD_LOGIC_1164.all;
ENTITY fulladder IS
PORT(A,B,C: IN STD_LOGIC;
        Sum,Carry: OUT STD_LOGIC);
```

```
END fulladder;
ARCHITECTURE structural OF fulladder IS
SIGNAL C1,S1,C2: STD_LOGIC;
COMPONENT Half_adder IS
PORT(I1, I2:IN STD_LOGIC;
      S,Ca: OUT STD_LOGIC);
END COMPONENT;

COMPONENT xor_gate IS
PORT(x, y:IN STD_LOGIC;
      z: OUT STD_LOGIC);
END COMPONENT;

BEGIN
INST_HA1:Half_adder PORT MAP(A,B,S1,C1);
INST_HA2:Half_adder PORT MAP(S1,C,Sum,C2);
INST_XOR:xor_gate PORT MAP(C1,C2,Carry);
END Structural;

CONFIGURATION config OF fulladder IS
FOR Structural
  FOR INST_HA1, INST_HA2:Half_adder
    USE ENTITY work.Half_adder(dataflow);
  END FOR;
FOR INST_XOR:Xor_gate
USE ENTITY work.Xor_gate(dataflow);
  END FOR;
END FOR;
END config;
```

Binding for Lower-level Configurations: Lower level configuration is useful when components which are to binded to the entity under design are binded using their own configurations, that is, let us say a full adder circuit is made up of two half adders and a XOR gate as components; but when the half adder is defined as an entity again it can be done in more than one ways by any style of modelling and can have its own configuration specification for the entity–architecture binding. So, when the full adder uses the half adder as a component and binding has to be specified for it, then any configuration of half adder can be binded to full adder instead of half adder architecture itself. This may cause nesting of configuration specifications with full adder configuration specification using configuration specification of half adder.

Let us see the following example in the coding form:

Code for Half_adder

```
LIBRARY IEEE;
USE IEEE.STD_LOGIC_1164.all;
ENTITY Half_adder IS
PORT(I1,I2: IN STD_LOGIC;
        S,Ca: OUT STD_LOGIC);
END Half_adder;

ARCHITECTURE dataflow OF Half_adder IS
BEGIN
S <= I1 XOR I2;
Ca <= I1 AND I2;
END dataflow;

ARCHITECTURE behave OF Half_adder IS
SIGNAL clk:STD_LOGIC;
BEGIN
PROCESS(clk)
IF clk'EVENT AND clk='1' THEN
S <= I1 XOR I2;
Ca <= I1 AND I2;
END IF;
END behave;
CONFIGURATION config_data OF Half_adder IS    -- configuration specification for dataflow
                                                 architecture and entity binding
FOR dataflow
END FOR;
END config_data;
CONFIGURATION config_behave OF Half_adder IS    -- configuration specification for
                                                   behavioural architecture and entity
                                                   binding
FOR behave
END FOR;
END config_behave;
```

Now, based on these configuration specifications of half adder and xor gate, configuration specification for binding components half adder and xor gate to entity full adder is:

Code for XOR gate configuration

```
LIBRARY IEEE;
USE IEEE.STD_LOGIC_1164.all;
```

```
ENTITY XOR_gate IS
PORT(x,y: IN STD_LOGIC;
      z: OUT STD_LOGIC);
END XOR_gate;

ARCHITECTURE dataflow OF XOR_gate IS
BEGIN
z <= x XOR y;
END dataflow;

CONFIGURATION XOR_data OF Half_adder IS      -- configuration specification for data-
                                                flow architecture and entity binding
 FOR dataflow
 END FOR;
```

Example 8.4 Example for component entity binding using lower level configurations

```
END XOR_data;
LIBRARY IEEE;
USE IEEE.STD_LOGIC_1164.all;

ENTITY fulladder IS
PORT(A,B,C: IN STD_LOGIC;
      Sum,Carry: OUT STD_LOGIC);
END fulladder;

ARCHITECTURE structural OF fulladder IS
SIGNAL C1,S1,C2: STD_LOGIC;
COMPONENT Half_adder IS
PORT(I1, I2:IN STD_LOGIC;
      S,Ca: OUT STD_LOGIC);
END COMPONENT;
COMPONENT xor_gate IS
PORT(x, y:IN STD_LOGIC;
      z: OUT STD_LOGIC);
END COMPONENT;

BEGIN
INST_HA1:Half_adder PORT MAP(A,B,S1,C1);
INST_HA2:Half_adder PORT MAP(S1,C,Sum,C2);
INST_XOR:xor_gate PORT MAP(C1,C2,Carry);
END Structural;
```

```
CONFIGURATION config_style2 OF fulladder IS
FOR Structural
   FOR INST_HA1, INST_HA2:Half_adder
     USE CONFIGURATION work.config_data;
  END FOR;
  FOR INST_XOR:Xor_gate
    USE CONFIGURATION work.xor_data;
  END FOR;
 END FOR;
END config_style2 ;
```

So, as seen in the example configurations config_data and XOR_data have been used, respectively, for component–entity binding for half adder and XOR gate, respectively.

8.3 INTRODUCTION TO PACKAGES

Packages are concepts introduced in VHDL allowing reusability and sharing of code across different design units by providing a modular storing mechanism. A package can be defined as a collection of components, functions, procedures, data types and other design units stored conveniently in design libraries so that any piece of code required by any design unit can be called and reused from the stored modules instead of required to be rewritten every time.

For example, XOR gate is defined and compiled as an entity . Now, a half adder, a full adder and a parity generator circuits need to be designed separately using structural style of modelling. All the three entities to be designed require XOR gate as a component. Instead of writing code for XOR gate for each entity under design separately, a user can write, compile and store XOR gate design in a package ABC and any entity which needs to use XOR gate can simply include the package ABC as shown in Figure 8.3 and use it in any number of times in any number of entities to be designed in different design libraries.

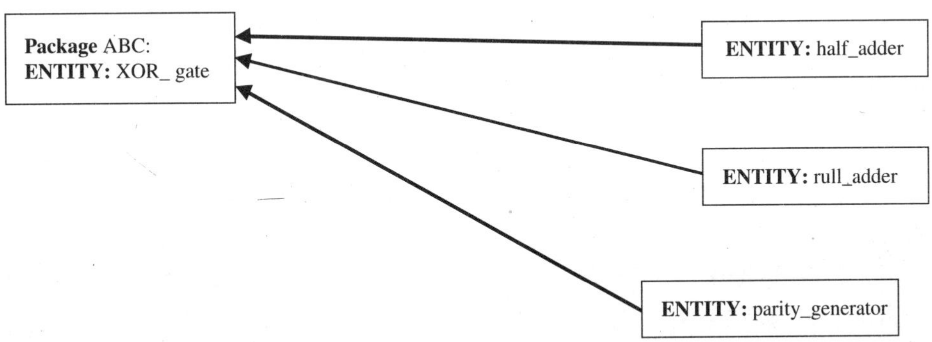

Figure 8.3 Packages Allow Reusability and Sharing of Code Across Different Design Units

8.4 WRITING A PACKAGE

Every package consists of two parts:

(i) Package declaration
(ii) Package body (optional)

SYNTAX:

```
PACKAGE name IS
  All declarations:
  Subprograms declaration;
  Type/Data Objects declarations;
  Component declarations;
  Constants/Deferred constants;
  Attributes;
  Aliases;
USE clauses;
END name;

PACKAGE BODY name IS (optional)
  Deferred constants initialization;
  Local decalartions, if any;
  Subprogram definitions;
END [name];
```

Package declaration defines the interface of the package. Every package is identified by a unique user-defined name. It is the declarative part of the package which can contain subprogram declaration, types, objects, components and so on, and all the declarations in the package body can be shared across different design units, even in different design libraries.

Package body is optional and is needed only for those packages which contain any deferred constants to be initialized, any subprograms to be defined or any local declarations. Package body shares the same name as package declaration, with which it is associated. If package does not have any deferred constants or subprograms no package body is needed. Also local declarations are allowed in a package body, that is, the scope of these items—signals, types sub-types and so on declared inside the package body, is limited only to the package body unlike the declarations in the package declarative section which have scope across different design units using it.

Let us see an example of simple package:

Example 8.5 A simple package

```
LIBRARY IEEE;
USE IEEE.STD_LOGIC_1164.all;

PACKAGE new_pack IS
TYPE word_size IS STD_LOGIC_VECTOR(7 downto 0);
```

```
TYPE mem_size IS ARRAY(0 TO 31 ) OF word_size;
END new_pack;
```

In Example 8.4, a package 'new_pack' is declared which contains declaration of a data type to create memory of 32 words with word size of 8 bits each. Since the package contains only type declaration there is no need for package body. Let us see an example to create package with declaration as well as package body.

Example 8.6 Package with package body

```
LIBRARY IEEE;
USE IEEE.STD_LOGIC_1164.all;
PACKAGE comp_pack IS
TYPE state IS (A,B,C,D);
FUNCTION edge_detect (SIGNAL x:STD_LOGIC) RETURN BOOLEAN;
CONSATNT m:INTEGER;
END comp_pack;
PACKAGE BODYcomp_pack IS
CONSTANT m:INTEGER:=5;                          -- deferred constant
FUNCTION edge_detect (SIGNAL x:STD_LOGIC) RETURN BOOLEAN IS
SIGNAL res:BOOLEAN;
BEGIN
IF x'EVENT AND x='1' THEN
res<= '1';
ELSIF x'EVENT AND x='0' THEN
Res<='0';
END IF;
RETURN res;
END edge_detect;
END comp_pack;
```

8.5 USING PACKAGE

The items stored in the form of packages can be reused anytime in any design unit by including the package containing the required items in the entity under design .VHDL gives the additional flexibility to include only the desired item from the package in an entity under design or the complete package with all its items included in the entity under design. This helps to save on the memory resources at the time of compilation and simulation by including only the required item from the design unit.

By default any package made and compiled is stored in the design library work, which is a default-user working library and any design unit which needs to use the part or whole of the package items need to include the package in the design unit, with USE clause having the following.

SYNTAX:

```
USE Library_name.Package_name.component_name;     -- includes only the named component
                                                    from the package
Or
USE Library_name.Package_name.subprogarm_name;    -- includes only the named subprogram
                                                    from the package
Or
USE Library_name.Package_name.all;                -- includes all the items from the package
```

Library_name is the name of the design library in which the package is stored. By default, it is a work library but can be any other user-defined library too, followed by the name of the package:*package_name* which has to be used, followed by the name of the component or item needed to be included or *all* if the whole package is to be included. USE should be included in the beginning of the design unit. Let us see an example for using the package:

Example 8.7 To design an up/down counter using packages

```
LIBRARY IEEE;
USE IEEE.STD_LOGIC_1164.all;
USE work.comp_pack.edge_detect;    -- including function edge_detect from package comp_pack
ENTITY up_down IS
PORT(clk,rst: IN STD_LOGIC;
        op: OUT STD_LOGIC_VECTOR(3 downto 0);
END up_down;
ARCHITECTURE behave OF up_down IS
SIGNAL intl:BOOLEAN;
BEGIN
PROCESS(clk,rst)
VARIABLE count:STD_LOGIC_VECTOR(3 downto 0);
BEGIN
  IF rst ='0' THEN
    count:="0000";
    IF intl=edge_detect(clk) THEN
      count := count + 1;
    ELSIF rst='1' THEN
      Count:= "1001";
    IF not(intl=edge_detect(clk)) THEN
      count := count -1;
      END IF;
  END IF;
```

```
END IF;
END PROCESS;
END behave;
```

SUMMARY

- Configurations are used to bind an entity with an architecture or a component with an entity.
- By default for every entity there exists a default configuration which binds the entity with its most recently compiled entity.
- Component to entity binding using configurations is applicable to structural style of modelling. It can be done in two ways: component to entity binding or in the form of lower-level configuration binding.
- Packages are collection of VHDL code allowing code reusability, code sharing and modular programming.
- Packages are stored as design units in design libraries and need to be included in an entity under design with USE clause.
- Package is comprised of package declaration and an optional package body.
- Package declaration and package body share the same name.
- Package body is required only for defining subprograms or for initialization of deferred constants.
- Package can be included across different design units either partially or as a whole package.

QUESTIONS

Objective Questions

1. The concept of configurations is based on ....

   a. early binding   b. late binding   c. static binding   d. none
2. .... Configurations bind component to an entity.

   a. Lower level   b. Upper level   c. Entity–architecture   d. none of these
3. Packages allow .... of code.

   a. portability   b. reusability   c. inheritance   d. extensibilty
4. Signals declared in a package are...........

   a. global   b. local   c local to entity   d. none of these
5. Deferred constants are constants, which are initialized ....

   a. along with its declaration in a package
   b. along with its declaration in a package body
   c. inside the entity using the package containing the constant
   d. anywhere when executed
6. STD_LOGIC_1164 is a package of Library IEEE. (True/False)
7. Package body is required only if the package contains ....

   a. subprograms   b. deferred constants   c. both   d. none

Descriptive Questions

1. What is a configuration? What are the advantages of using configuration in a VHDL design?
2. Write short notes on:
   a. Configurations b. Packages
3. What is the need for configurations in VHDL? Explain with the help of an example.
4. Explain configuration with a special reference to structural modelling.
5. Packages support the concept of reusability. Justify.
6. Write down the truth table, entity declaration and architectures—dataflow and behavioural for a 4-bit comparator. Define a default configuration for binding entity to its dataflow architecture.
7. Write down the configuration specification for full adder (MDU).
8. Write a function to add 2-bit vectors with range 7 DOWNTO 0 in a package contained in a work library.
9. Write a VHDL subprogram to implement a stack memory inside a package. Call this subprogram inside an entity ALU which performs the arithmetic operations on integers. Use the stack subprogram for push and pop operations.

ADVANCED PROGRAMMING CONCEPTS

9

CHAPTER OBJECTIVES

- To resolve multiple driver signals
- Learn about different attributes
- To create libraries
- Use of qualified expressions

9.1 MULTIPLE DRIVERS

Every output signal created has associated with it driver(s) which holds its values.

For example:

```
s<= a XOR b;
```

In this signal s is driven by the expression value computed by '**xor**' operation of '**a**' and '**b**'. But at times, situation may arise where the same signal has more than one driving statements associated with it. This gives rise to following situations that are encountered in VHDL:

- Inside a process
  s<= 1 after 3 ns, 4 after 6 ns, 12 after 20 ns
- Between two or more processes

```
P1: process( )
    Begin
    ........
    ........
    s<= a;
    ...........
End Process P1;
P2: : process( )
    Begin
    ........
    ........
```

```
    s<= not(b);
    ............
End Process P2;
```

- More than one driving expressions with different input signals involved

```
s<= a or b;
s<= c;
```

- Concurrent signal select statements

```
S<= '1' when a= '1',
    '0' when a= '0';
```

The problem of multiple drivers exists when a decision has to be taken regarding the value to be assigned if more than one assignment is made to the same target signal. So in the above listed cases, in Case 3 where more than one concurrent signal assignments are made to signal 's', a decision is to be made as which is the effective value of signal 's'. This requires writing of a user-defined resolution function which will resolve the conflict between two values that are to be assigned to the same output. The resolution function is written just like any other user-defined function except that at function call no arguments need to be given and function name is part of signal declaration for a signal having multiple drivers. The function contains user-specified logic to resolve the conflict in the assignment of the signal which is mostly taken as wired OR or wired AND, though it can be any user-specified logic. If open-collector or open-drain output drivers are used with pull-up load on a signal, then the signal can be pulled low by any driver and pulled high by load only when all drivers are off. This is called wired-or or wired-and connection. If tri-state drivers are used only one driver which drives the signal is active at a time. Figure 9.1 explains the logic of using resolution function. Let's say output signal has 3 driver sources v1, v2, v3. Resolution function uses to logic to assign only one value to output signal at a time so that no conflict occurs due to multiple input values for the output signal.

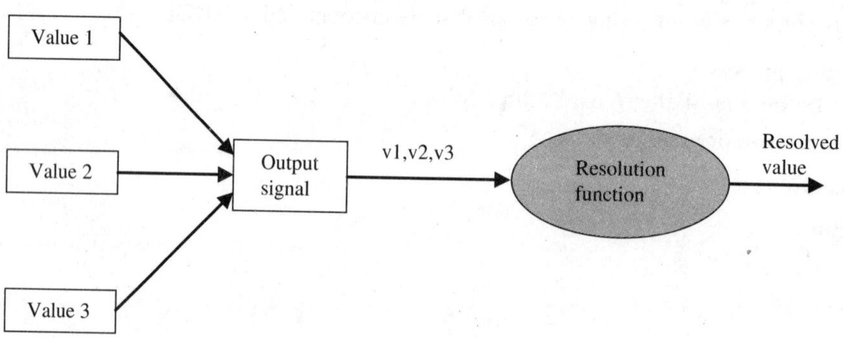

Figure 9.1 Resolution Function

Example 9.1 Resolution function

```
LIBRARY IEEE;
USE IEEE.STD_LOGIC_1164.all;
```

```
ENTITY fnc_main IS
PORT(a,b,c: IN BIT; s: OUT BIT);
END fnc_main;
ARCHITECTURE fnc_main OF fnc_main IS
FUNCTION r_fnc( SIGNAL drivers: BIT_VECTOR) RETURN BIT IS
  VARIABLE res:BIT:='0';                                -- FUNCTION DEFINITION
BEGIN
FOR i IN drivers'RANGE LOOP
IF drivers(i)='1' THEN
  res:='1';
ELSE
res:='0';
END IF;
END LOOP;
RETURN res;
END r_fnc;
SIGNAL s1:r_fnc BIT:='0';
BEGIN
s1<= a ;
s1 <= NOT (b);
s<=s1;
END fnc_main;
```

Example 9.1 shows the use of resolution function defined by user based on WIRED_OR logic. Resolution function is used by the simulation tool to resolve the driver conflict and is ignored by most of the synthesis tools. It should always be independent of the order of its inputs. It is always a single input function where the single input is the array of its drivers.

9.2 ATTRIBUTES

Attributes are the characteristics of the data depending on its type and class. They can be predefined or user-defined. They are important as they help to reduce the code complexity. Predefined attributes are broadly classified into five categories as follows:

- Value Attributes: They return a simple value.
- Function Attributes: They take an input argument and make a function call to invoke the attribute.
- Signal: They are applied on signals and create new signals in turn.
- Type: They are applied on a type and return.
- Range: They are applicable to a range of data types.

9.2.1 Value Attributes

Classification of value attributes is as shown in Figure 9.2:

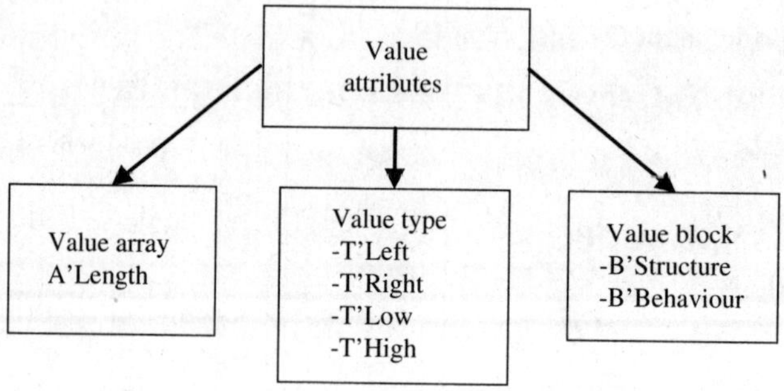

Figure 9.2 Value Attributes

Value Array Attribute: A'Length is the only value array attribute which gives the number of values in an array range.

Value Type Attribute: Value type attributes give the bounds of the type which can be any physical, enumeration, integer or floating type or subtype.

Let T denote any physical, enumeration, integer or floating type or subtype. The following attributes are defined in VHDL:

- T'Left: Left bound of a type or subtype
- T'Right: Right bound of a type or subtype
- T'Low: Lower bound of a type or subtype
- T'High: Upper bound of a type or subtype
- T'Ascending: Returns a Boolean value true if the range of type 'T' is **TO** and false if it is **DOWNTO.**

Value Block Attribute: Value block attributes are used to check the way the design has been modelled. If B denotes a block or architecture name then

- B'Structure: Returns true if B contains component instantiation statements or passive processes, else false.
- B'Behaviour: Returns true if block does not have any component instantiation.

9.2.2 Function Attributes

Function attributes use an input argument and a function call is made to invoke the function attribute. Function attributes act on signals, array and types and are classified accordingly as shown in Figure 9.3.

Function Type Attributes

- T'Pos(X): Gives position number of 'X' in 'T'
- T'Val(X): Gives a value corresponding to position number indicated by 'X'
- T'Succ(X):Gives value in Type 'T' next to 'X'
- T'Pred(X): Gives value in type 'T' preceding 'X'

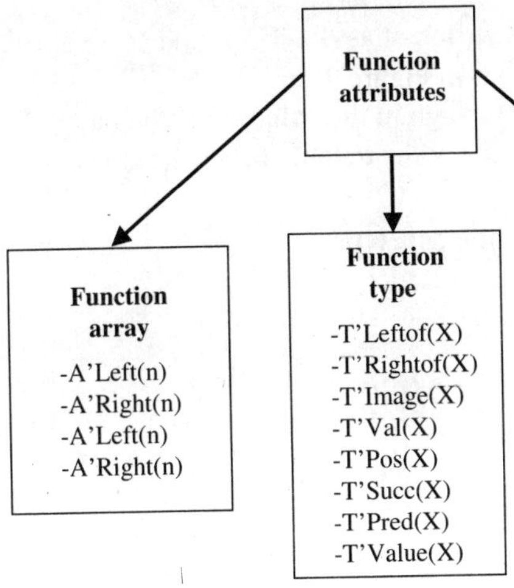

Figure 9.3 Function Attributes

- T'Leftof(X): Gives value in type 'T' which is in the left of 'X'
- T'Rightof(X): Gives value in type 'T' which is in the right of 'X'
- T'Image(X): The string representation of 'X' in 'T'
- T'Value(X): Value of type 'T' whose string representation is 'X'

Function Array Attributes

- A'Left(n): Gives value to left of n in 'A'
- A'Right(n): Gives value to right of n in 'A'
- A'Low(n):Gives lower bound of index range 'n'
- A'High(n): Gives upper bound of index range 'n'

If not specified n=1 by default.

Note: For ascending ranges A'Left(n) is the same as A'Low(n) and A'Right(n) is the same as A'High(n) whereas for descending ranges A'left(n) is the same as A'High(n) and A'right(n) is the same as A'Low(n).

Function Signal Attributes: If S is a signal, then

- S'EVENT: Returns true if an event occurred on signal S in the current delta.
- S'ACTIVE: Returns true if signal S is active in the current delta. It is important to understand the difference between an event on a signal and a signal that is being active. A signal is said to be active when a new value is assigned to the signal, even if the new value is the same as the old value; while an event is said to have occurred only if the new value is different from the old value.
- S'LAST_EVENT: Returns time elapsed since the last event on signal S.
- S'LAST_ACTIVE: Returns time elapsed since the last time signal was active.
- S'LAST_VALUE: Returns the value of S, before the last event.

Note: Attribute 'Active and 'Event are different as 'ACTIVE and 'LAST_ACTIVE are invoked when a transaction takes place while 'EVENT and 'LAST_EVENT are invoked when an event occurs. Transaction is the change in the value of the signal or in or inout port even when no change in the value occurs at the output port or the signal.

Example 9.2 shows all the attributes used of a data type.

Example 9.2 To show TYPE attributes

```
LIBRARY IEEE;
USE IEEE.STD_LOGIC_1164.all;
ENTITY p_attri IS
PORT (i1:OUT INTEGER;
      i2,i3:OUT BOOLEAN );
END p_attri;
ARCHITECTURE behave OF attri IS              -- behavioural architecture model of the entity
TYPE attrib IS ( a,b,c,d,e);
SUBTYPE s_attrib IS attrib RANGE e DOWNTO c;
SIGNAL i4,i5,i6,i7,i8,i9,i10,i11,i12: attrib;
BEGIN
  i1<=attrib'Pos(c);
  i2<=attrib'Ascending;
  i3<=s_attrib'Ascending;
  i4<=attrib'Left;
  i5<=attrib'Right:;
  i6<=attrib'Low;
  i7<=attrib'High;
  i8<=attrib'Val(4);
  i9<=attrib'Succ(b);
  i10<=attrib'Pred(b);
  i11<=attrib'Image(b);
  i12<=attrib'Value("b");
END behave;
OUTPUT:
i1 = 2
i2=true
i3 =false
i4 =a
i5=e
i6=a
i7=e
```

```
i8=e
i9=c
i10=a
i11= "b"
i12= b
```

*Image and value attributes are not synthesizable with Xilinx.

Example 9.3 shows use and result of all the attributes defined for ARRAYS.

Example 9.3 Array attributes

```
LIBRARY IEEE;
USE IEEE.STD_LOGIC_1164.all;

ENTITY p_attri IS
PORT (i1,i2,i3,i4,i7:OUT INTEGER;
      i8:OUT BOOLEAN );
END p_attri;

ARCHITECTURE behave OF p_attriIS          -- behavioural architecture model of the entity
  SIGNAL Attrib,i5,i6: STD_LOGIC_VECTOR(7 DOWNTO 0)
 BEGIN
  i1<=attrib'Low;
  i2<=attrib'High
  i3<=s_attrib'Left;
  i4<=attrib'right;
  i5<=attrib'Range:;
  i6<=attrib'Reverse_range;
  i7<=attrib'Length;
  i8<=attrib'ascending;
END behave;
OUTPUT:
i1 = 0
i2=7
i3 =7
i4 =0
i5=7 downto 0;
i6=0 to 7
i7=8
i8=FALSE
```

Example 9.4 shows use of function signal attributes S'EVENT and S'LAST_VALUE with help of assert statement to check set-up time violation for a signal.

Example 9.4 Function signal attributes

```
LIBRARY IEEE;
USE IEEE.STD_LOGIC_1164.all;
ENTITY fnc_attrib IS
PORT (d,clk:IN STD_LOGIC;
      q:OUT STD_LOGIC );
END fnc_attrib;

ARCHITECTURE behave OF fnc_attrib IS          -- behavioural architecture model of the entity
  CONSTANT setup_time:TIME:=10 ns;
 BEGIN
  PROCESS(d,clk)
  BEGIN
  IF clk'EVENT AND clk='1' AND clk'LAST_VALUE='0' THEN
     ASSERT (d'LAST_EVENT >= setup_time)
        REPORT "set up condition violated"
        SEVERITY ERROR;
     q<=d;
  END IF;
  END PROCESS;
END behave;
```

Signal Attributes

- S'DELAYED(T): Returns a signal of type same as S but delayed by specified time T.
- S'STABLE(T): Returns Boolean true or false. True if the reference signal had no events for a specified time T. It has the same function as S'Event but S'Event is more efficient in memory and speed as it does not create a new signal.
- S'QUIET[(T)]: Returns the Boolean True/False, True if the reference signal had no transactions/events for a specified time.
- S'TRANSACTION: Returns a BIT value that toggles its value for every transaction or event that occurs on the reference signal.

Function attributes for signals and signal attributes are different. Signal attributes create new signals invoking events when used in concurrent statements whereas function attributes for signals do not create new signals; they only return the value for the attribute. So, signal attributes can be used wherever a signal is expected as a sensitivity list whereas function signal attributes cannot be used as independent signals. Example 9.5 shows the use of signal attribute S'TRANSACTION.

Example 9.5 Signal attributes

```
LIBRARY IEEE;
USE IEEE.STD_LOGIC_1164.all;

ENTITY sig_attrib IS
```

```
PORT( a:IN STD_LOGIC;
      b:OUT STD_LOGIC );
END sig_attrib;
ARCHITECTURE behave OF sig_attrib IS      -- behavioural architecture model of the entity
  BEGIN
   PROCESS(a'Transaction)
  BEGIN
    IF a'TRANSACTION='1' THEN
      b<= NOT (a);
    END IF;
  END PROCESS;
END behave;
```

9.2.3 Type Attributes

There is only one type attribute T'Base where T denotes a physical, enumeration, floating or integer type.

T'Base: returns base type of T

For example,

TYPE T is state (A,B,C,D,E)

So, T'Base returns state.

9.2.4 Range Attributes

There are two range attributes which work with constrained array types alone and has optional parameter n, which denotes index range.

T'Range: returns the range in the order specified

T'Reverse_range: returns the range in the reverse order.

SIGNAL a: **STD_LOGIC_VECTOR** (0 **TO** 7);

a'range returns 0 to 7

a'reverse_range returns 7 down to 0

User-defined Attributes: User-defined attributes are used to attach information to VHDL constructs. It has a declaration part and a specification part. Declaration is

ATTRIBUTE Attributename: **TYPE**;

Here Attributename is the user-defined attribute and Type indicates the type of data attribute will return. This is attribute declaration which can be done in package, architecture body (declarative section).

Attribute specification:

ATTRIBUTE Attributename **OF** Constructname: Class is expression/value;

Here attribute name is the same user-defined name of attribute as declared in attribute declaration. Constructname is the VHDL construct for which the attribute is to be defined. It can be the name of an entity, architecture, type, configuration, component, label, signal, variable, constant, subprogram or a subtype. Class is the entity, architecture, type, configuration, component, label, signal, variable, constant, subprogram or a subtype. Expression/value is the expression or value of the attribute which is expected when the attribute is invoked.

Once the attribute is declared and specified it can be called or used anywhere in the program like predefined attributes as VHDL constructname' attributename; Example 9.6 shows the use of user-defined attribute. It defines an attribute set-up for signals which can be used as additional characteristic of any signal defined in the same architecture.

Example 9.6 User-defined attributes

```
LIBRARY IEEE;
USE IEEE.STD_LOGIC_1164.all;

ENTITY user_attrib IS
PORT (a,b:IN STD_LOGIC;
        c:OUT STD_LOGIC);
END user_attrib;

ARCHITECTURE behave OF user_attrib IS          -- behavioural architecture model of the entity
  ATTRIBUTE setup_time :TIME;
  ATTRIBUTE setup_time OF a,b:SIGNAL IS 20ns;
  SIGNAL time_chk:TIME;
  BEGIN
  PROCESS(a,b)
  BEGIN
  IF a'LAST_EVENT >a'setup_time THEN
    IF b'LAST_EVENT >a'setup_time THEN
   c<= a AND b;
     END IF;
  END IF;
  END PROCESS;
END behave;
```

9.3 LIBRARIES

The design unit coded in VHDL is always stored in a design library. A design library refers to the storage of VHDL descriptions in the file system of the host environment. The storage format is tool implementation-specific, but the user has to always work in a current working library known as a 'WORK' library. There can be only one working library in VHDl suite. An arbitrary number of design libraries may be specified, containing diff erent constructs which can be included when and where needed by including the library in the design unit using 'Use Library' clause. The VHDL source is present in an ASCII file called the design file. The design file is an ASCII file containing the VHDL source. A design library consists of a number of compiled design units. Design units are further classified as

1. **Primary units:** These units allow items to be exported out of the design unit. Primary design units in VHDL are

- entity declaration
- package declaration
- configuration declaration

2. **Secondary units:** These units do not allow the items declared within them to be exported out of the design unit, that is, these items cannot be referenced in other design units. Secondary units in VHDL are

- architecture body
- package body

There can be only one primary unit with a given name in a single design library where as secondary units associated with different primary units can have identical names in the same design library. And, more importantly, a secondary unit may have the same name as its associated primary unit. Secondary units must coexist with their associated primary units in the same design library. Even though a configuration declaration is a primary unit, it must reside in the same library as the entity declaration to which it is associated. Most of the present day tools allow creating of user libraries or even moving the design units from one library to another.

To use user-created libraries, 'Library' and 'use Library' clause are to be used.

For Example:

IF a library 'my work' is created by the user which contains the entity and architecture of And gate and Xor_gate as

```
ENTITY and_g IS
PORT(a,b:IN STD_LOGIC;
      c :OUT STD_LOGIC);
END and_g;

ARCHITECTURE behave OF and_g IS
  BEGIN
c<=a AND b;
END behave;

ENTITY xor_g IS
PORT(x,y:IN STD_LOGIC;
      z :OUT STD_LOGIC);
END and_g;

ARCHITECTURE behave OF xor_g IS
  BEGIN
z<=a XOR b;
END behave;
```

Now to use let us say And _gate in the current design we need to include the entity and_g in our design as

```
LIBRARY IEEE;
USE IEEE.STD_LOGIC_1164.all;
```

LIBRARY my_work;
USE my_work.and_g;

9.4 ALIAS

Alias declares an alternate name for all or part of the VHDl object. It is useful as a convenience of shorthand notation or as a mechanism to refer to same name in different context.

SYNTAX:

ALIAS name [: **TYPE**] **IS** Vhdl_object/construct [**SIGNATURE**];

Where ALIAS is the keyword, name is the identifier or the user name to be given as alias, TYPE is the data type of the alias and vhdl object or construct can be signal, variable, constant, a function, a type or a literal.

TYPE is the optional required for alias created for a VHDL object alone, whereas for non-object alias no type information is required. Signatures are required for alias created for a subprogram or an enumeration literal. Aliases are useful in particular to the reference part of array.

FOR EXAMPLE:

SIGNAL a:**STD_LOGIC_VECTOR**(7 **DOWNTO** 0);
ALIAS a1:**STD_LOGIC_VECTOR**(3 **DOWNTO** 0) **IS** a(7 **DOWNTO** 4);

For non-object alias,

For enumerated type,

ALIAS zero **IS** **FALSE** [**RETURN BOOLEAN**];

Name — Value — Type

9.5 SIGNATURES

Signatures distinguish between overloaded subprograms and enumeration literals based on their parameter and result type. A signature may be used in an attribute name, entity designator or alias declaration.

Typemark, type_mark,………,type_mark return type_mark.

Parameter types in their defined order

Signatures are used immediately after a subprogram or an enumeration literal name. When used with enumerated types, a type is treated as a function with no parameters.

FOR EXAMPLE:

TYPE bit_val IS ('0', '1');
TYPE four_val IS('U', '0', '1', 'Z');
SIGNATURE [**RETURN** bit_val];
ALIAS zero is '0' [**RETURN** bit_val];

9.6 QUALIFIED EXPRESSIONS

Qualified expressions are used for defining the type of an expression involving ambiguity. It is used for exact type matching in case of overloaded subprograms. It performs no type conversion but helps to overcome the ambiguity in case of overloaded functions and procedures. Also, qualified expressions can be used to assign values to aggregate when the range of aggregates is likely to be vary.

SYNTAX:

TYPE '(Expression/Val);

Let us take an example,

If in CPU design both 8-bit as well as 16-bit data are allowed for the rest option and if CPU has to initialize all data bits to zeros, then to do this one way is to check if data is 8 bits or 16 bits and then use direct assignment as:

```
Data8<= "00000000";
Data16 <= "0000000000000000";
```

Another way is using qualified expressions as

```
Data_rst<= data8'(others <= '0');
```

This is another use of qualified expressions, to determine the width of the output signal and in accordance to that change the width of the input signal or source signal.

9.7 GENERATE STATEMENT

Generate is a concurrent statement that is used to conditional and repetitive structures during the elaboration phase of simulation. It is used for repeating concurrent statements for a predefined number of times or till a condition is satisfied. There are two forms of generate statement as

- FOR generate scheme
- IF generate scheme

FOR generate allows repetitive/looping structures in the concurrent style of modelling.

SYNTAX:

```
Label: FOR identifier IN RANGE GENERATE
Concurrent statements
..........
............
.........
END GENERATE Label;
```

IF generate allows conditional looping.

SYNTAX:

```
Label: IF condition GENERATE
Concurrent statements
......................
...................
END GENERATE Label;
```

Label is compulsory for to be used with a generate statement. Both forms of generate can be nested to any level. Range and condition included in FOR generate and IF generate, respectively, have to be static, which implies they cannot be signals. Example 9.7 shows use of generate statements as concurrent iterative constructs.

Example 9.7 Example of generate statements

```
LIBRARY IEEE;
USE IEEE.STD_LOGIC_1164.all;
ENTITY rippadd IS
PORT (a,b: IN BIT_VECTOR(0 TO 7);
      cin: IN BIT;
    S:OUT BIT_VECTOR(0 TO 7 );
  Cout:OUT BIT);
END rippadd;
ARCHITECTURE gen OF rippadd IS
  COMPONENT fulladd IS
 PORT(A,B,Cin:IN BIT;
        Su,Ca:OUT BIT);
 END COMPONENT;
 COMPONENT halfadd IS
 PORT(a,b:IN BIT;
S,ca:OUT BIT);
END COMPONENT;
SIGNAL c:BIT_VECTOR(0 TO 7);
BEGIN
 row(0) <= "0000" & ip;
 Gen_add: FOR i IN 0 TO 7 GENERATE
 Lower_bit: IF i=0 GENERATE
          U1:halfadd PORTMAP(A(i), B(i),S(i),C(i));
        END GENERATE Lower_bit;
 Upper_bit: IF i>0 GENERATE
        U2:fulladd PORTMAP (A(i),B(i),C(i-1),S(i),C(i));
       END GENERATE Upper_bit ;
  END GENERATE Gen_add;
    Cout<=C(7);
END gen;
```

9.8 NULL TRANSACTIONS

VHDL facilitates the modelling of outputs which can be turned off if no input value is connected. NULL transaction allows an input to be disconnected if no input is provided at source. For example,
a<= null after 5 ns;

SUMMARY

- Signals with multiple sources need to be resolved.
- Resolution functions are user-defined functions needed for driver resolution.
- Attributes are used to attach additional information to VHDL constructs.
- There are predefined and user-defined attributes.
- Predefined attributes are classified in five categories: value, function, type, signal and range.
- User-defined attributes give user flexibility to specify additional characteristics to VHDL constructs.
- Libraries specify a storage area for VHDL programs.
- There is only one current work library in VHDL. Else, predefined libraries containing inbuilt VHDL constructs exist in VHDL as well as user-defined libraries can be created.
- Alias allows an alternate name to be given to all or part of a VHDL object. It is useful particularly for arrays.
- Qualified expressions return specified data type for given expression and help in type matching.
- Generate statements allow repetitive and conditional constructs in concurrent modelling.
- Generate statements exist in two forms: IF generate and FOR generate.
- Null transactions are used to disconnect the input source from a target.

QUESTIONS

Descriptive Questions

1. What are attributes?
2. Write a note on all types of attributes.
3. What is the use of null transaction?
4. What is the difference between function signal attributes and signal attributes? Give an example.
5. What do signatures signify in VHDL? Where are they used?

Program Exercise

1. Write a resolution function based on wired OR logic for a target signal having four inputs at its source.
2. Write a program using IF generate for full adder design.
3. Write a code for an ALU design which has an 8-bit data for a function selection of which 4 bits are meant for function opcode for arithmetic and logical functions. Use alias to refer to different parts of data.
4. Write a program for a shifter using generate statements.
5. Write a program to use signal attribute.
6. Define a priority encoder. Can it be designed using generate statements?

TESTBENCHES

10

CHAPTER OBJECTIVES

- To know what is testbench
- Writing a testbench
- Verification of VHDL designs using testbenches
- Using test vectors in files for testbenches

10.1 INTRODUCTION

Verification of the circuit functionality is done as the first step once complete VHDL code for circuit has been coded. Verification of design without actual hardware in place is done using simulation. For simulating any design two things are required: design to be tested and stimulus to inputs. Two approaches are used in VHDL for simulation: one is assigning input values in simulation window and getting the output waveforms and the other is using testbenches.

Testbench is different from VHDL simulation of the circuit, which uses run-time assignment of input values. In case of testbench stimulus needed to drive the circuit is defined in the testbench itself instead of instantiating the inputs in the simulator window. This gives an advantage as test-bench applies stimulus to the design in the form of test vectors or stored inputs (either in testbench model or as an ASCII file) which can be used to analyze the results, or compare the results of two simulations or compare the actual output against the expected output. The purpose of a testbench is to verify the functionality of a developed model or package. Figure 10.1 shows a typical testbench structure. A testbench is a diffferent design unit separated from the design to be verified and placed in a design library that is separated from the circuit design itself.

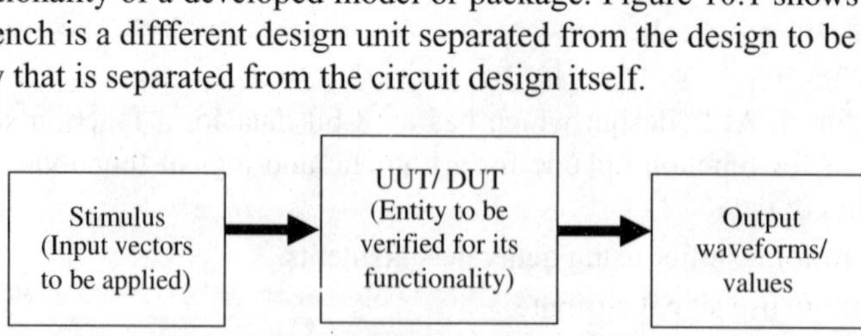

Figure 10.1 Typical Testbench

10.1.1 Advantages of Using Testbenches

Using testbenches for verification of circuit functionality has several advantages for the simpler approach of assigning values in simulation window as follows:

1. In a testbench input vectors are stored either as a part of the testbench file or as a TEXT (ASCII) file.
2. Testbench simulation is suitable for complete design test as well as individual component tests using the same values of inputs from the component to the design level. This gives a methodological approach to design verification.
3. The outputs generated can be observed as waveforms or can be stored as values for comparison against expected values or for later analysis or to compare two simulations.
4. If any modification is made to design test vectors can be easily applied to the new design.
5. Testbench approach is useful for post-fit verification as well.
6. It is very useful for analysis specially to analyse circuit behaviour with respect to particular input without changing the other input values.
7. It gives an idea about the timing requirements of the circuit under test.

10.2 WRITING A TESTBENCH

When writing a testbench, the design for which the testbench has to be written is always known as 'Unit Under Test (UUT) or 'Design Under Test (DUT)'. DUT is included as a component in the testbench model. Testbench is written as an entity just like any other VHDL model with an architecture body which includes DUT as a component. But as testbench entity does not have any interface the entity declaration is empty with no generic or port declarations as follows:

SYNTAX:

```
ENTITY tb IS
END tb;
ARCHITECTURE tb OF tb IS
COMPONENT DUT1
  PORT (port_names : MODE TYPE);
 END COMPONENT;
Local signals;
BEGIN
      Generating test_vectorss;
      L1: DUT1 PORT MAP (association list );
      applying test vectors to DUT;
     [ verify or compare or store results;]
END tb;
```

The entity declaration has no ports as the testbench itself does not have any inputs or outputs. It just uses the test vectors which are applied to the UUT/DUT and having an architecture body without an entity declaration is illegal. Therefore, we have empty entity declaration. Architecture part has a component which is DUT with the same name and matching ports. In the executable part the first thing is to generate the test vecors. It can be a repetitive signal or a constant array of values. Once the test vectors (values to be applied as stimulus to inputs) are available the next thing is to apply these to DUT and finally observe the output or to write the result to a file or to report errors. Monitoring of output values in the testbench is optional. The output can simply be observed as waveform also as in simulation windows. If the output is to be compared against the expected values available, a set of expected values for output parameters is required. This set of expected values is known as intent model.

Based on this we can have two main types of testbenches:

- Testbench having test vector values in the testbench model
- Importing test vector values from a file

Let us see a simple testbench which simply uses test vectors for input stimulation in Example 10.1.

Example 10.1 Testbench for 4:1mux

Consider an entity mux, 4:1 mux with four inputs a,b,c,d, two select inputs which are vector 'sel' (1 DOWN TO 0) and has output y.

```
LIBRARY IEEE;
USE IEEE.STD_LOGIC_1164.all;

ENTITY tb_mux IS
END tb_mux;

ARCHITECTURE tb OF tb_mux IS
COMPONENT mux IS
PORT( a,b,c,d: IN STD_LOGIC;
      sel: IN STD_LOGIC_VECTOR(1 DOWNTO 0);
      y: OUT STD_LOGIC);
END COMPONENT;

SIGNAL a,b,c,d,y: STD_LOGIC;
SIGNAL sel: STD_LOGIC_VECTOR(1 DOWNTO 0);

BEGIN
sel <= "00", "01" AFTER 20 ns, "10" AFTER 40 ns, "11" AFTER 60 ns, "00"AFTER 80ns;
a<= '0' AFTER 10 ns, '1' AFTER 20 ns;
b<= '0' AFTER 30 ns, '1' AFTER 40 ns;
c<= '0' AFTER 50 ns, '1' AFTER 60 ns;
d<= '0' AFTER 70 ns, '1' AFTER 80 ns;
M1: mux PORT MAP(a=> a, b=> b, c=> c,d=> d,sel=> sel,y=> y);          -- instantiation
END tb;
```

Consider an entity up-counter

```
LIBRARY IEEE;
USE IEEE.STD_LOGIC_1164.all;
USE IEEE.STD_LOGIC_UNSIGNED.all;
ENTITY counter IS
PORT( clk,rst: IN STD_LOGIC;
      count: INOUT STD_LOGIC_VECTOR(3 DOWNTO 0));
END counter;
ARCHITECTURE behave OF counter IS          -- behavioural architecture model of the entity
BEGIN
  P1: PROCESS(rst,clk)                     -- declarative part of process
      BEGIN
        IF rst= '0' THEN
           count<= "0000";
        ELSIF clk'EVENT AND clk='1' THEN
           count<= count + 1;
        ELSE
           count <= count;
        END IF;                            -- executable part of process
  END PROCESS P1;
END behave;
```

Consider another testbench (Example 10.2) for an up-counter which gives input test values.

Example 10.2 Testbench for a 4-bit up-counter

```
LIBRARY IEEE;
USE IEEE.STD_LOGIC_1164.all;
ENTITY tb_cnt IS
END tb_cnt;
ARCHITECTURE tb OF tb_cnt IS
COMPONENT counter IS
PORT( clk,rst: IN STD_LOGIC;
      count: INOUT STD_LOGIC_VECTOR(3 DOWNTO 0));
END COMPONENT;


SIGNAL clk ,rst: STD_LOGIC;
SIGNAL count: STD_LOGIC_VECTOR(3 DOWNTO 0);
TYPE vector IS RECORD
```

```
        rst : STD_LOGIC;
        clk: STD_LOGIC;
        count: STD_LOGIC_VECTOR(3 DOWNTO 0);
 END RECORD;
 TYPE test_array IS ARRAY(NATURAL RANGE<>) OF vector;
 CONSTANT testing_val: test_array:=
 (( rst=> ‘0’, clk=> ‘1’, count–> “0000”),
  (rst=> ‘0’, clk=> ‘0’, count=> “0000”),
  (rst=> ‘1’, clk=> ‘1’, count=> “0000”),
  (rst=> ‘0’, clk=> ‘1’, count=> “0001”),
  (rst=> ‘0’, clk=> ‘0’, count=> “0001”),
  (rst=> ‘0’, clk=> ‘1’, count=> “0010”),
  (rst=> ‘0’, clk=> ‘0’, count=> “0010”),
  (rst=> ‘0’, clk=> ‘1’, count=> “0011”),
  (rst=> ‘0’, clk=> ‘0’, count=> “0011”),
  (rst=> ‘0’, clk=> ‘1’, count=> “0100”),
  (rst=> ‘0’, clk=> ‘0’, count=> “0100”),
  (rst=> ‘0’, clk=> ‘1’, count=> “0101”),
  (rst=> ‘0’, clk=> ‘0’, count=> “0101”),
  (rst=> ‘1’, clk=> ‘0’, count=> “0000”));
BEGIN
U1: counter PORT MAP(clk=> clk, rst=>rst, count=> count);          -- instantiation
P1: PROCESS
      VARIABLE t1: vector;
      BEGIN
            FOR i IN testing_val’RANGE LOOP                      -- applying test vectors
                  t1:= testing_val(i);
                  clk<= t1.clk;
                  rst<= t1.rst;
            WAIT FOR 15 ns;
                  ASSERT count /= t1.count
                        REPORT “wrong count”
                  SEVERITY ERROR;
            END LOOP;
      END PROCESS;
END tb;
```

The waveforms results for testbench written for 4:1 mux are shown in Figure 10.2.

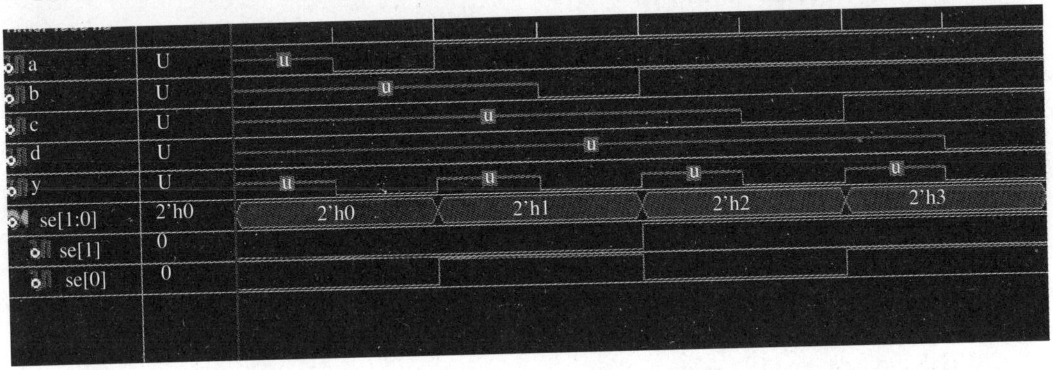

Figure 10.2 Simulation Waveforms for Multiplexer Testbench

Testbench can also use stimulus for inputs from a text. Even the output generated can be written to a file. These involve use of READ and WRITE procedures for reading a file or writing output to a file.

READ(filename, data_name)
WRITE(filename, data_name)

Example: Testbench reading test vectors from the file
Let the input file be of the form: File_name= inp_test

Rst	clk	count
0	0	0000
0	1	0000
1	1	0000
0	0	0001
0	1	0001
0	0	0010
0	1	0010
0	0	0011
0	1	0011
0	0	0100
0	1	0100
0	0	0101
0	1	0101
1	0	0000

```
LIBRARY IEEE;
USE IEEE.STD_LOGIC_1164.all;

ENTITY tb_cntl IS
END tb_cntl;
```

```
ARCHITECTURE tb OF tb_cnt1 IS
COMPONENT counter IS
PORT( clk,rst: IN STD_LOGIC;
      count: INOUT STD_LOGIC_VECTOR(3 DOWNTO 0));
END COMPONENT;
SIGNAL clk ,rst: STD_LOGIC;
SIGNAL count: STD_LOGIC_VECTOR(3 DOWNTO 0);
BEGIN
 U1: counter PORT MAP(clk=> clk, rst=>rst, count=> count);          -- instantiation
 P1: PROCESS
     FILE inp_test: text IS IN "c:\ test";
      VARIABLE line_rd: LINE;
     VARIABLE nxt_sg: BOOLEAN;
     VARIABLE ch: BOOLEAN;
     VARIABLE trst, tclk: STD_LOGIC;
     VARIABLE tcount: STD_LOGIC_VECTOR(3 DOWNTO 0);
     BEGIN
          WHILE NOT endfile(inp_test)LOOP
               READLINE( inp_test, line_rd);
               READ(line_rd,ch,nxt_sg);
               IF NOT nxt_sg OR ch/=HT THEN
                    NEXT;
               END IF
               READ(line_rd,trst,nxt_sg);
               NEXT WHEN NOT nxt_sg;
               READ(line_rd,ch);
               READ(line_rd,tclk);
               READ(line_rd,ch);
               READ(line_rd,tcount);
          WAIT FOR 20 ns;
          rst => trst;
          clk=> tclk;
    ASSERT count /= tcount
                   REPORT "wrong count"
              SEVERITY ERROR;
   END LOOP;
   END PROCESS;
END tb;
```

If the output is to be stored in a text file, on similar lines as the above example the write procedure can be invoked. File should be opened in OUT mode and using write procedure values can be written to file.

10.3 WAVEFORM GENERATION

At times using testbenches waveforms need to be generated either as a reference signal or as a stimulus signal. In design of synchronous signals in addition to input vectors the clock signal is required which needs to be generated in the testbench itself. So, waveform generation is an important part of a testbench. Waveform generated can be as follows:

- A pulse train with a uniform repetitive pattern
- Waveform with varying on–off time

Repetitive pulse can be generated simply by using a concurrent signal assignment signal with a constant on-off period specified using 'after' clause.

For Example:

Let 'A' be the signal to be generated of type 'BIT'

;

a **NOT** (A) **AFTER** 10 ns;

The uniform pulse waveform generated for this statement is shown in Figure 10.3.

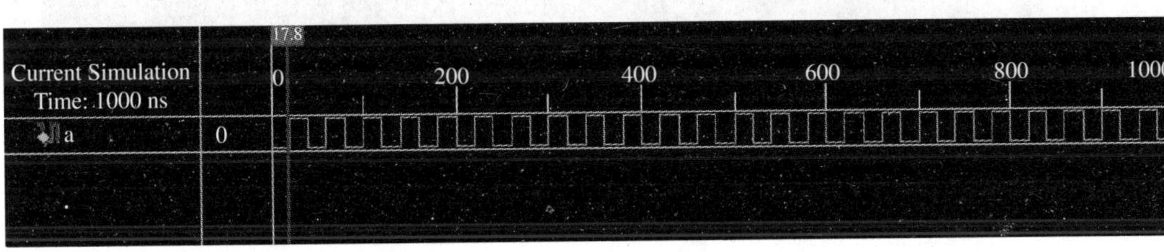

Figure 10.3 Generated Waveform for Repetitive Uniform Pulse

Sometimes, a synchronous system requires signals such as reset or preset clear which may be repetitive but have non-uniform on–off periods. To generate a pulse waveform with duty cycle other than 50 per cent (different on–off periods), the time period for pulse high and pulse low has to be separately specified as:

```
LIBRARY IEEE;
USE IEEE.STD_LOGIC_1164.all;

ENTITY pulse_gen IS
        PORT( pul: OUT STD_LOGIC);
END pulse_gen;

ARCHITECTURE behave OF pulse_gen IS        -- behavioural architecture model of the entity
BEGIN
  P1: PROCESS
```

```
    CONSTANT pul_period: TIME:= 100 ns;
    CONSTANT pul_high: TIME := 60 ns;
BEGIN
LOOP
pul<= '1';
WAIT FOR pul_high;
Pul<= '0';
WAIT FOR pul_period- pul_high;
END LOOP;
END PROCESS;
END behave;
```

This code generates a pulse waveform with non-uniform duty cycle as shown in Figure 10.4.

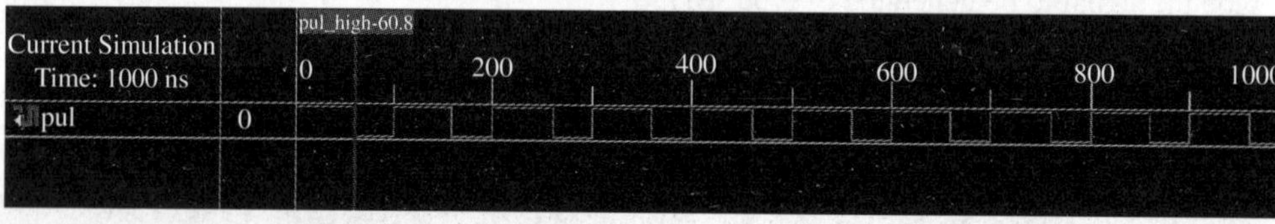

Figure 10.4 Waveform Showing Pulse with a Duty Cycle of 60%

SUMMARY

- Testbenches are used for the functional verification of the design.
- Testbenches are used to test vector generation, output monitoring and analysis of results.
- Entity for which the testbench is written is known as DUT or UUT.
- Testbench is written as VHDL entity which does not have any ports.
- A testbench essentially has test vectors, instantiation of DUT to testbench signals and optional output monitoring.
- Testbench can generate the test input values in the testbench architecture itself or can read values from a file.
- Output of testbench can be observed in simulation window or it can be stored in a file for analysis/comparison.

QUESTIONS

Objective Questions

1. Testbench is used to test.............. of design.

   a. architecture   b. syntax   c. functional behaviour   d. validity

2. Testbench is different from simple simulation in ....

   a. its treatment of inputs and outputs   b. its treatment of inputs

   c. its treatment of outputs   d. none of these.

3. For which of the following the entity declaration has no ports defined: ....
   a. testbench b. package c. component d. none of these
4. A testbench architecture is comprised of a ...., DUT AND output monitor.
   a. stimulus b. input c. signals d. components
5. Test vectors can be applied from a file. Allowed file types are ....
   a. ASCII, text, rtf,xls b. Text and ASCII c. all files d. none of these
6. Duty cycle is a measure of signal time for which it remains high. (True/False)
7. Testbenches use .... style of modelling.
   a. structural b. behvaioural c. mixed d. dataflow
8. For a synchronous system at least .... stimulus signals are required.
   a. 3 b. 2 c. 1 d. 4

Descriptive Questions

1. What is a testbench? Describe the architecture of a typical testbench.
2. What are the advantages of using a testbench?
3. What are intent models? Where are they used?
4. Write a simple testbench for the priority encoder.
5. How can procedures be used for stimulus generation? Explain with an example of D flip-flop with a synchronous reset.
6. Write a VHDL testbench for a positive edge detector. Use an assert statement to check for violation of set-up time to filter out glitches.
7. Is there any difference in writing testbench for a combinational circuit and a sequential circuit? If yes, what?
8. Design a generic n-bit up-counter and verify its functionality using testbenches which initiates the test vectors in the testbench itself. Also monitor the output against the expected output.
9. Design a successive approximation register in VHDL. To verify its functionality write a VHDL testbench.
10. Can clock signal be generated using testbench. If yes how? Extend the code to generate a clock with 40 per cent duty cycle.

CIRCUIT SYNTHESIS

11

CHAPTER OBJECTIVES

- What is synthesis
- Significance of synthesis in design flow
- Differences between synthesis and simulation
- Steps to synthesize a circuit
- Synthesizing combinational and sequential circuits
- Modelling and synthesis of finite state machines

11.1 INTRODUCTION

Synthesis of circuit is defined as a process of generating netlist from a circuit design model. Synthesis means 'to generate'. It is a step to generate circuit hardware schematics. VLSI design flow is revisited here (Figure 11.1), explaining the role of synthesis in digital design flow. In VHDL design flow, synthesis is performed for hardware generation after simulation verifies the logic functionality of circuit under design and is ready for its hardware implementation. Synthesis outcome maybe, gate-level schematic diagram from its RTL-level description generally obtained by pre-synthesis simulation. Synthesis is mapping of code in VHDL to circuit in terms of components comprising the circuit. It also gives information on important circuit parameters as propagation delays and critical path delays. In fact, the optimization of circuit design parameters is a part of the synthesis itself. The generated hardware is optimized using a user-specified target technology supported by VHDL synthesis module in terms of speed, area and power for the given family.

Circuit design flow is revisited here for an understanding role of synthesis in VHDL design process. Synthesis follows after functional verification of the circuit design which is done by simulation of the VHDL model. The result of simulation is RTL design which serves as the input for synthesis. This register transfer description of the circuit is then converted into gates schematic by the synthesis compiler. Since hardware components are generated in synthesis it requires the user to specify what target technology will he/she be using for implementing the circuit design in hardware. Synthesis compiler then optimizes the generated hardware with respect to speed, power and area considerations for the target technology.

Synthesis is a very important step in hardware implementation of the circuit. It is synthesis which translates code into equivalent hardware components. This consequently places many restrictions on the modelling style for the circuit under design as each and every statement defined in VHDL may not generate an equivalent hardware. So, a part of VHDL constructs are not synthesizable and may restrict the coding options to include only synthesizable constructs. The main reason for this is developing VHDL as a simulation language and not as a synthesis language. Moreover, synthesis is technology specific because it deals with hardware. Target technology for hardware implementation and optimization parameters need to be specified by the user. This too places restrictions on the portability of VHDL code across different synthesis tools.

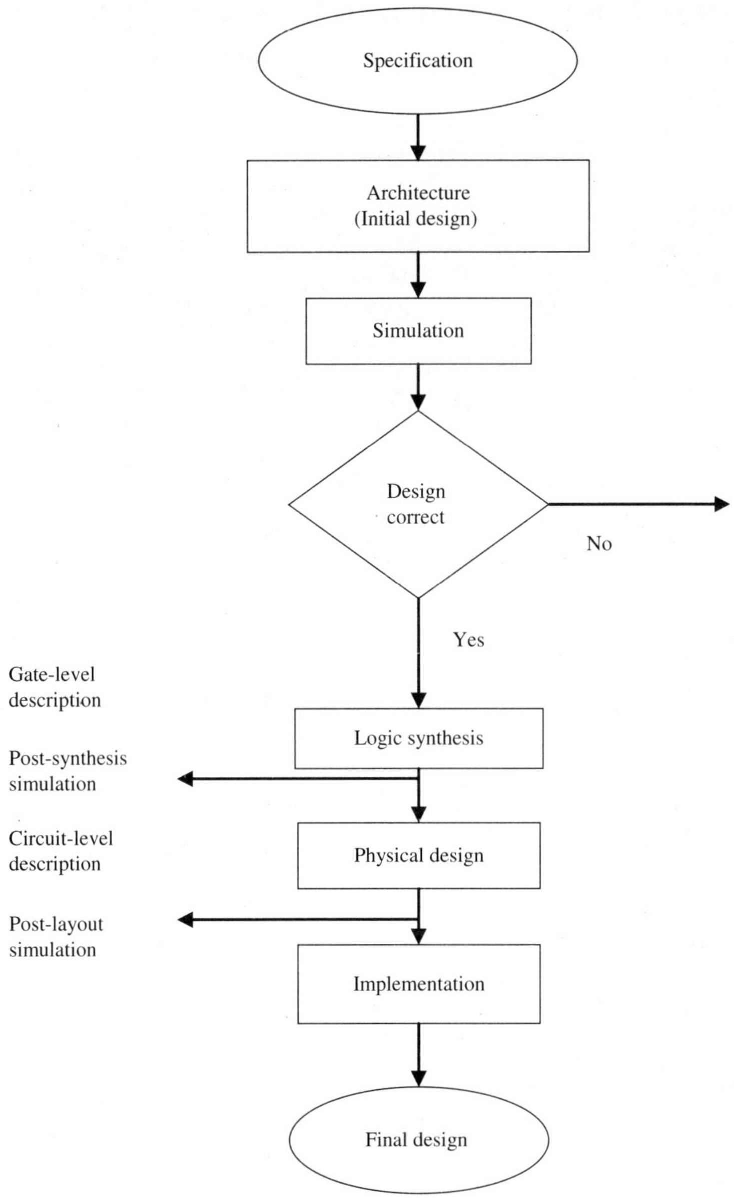

Figure 11.1 Design Flow

11.2 DIFFERENCES BETWEEN SIMULATION AND SYNTHESIS

Simulation and synthesis are complementary processes. Simulation is verification of VHDL code design of the circuit. It is the preliminary step to verify the correct functioning of the circuit in terms of waveforms showing output and input signal variations. It is the best way to save on hardware and still check the functional behaviour of the circuit without any hardware. Only if the simulation results indicate correct results as expected, does the designer proceed onto the implementation of design. Generally, simulation is an integral step at any stage of design as seen in the design flow, but most importantly two types of simulations are necessary in the design flow as follows:

- RTL simulation
- Functional simulation

RTL simulation is pre-synthesis simulation where the VHDL code for circuit under design is verified for correct results. This is a designer's first step to begin with a design. RTL simulation requires two inputs: VHDL code for circuit or entity under design and stimulus to drive the inputs. Here the code for the entity under design is compiled to check for syntax errors and then simulated using test values to verify that the design functions correctly. The flow diagram below shows the RTL simulation process. At this stage circuit is treated internally as register transfer logic, that is, flow of data between registers. The stimulus to inputs is given externally as a direct value in the simulation itself or using a testbench or in rare cases, for small designs input can be self-stimulating. The output of RTL simulation stage is functionally tested code which is then given to synthesis tool for schematic generation and further implementation.

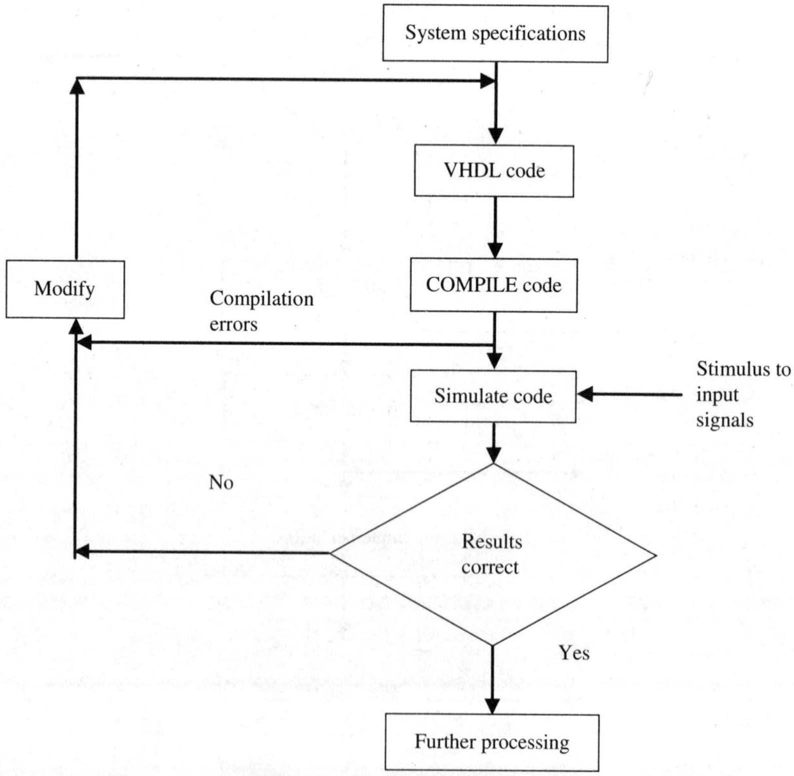

Figure 11.2 RTL Simulation Flow

Functional simulation on the other hand refers to post-synthesis simulation after a required optimization is performed and constraints applied. Its main function is not just to verify the functionality of design alone as done in RTL simulation, but to see the timing constraints, critical path analysis, total interconnect delays, and whether the functional simulation is performed. The wire-load models give estimates of interconnect delays, which help to compute timing values for propagation across the optimized design. In fact, this information is given to place and route which optimizes the design with a new place and route optimizations to meet the timing constraints. Some tools provide for static time analysis at this stage while others use another stage of functional simulation after place and route to see the timing constraints.

So, in simulation from the given VHDL coding of circuit which can be any level of abstraction or can be block diagram or FSM representation, its functionality is verified using waveform and timing analysis. Synthesis on the other hand, generates circuit as a netlist of gates from the given VHDL code.

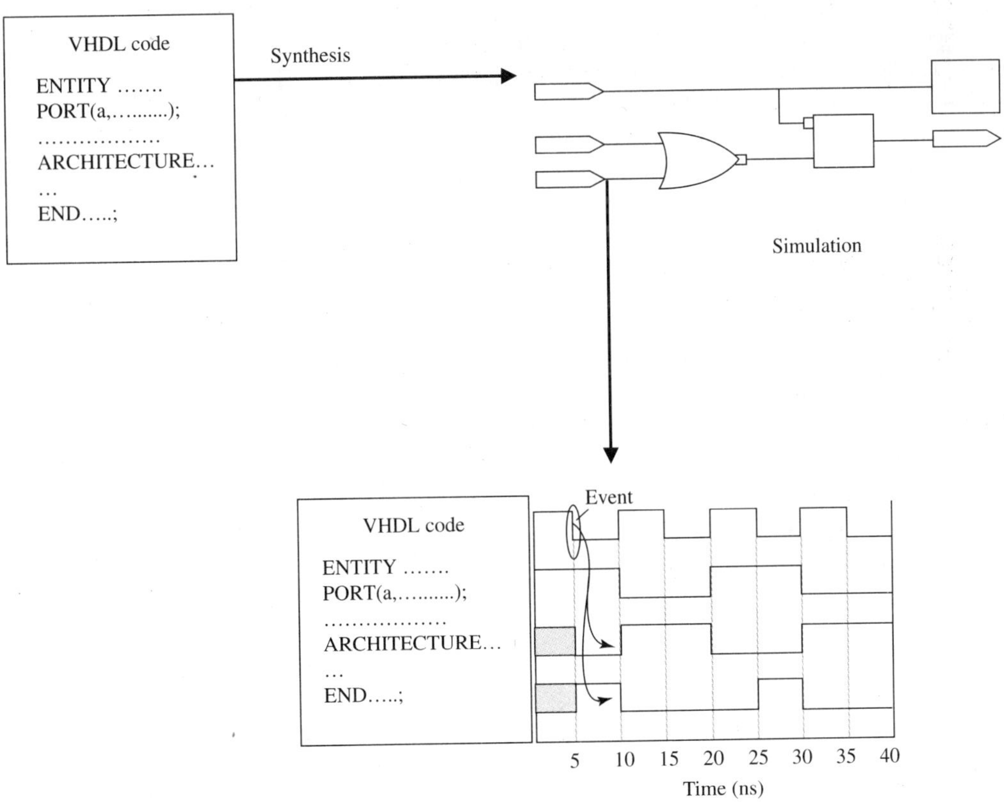

Figure 11.3 Synthesis vs Simulation

11.3 SYNTHESIS FLOW

Synthesis is the process of generating gate-level schematics or RTL schematics for an entity under design.

Synthesis consists of three phases such as

- Logic schematic
- Optimization
- Technology Mapping

The input to synthesis is VHDL code for circuit, the library of technology used and area as well as timing constraints. First of all, the generic (technology independent) schematic is generated based on the VHDL code input. This is synthesizing the circuit in terms of Boolean expressions. The next step is optimization. The output of synthesis without optimization is RTL schematic of the circuit. Most of the synthesis tools provide for both synthesis and optimization together. Optimization is circuit layout satisfying the area and timing constraints for a given technology. It requires the family of FPGA/CPLD to be specified so as to achieve optimization for the given family. The RTL schematic along with the library files from the family of FPGA/CPLD specified by the user and area and speed constraints serve as the input for optimization process. The outcome is gate-level circuit with given library components satisfying the user-specified speed and area constraints. The final phase of synthesis is to determine how to realize the circuit in hardware technology. This is technology-mapping phase outcome of which is optimized gate-level schematic of circuit in the given technology that is ready for implementation. Also, apart from the gate schematic, netlist and error log file are generated. A detailed synthesis report is generated. Synthesis does not check for logical errors but it checks only for syntactical errors and discrepancies in timing constraints for the given family.

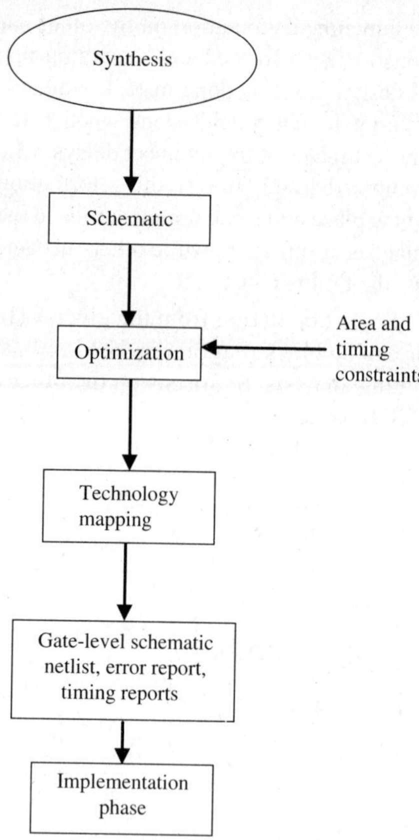

Figure 11.4 Design Flow for Synthesis

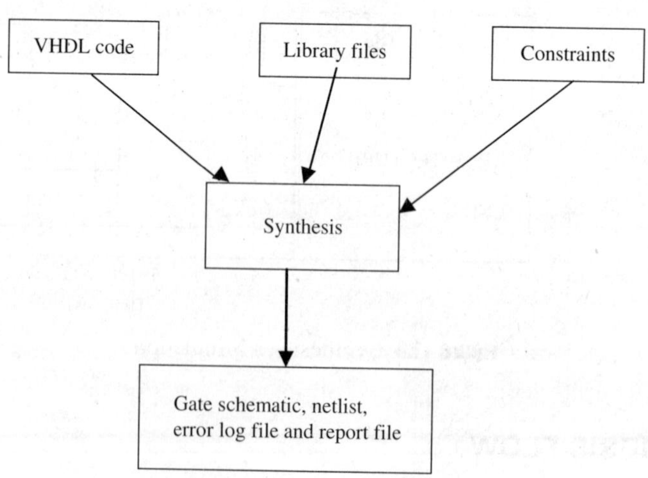

Figure 11.5 The Synthesis Process

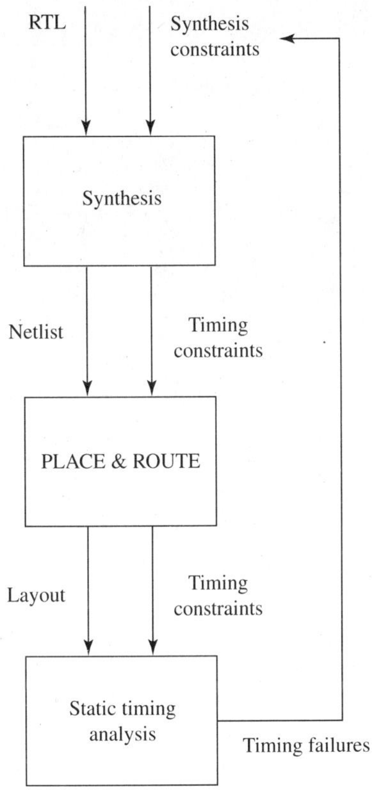

Figure 11.6 Timing Constraints in Synthesis

So, synthesis tool acts just like a compiler. VHDL code is the code of circuit to be synthesized. Synthesis coverts circuit code from high level of abstraction to lower level of abstraction closest to circuit. Since it produces gate-level circuit it is closest to structural style of modelling which describes the circuit in terms of components. The netlist is a representation of circuit in terms of components and nets connecting them in a text format. Error log file lists if any synthesis errors are generated such as unsynthesizable constructs as wait statement, syntax errors, and so on. The report file is a detailed information on how the synthesis is performed, system hierarchy, time taken and circuit analysis. Most of the synthesis tools allow the user to achieve speed versus area optimization. If the user does not specify any optimization constraints, the synthesis tool picks up the default values for the given family and optimizes the design both for speed and area partially. Optimization preferences determine the topology of the circuit. Generally, if the speed is more or fast for optimization the area required will be more and vice versa. This is because, in general, for fast designs more of parallelism is incorporated in the structure. More parallel structures require more area. Most of the synthesis tools use wire-load models for computing the interconnect delays to achieve speed optimizations. These estimates are then used to achieve gate-level speed optimization.

Synthesis has a very important role to play in the final design because of the effect on timing constraints. As seen in the following figure, after the synthesis any timing violations are checked for by placement and route tool if any discrepancies are found. Timing violations are reports to synthesis tools for modifications. So, choosing the speed and area optimizations are to be carefully done. Due to timing failures, fast optimization may result in slower design.

The following example shows the coding for ripple adder:

```
LIBRARY IEEE;
USE IEEE.STD_LOGIC_1164.ALL;
USE IEEE.STD_LOGIC_ARITH.ALL;
USE IEEE STD_LOGIC_UNSIGNED.ALL;

ENTITY word_adder IS
  PORT (a : IN STD_LOGIC_VECTOR(31 DOWNTO 0);
    b : IN STD_LOGIC_VECTOR(31 DOWNTO 0);
      c_in: IN STD_LOGIC;
      s : OUT STD_LOGIC_VECTOR(31 DOWNTO 0);
      c_out : OUT STD_LOGIC;
    ovfl : OUT STD_LOGIC);
  END word_adder;
```

```
ARCHITECTURE ripple OF word_adder IS
BEGIN
add_op: PROCESS (a, b, c_in)
   VARIABLE result : STD_LOGIC_VECTOR(31 DOWNTO 0);
   VARIABLE intern : STD_LOGIC;
   VARIABLE carry_out, carry_in : STD_LOGIC;
BEGIN
    carry_out := c_in;
    FOR index IN 0 to 31 LOOP
       carry_in := carry_out;                                    -- of previous bit
       intern := a(index) XOR b(index);
       result(index) := intern XOR carry_in;
       carry_out := (a(index) AND b(index)) OR (carry_in AND intern);
    END LOOP;
    ovfl <= carry_out XOR carry_in;
    s <= result;
    c_out<= carry_out;
 END PROCESS add_op;
END ripple;
```

Now, synthesizing the circuit for fast implementation generates the following results:

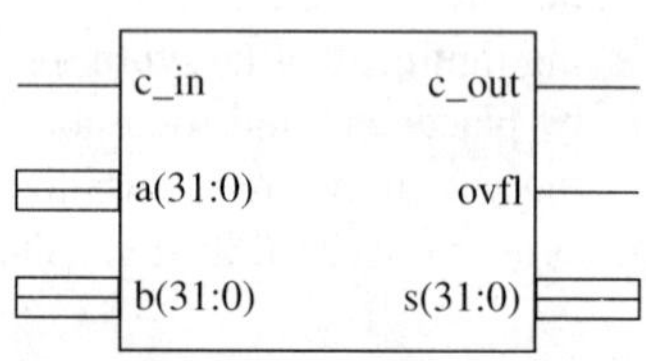

Figure 11.7 31-bit Adder

Final Results	
Optimization Goal	: Speed
Cell Usage :	
# BELS	: 206
# AND2	: 94
# OR2	: 47
# XOR2	: 65
# IO Buffers	: 98
# IBUF	: 65
# OBUF	: 33

CPU : 9.06 / 9.67 s | Elapsed : 9.00 / 10.00 s
Now optimizing for area

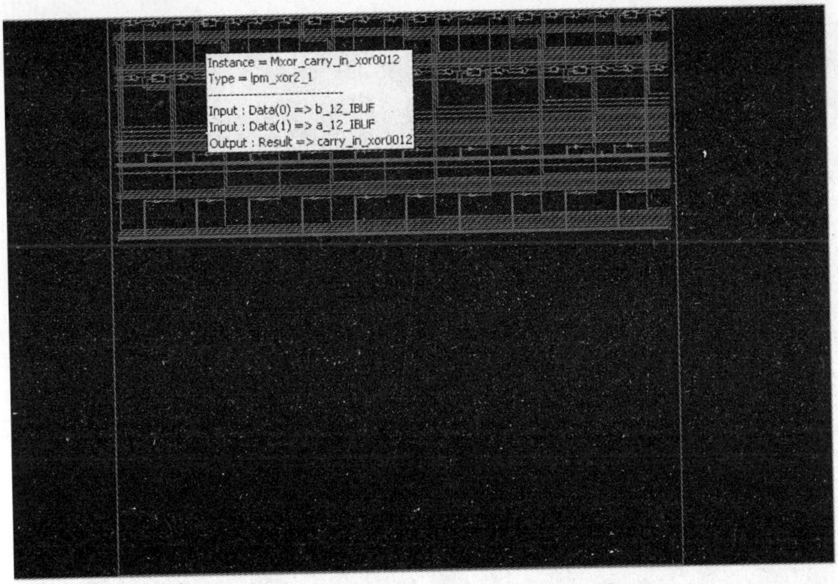

Figure 11.8 Technology Schematic for 31-bit Ripple Adder

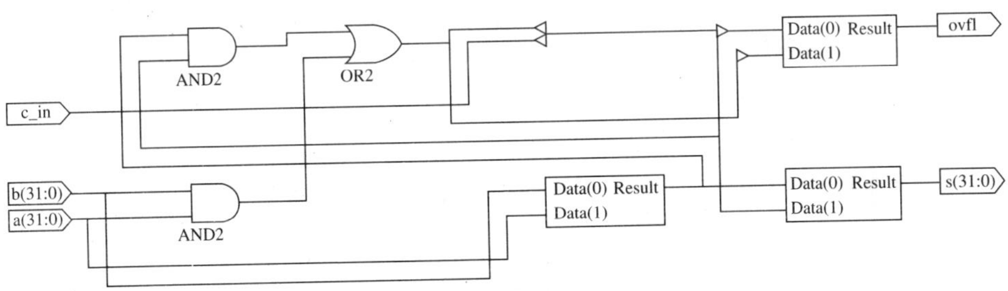

for 31-bit ripple adder

Figure 11.9 RTL Schematic

Final Report*	
Final Results	
Optimization Goal	: Area
Cell Usage :	
# BELS	: 161
# AND2	: 64
# OR2	: 32
# XOR2	: 65
# IO Buffers	: 98
# IBUF	: 65
# OBUF	: 33

CPU : 6.05 / 7.89 s | Elapsed : 7.00 / 8.00 s

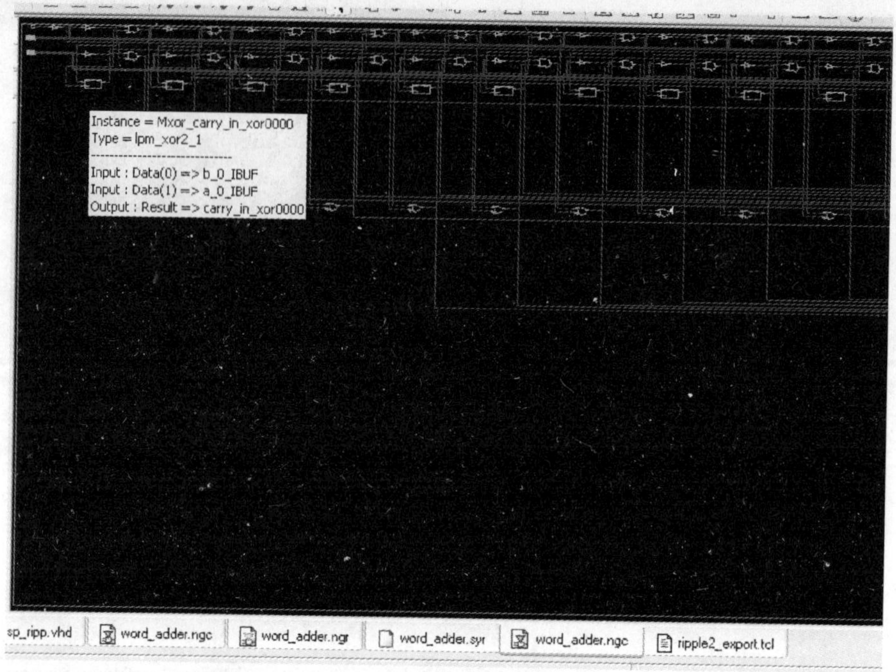

Figure 11.10 Technology Schematic for Optimized Area

As seen from the reports of synthesis though the area optimization achieves its goal significantly well with 161 components against speed optimization which uses 206 gates. the speed goal is not achieved although synthesis is optimized for a speed time taken is 9.00 s against 5.00 s for area-optimized circuit . So, the design instead of becoming faster has slowed down. The reason for this is explained by the banana curve. The relation between time and area is given by banana curve, in general, a design which is optimized for area will be small but slow and the design optimized for speed will be fast but large. But in the above case due to wrong timing constraints. Now, if we modify our constraints and optimize for speed the following results are achieved.

Final Results	
Optimization Goal	: Speed
Cell Usage :	
# BELS	: 206
# AND2	: 94
# OR2	: 47
# XOR2	: 65
# IO Buffers	: 99
# IBUF	: 65
# OBUF	: 34

CPU : 5.26 / 5.67 s | Elapsed : 6.00 / 6.00 s

So, timing constraints have a very important role to play in synthesis. In fact, all synthesis tools are based on optimization in relation to timing constraints. These constraints can be specified by the user or the default constraints for the given technology can be picked up from the technology library.

11.4 SYNTHESIZING DATA TYPES

Understanding the treatment of data types in synthesis tool is very important as it determines the size of the data objects that are used in design.

Std_logic: Logic values corresponding to states '1' , 'H' , '0' and 'L' can be used in any manner. Some synthesis tools do not accept the uninitialized logic state 'U'. Logic states 'Z' , 'X' , 'W' and '-' can be used in signal and variable assignments. 'Z' is synthesized to three-state logic. The states 'X' , 'W' and '-' are treated as unknown or don't care values. The values 'Z' , 'X' , 'W' and '-' may be used in the conditional clauses such as the comparison in an 'if' or case statement. However, some synthesis tools will ignore them and only match the surrounding '1' and '0' bits. Consequently, a synthesized design may behave differently from the simulation if a stimulus uses 'Z' , 'X' , 'W' or '-' . So it is always preferable to use IEEE synthesis packages STD_MATCH function for comparisons.

Integer Types: Integer types are treated internally in bit representation. For signed integers, the given range of integers is represented in 2's complement form with number of bits equal to minimum number of bits required to code the largest inter value in the specified range. For unsigned it is simple with bits representing the integer value with a minimum number of bits again determined by the largest value in the range.

Enumeration Types: Enumerated data types are treated in a similar fashion as integers where the largest integer value to be represented for given type is highest position value in the enumerated type.

FOR EXAMPLE:

TYPE week IS (Mon, Tue, Wed, Thur, Fri, Sat, Sun).

In this starting from 0 to Monday, 7 is assigned to Sun internally to represent 7 in a bit format minimum 3 bits are needed.

Arrays and Vectors: Single as well as multidimensional types are supported by synthesis. Arrays represent collection of wires or flip-flops. Array values can be accessed using the index value with an array name where index can be a constant value or a non-constant value.

11.5 SYNTHESIS OF COMBINATIONAL LOGIC

Combinational circuit is a circuit containing no memory element. So, most of the signals are modelled as wires or level-sensitive latches in the hardware logic. Most of the times a combinational circuit is synthesized in the form of registers, until and unless restricted by the user. VHDL coding for combinational circuits uses concurrent signal assignment statements or a process statement which is level sensitive. Level-sensitive process means whose signals in the sensitivity list are not using their attributes in the process. Moreover, synthesizable process for a combinational logic should have signals whose value is read inside the process as a part of sensitivity list and all signal assignment statements are covered by the conditional, combinational statements.

Example 11.1

```
LIBRARY IEEE;
USE IEEE.STD_LOGIC_1164.all;
ENTITY comb_sens IS
PORT(a,b,clk: IN STD_LOGIC;
        c: OUT INTEGER);
END comb_sens;

ARCHITECTURE comb_sens OF comb_sens IS
 BEGIN
PROCESS(a,clk)
BEGIN
IF Clk'EVENT AND Clk='1'THEN                 -- not a level-sensitive process
c<= a AND b;                                 -- b is not part of sensitivity list
  END IF;
END PROCESS

END comb_sens;
```

In the above example, the process is not a level-sensitive process as clk signal is using attribute 'EVENT' and hence, the synthesizer will generate a warning also. For a combinational logic all signals to be read (in this case, a, b and clk) should be included in the sensitivity list.

Synthesizing Assignment Statement: Signal assignment statement is synthesized as b<=a;

Here signal on right hand side drives the output 'b'. The output of assignment statement can represent a wire, a flip-flop, bus depending on the context in which signal assignment is made.

In the above example, simple assignment statement represents a wire.

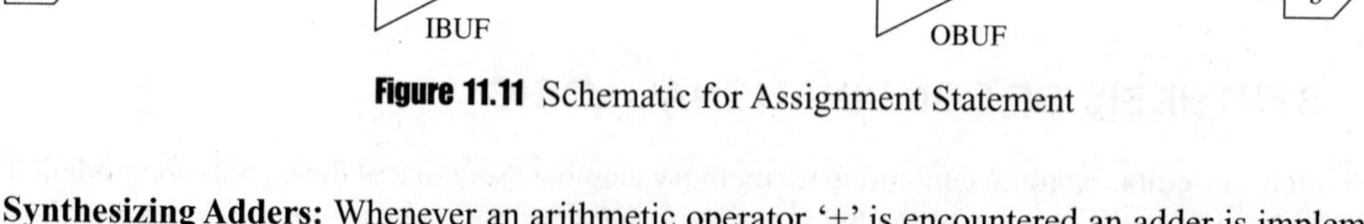

Figure 11.11 Schematic for Assignment Statement

Synthesizing Adders: Whenever an arithmetic operator '+' is encountered an adder is implemented by the synthesizer to perform the operation. Though for bit_vectors an addition operation is not defined an overloaded addition operator can be used while for std_logic_vectors. It is defined for unsigned/signed types and needs to include library file IEEE.std_logic_unsigned all package. Else, addition is defined for bits and integers. In case of integers, an internal adder of 32 bits is generated, so as a save on time, it is always better to specify the range of integers in case the integers to be added can be represented using bits less than 32.

Example 11.2 Addition of integers

```
LIBRARY IEEE;
USE IEEE.STD_LOGIC_1164.ALL;
USE IEEE.STD_LOGIC_ARITH.ALL;
USE IEEE STD_LOGIC_UNSIGNED.ALL;

ENTITY adder IS
PORT (a,b : IN INTEGER;
        sum : OUT INTEGER);
END adder;

ARCHITECTURE add OF adder IS
BEGIN
Sum <= a + b;
  END add;
```

This will be implemented by the synthesizer as

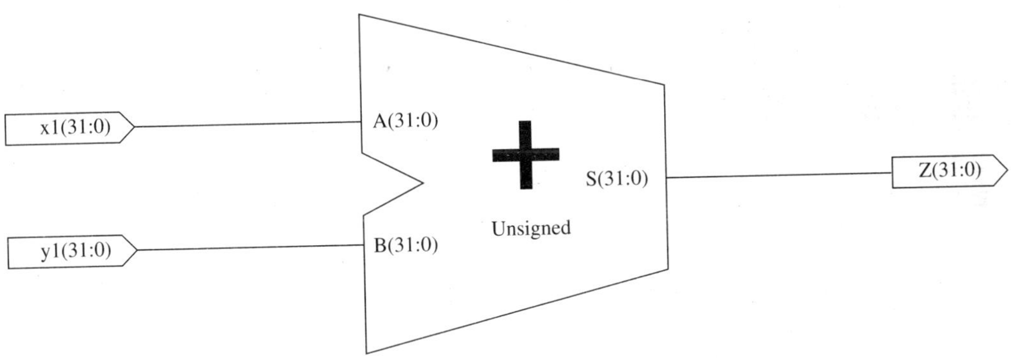

Figure 11.12 Integer Addition

Adders in VHDL can be synthesized for adding n bit numbers. If the number of operands are more than two, it is always advantageous to use parenthesis as it helps to make the synthesis faster because at a time only two operands are considered.

For Example:

For sum <= a + b+ c +d;

Will be evaluated as x = a+b

y = x +c

sum = y + d; so three addition operations are performed. Using parenthesis,

sum <= (a+b) + (c+d)

will generate the same 3 addition operations faster because of lower stacking values.

Synthesizing Multiplexers: Multiplexers can be synthesized using case, when select statement or IF-elseif ladder. All the three produce similar results.

Example 11.3 Multiplexer

```
LIBRARY IEEE;
USE IEEE.STD_LOGIC_1164.ALL;
USE IEEE.STD_LOGIC_ARITH.ALL;
USE IEEE STD_LOGIC_UNSIGNED.ALL;

ENTITY Mux IS
 PORT
(ip:IN BIT_VECTOR(3 DOWNTO 0); sel: IN BIT_VECTOR(1 DOWNTO 0); s: OUT BIT);
END Mux4;
ARCHITECTURE syn of Mux is
BEGIN
PROCESS (sel, i)
BEGIN
CASE sel IS
WHEN "00" => s <= i(0);
WHEN "01" => s <= i(1);
WHEN "10" => s <= i(2);
WHEN "11" => s <= i(3);
END CASE ;
END PROCESS ;
END syn;
```

This code is synthesized as in time 9.99 ns and CPU time 10.28 s:

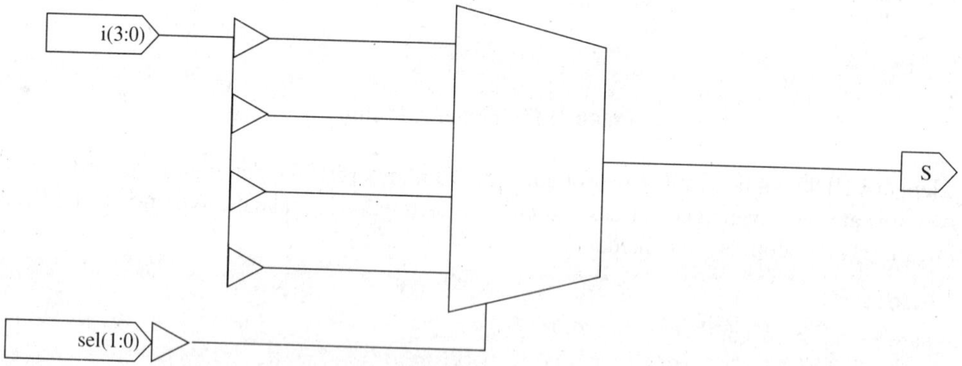

Figure 11.13 Mux Schematic

Using when select statement synthesis of mux is with datapath time 9.99ns and CPU time 10.53 sec.

So, the use of case or concurrent conditional assignment statement does not make much of a difference. Synthesizing of decoders is similar to that of the multiplexers as the decoders can be realized using with-select-when construct or case statement using level-sensitive process.

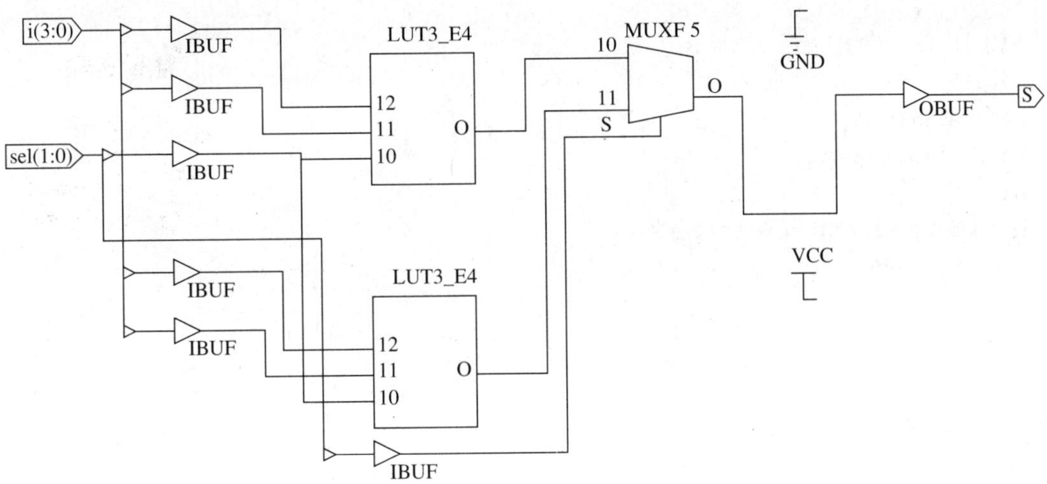

Figure 11.14 Schematic for Mux with Select Statement

11.6 SYNTHESIZING SEQUENTIAL LOGIC

Sequential circuit is a combinational circuit with memory elements. Sequential circuits have a feedback path which requires memory as an output is a function of both present and past inputs. In hardware, the sequential circuit VHDL objects can be mapped to wires, flip-flops or latches. Mapping of an object done by wire or flip-flop depends on the logic in which the object is used. Generally, with clock, data objects generate flip-flops. When it comes to synthesis of sequential circuits, a major impact is made by use of signals and variables.

Sequential logic uses an edge-sensitive process, that is, a process statement with sensitivity to the edge of a signal. This can be realized using sensitivity list and test for at least one signal in the sensitivity list for 'EVENT' attribute and positive or negative edge-level condition. It is also possible to use 'wait' until statement without ant sensitivity list. The other forms of wait are not synthesizable. Since the sequential logic requires memory, past values as well as present values are needed. This is possible for input signals not contained in the sensitivity list of the process but being used in the processor using a data object before being assigned a new value or using multiple wait statement for separate control paths.

Variables have a characteristic of taking the value immediately on an assignment without any delay. So, most of the times, variables are mapped in hardware to wires except for a typical case when variables are used before their assignment; in this case, variables generate flip-flop as they are using their previous value and a new assignment is not yet made. So, memory is required which uses flip-flops.

Example 11.4 Variables generating flip-flops

```
LIBRARY IEEE;
USE IEEE.STD_LOGIC_1164.ALL;
ENTITY var IS
  PORT
(ip,clk:IN BIT; s: OUT BIT);
END var;
```

```
ARCHITECTURE syn of var is
BEGIN
PROCESS (clk)
VARIABLE a,b: BIT;
BEGIN
IF clk'EVENT AND clk='1' THEN
s<= a XOR b;;
END IF;
a:= ip;
b:=a;
END PROCESS;
END syn;
```

This code generates the following schematic which has a flip-flop.

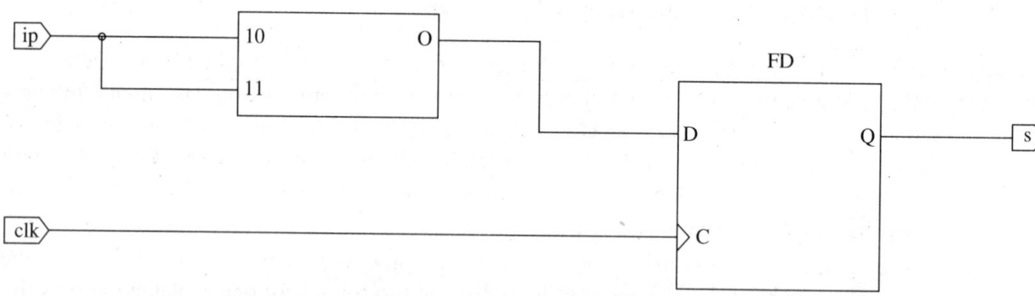

Figure 11.15 Flip-Flop Schematic

Now, if the same code is written as

Example 11.5 Variable use in synthesis

```
LIBRARY IEEE;
USE IEEE.STD_LOGIC_1164.ALL;
ENTITY var IS
  PORT
 (ip,clk:IN BIT; s: OUT BIT);
END var;
ARCHITECTURE syn of var is
BEGIN
PROCESS (clk)
VARIABLE a,b: BIT;
BEGIN
```

```
a:= ip;
b:=a;
IF clk'EVENT AND clk='1' THEN
s<= a XOR b;;
END IF;
END PROCESS;
END syn;
```

The same code when synthesized with a variable assignment before its uses is synthesized as simple wire is shown as follows:

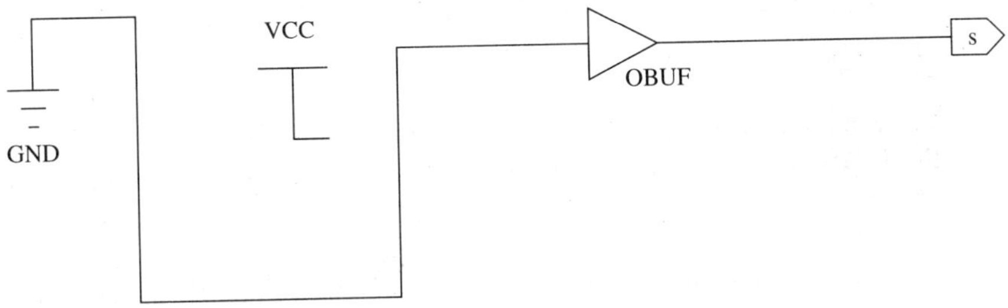

Figure 11.16 Technology Schematic for the Above Code

Synthesizing Flip-flops: For edge-sensitive process having clock is always synthesized as flip-flop.

Example 11.5 Flip-flop with synchronous reset

```
LIBRARY IEEE;
USE IEEE.STD_LOGIC_1164.ALL;
ENTITY ff_r IS
PORT (Clk, Reset, D : IN STD_ULOGIC;
Q : OUT STD_ULOGIC);
END ff_r;
ARCHITECTURE syn OF ff_r IS
BEGIN
PROCESS (Clk)
BEGIN
IF (Clk'EVENT AND Clk = '1') THEN
IF (Reset = '1') THEN Q <= '0';
ELSE Q <= D;
END IF;
END IF;
END PROCESS ;
END syn;
```

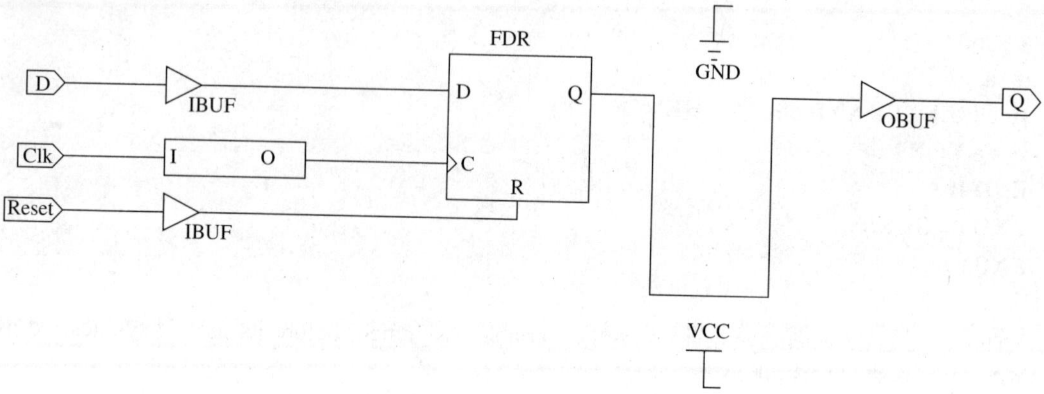

Figure 11.17 Schematic for Flip-Flop with Synchronous Reset

Note that the reset has not been included in the sensitivity list as change in its value does not affect the working of flip-flop until the clock has a rising edge. But for asynchronous rest flip-flop it is necessary to include the reset in the sensitivity list.

The same code will be synthesized as a latch, if the output signal is not assigned a value in every conditional statement branch and there is not an edge detection. If in a PROCESS one has not defined what to do in all possible cases then a latch is necessary to generate the functionality— even if no clock is defined. If one has defined what to do in all possible cases and there is no defined clock then the combinational circuitry will be built.

Example 11.6 Generating a latch

```
LIBRARY IEEE;
USE IEEE.STD_LOGIC_1164.ALL;
ENTITY ff_r IS
PORT ( Reset, D : IN STD_ULOGIC;
Q : OUT STD_ULOGIC);
END ff_r;
ARCHITECTURE syn OF ff_r IS
BEGIN
PROCESS (reset)
BEGIN
IF (Reset = '1') THEN
Q <= '0';
END IF;
END PROCESS ;
END syn;
```

This code does not generate a flip-flop but a latch as shown

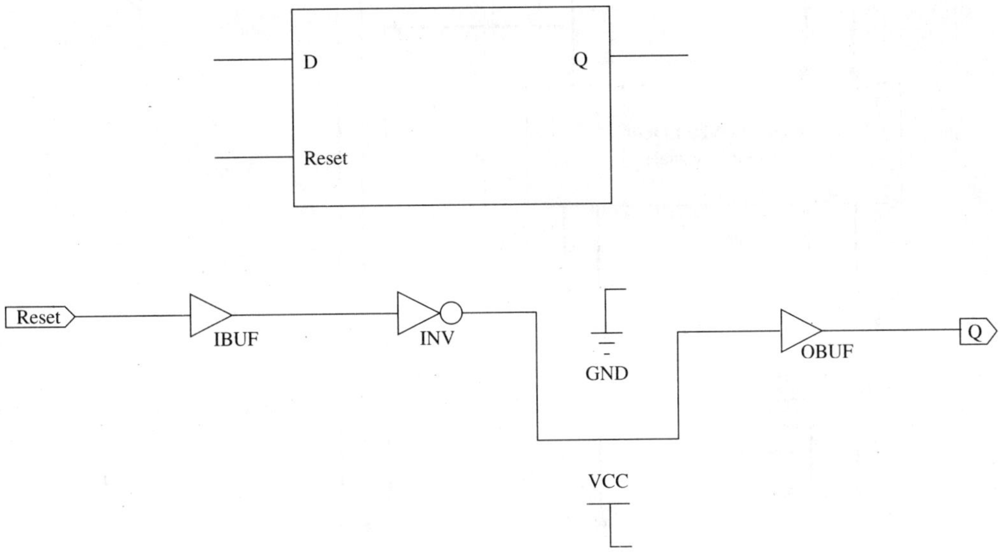

Figure 11.18 Schematic Synthesis for the Earlier Code

11.7 MODELLING FINITE STATE MACHINES

State machines are sequential logic systems in which the output is a function of present as well as past inputs. Further, a state machine can be Mealy machine or a Moore machine. Moore machine has an output as a function of only the present state and does not depend directly on the current inputs. On the other hand, Mealy machine has an output as a function of present state and current inputs.

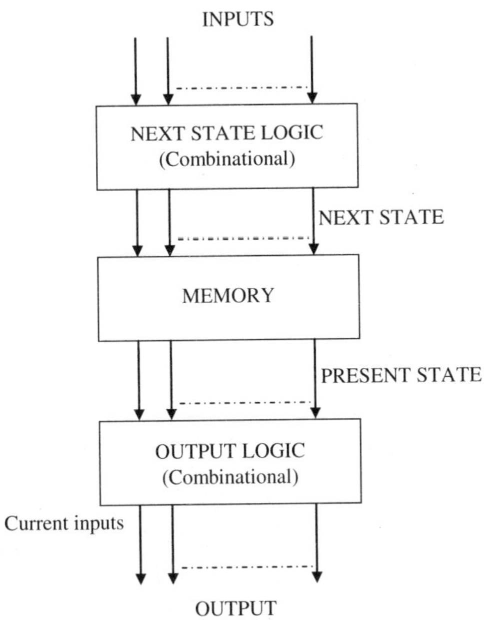

Figure 11.19 Moore Machine

Modelling of state machines has certain typical requirements. Firstly, designer has to decide the form of the state machine—Mealy or Moore. Next, to represent the state of state machine enumerated data type is required. Signal of type this is used to access the present state. Multiple processes are required. Three process, namely, for-updating the next state using next state logic, another for updating present state and the third one for updating of outputs. Each process further uses case statement (if-else could be used, but case is more efficient). In case a single process is used for state machine design, buffered outputs are used by synthesis tools to store the past values required. Case statement used in finite state modelling does not have an entry for others case as its inclusion would bring in extra logic for unused states. It is always better to use don't cares for unused states of FSM (finite state machine).

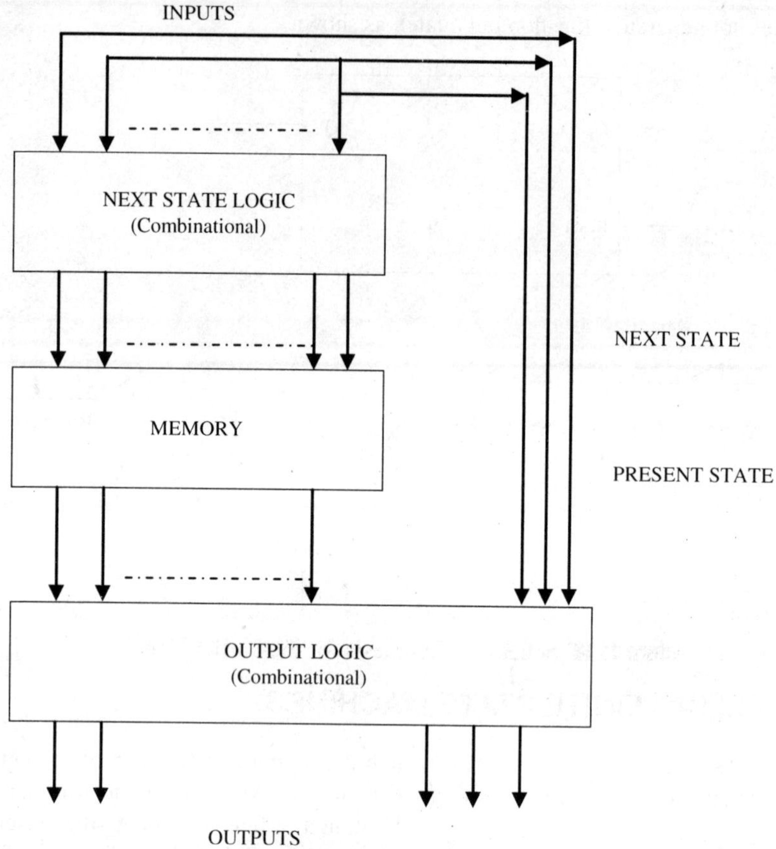

Figure 11.20 Mealy Machine

Next in the transition process if any assignments for output state or next state are missed under certain conditions where they do not change, an error is generated by a synthesis tool. So, it is necessary to make assignments to output and next state signals under all conditions.

Example of State Machine: Consider a state machine with the following state diagram. It has three states A,B and C. The input is w and the output is z.

Example 11.7 FSM coding

```
LIBRARY IEEE;
USE IEEE.STD_LOGIC_1164.ALL;
ENTITY fsm IS
PORT ( Reset, w,clk : IN STD_ULOGIC;
z : OUT STD_ULOGIC);
END fsm;
```

```
ARCHITECTURE behave OF fsm IS
TYPE state IS (A,B,C);
SIGNAL y:state;
BEGIN
 PROCESS (reset,clk)
BEGIN
IF (Reset = '0') THEN
 y <= A;
ELSIF (clk'EVENT AND clk='1') THEN
 CASE y IS
WHEN A=>
   IF w='0' THEN
      y<=A;
   ELSE
      y<=B;
   END IF;
WHEN B=>
   IF w='0' THEN
      y<=A;
ELSE
      y<=C;
END IF;
```

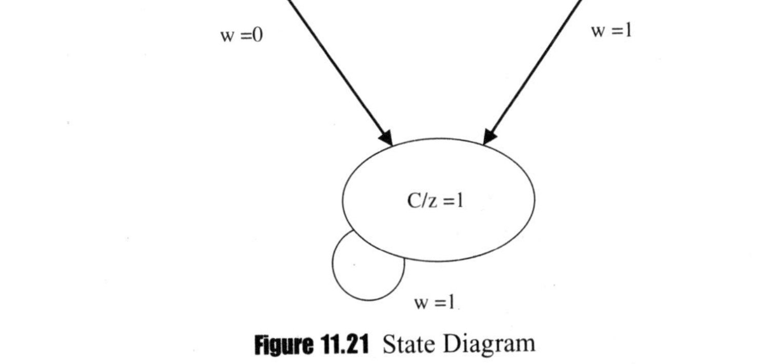

Figure 11.21 State Diagram

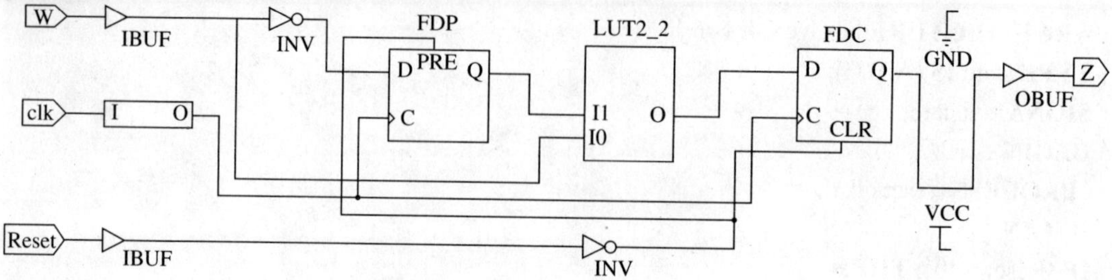

Figure 11.22 Schematic for FSM

SUMMARY

- Synthesis is converting higher level of abstraction to lower level of abstraction.
- Synthesis and simulation are complementary processes.
- Synthesis takes the input as RTL schematic or VHDL code and generates gate-level schematics.
- Synthesis tools perform optimization along with synthesis.
- Synthesis optimization is for area and speed. Generally, for fast-circuit implementations area is large and vice versa.
- VHDL is basically not a synthesizing language but a simulation language. So, a simulated code is not necessarily synthesizable.
- Not all VHDL constructs are synthesizable in VHDL.
- Combinational logic is synthesizable with level-sensitive process and sequential logic is synthesizable for edge-sensitive process.
- Finite state machine synthesis uses multiple processes and case statement.

QUESTIONS

1. Explain the role of synthesis in digital design and implementation.
2. Explain the complete synthesis flow.
3. What is static time analysis?
4. Draw the comparison between simulation and synthesis.
5. What are FSMs? Explain the difference between mealy and moore machines.
6. Discuss the differences between combinational and sequential circuits. How do these differences affect the synthesis process?

12 EMBEDDED DESIGN USING VHDL

CHAPTER OBJECTIVES

- To understand what is an embedded system
- Architecture of embedded system
- VHDL coding for each of the components of embedded system
- How to achieve design optimizations
- Limitations of VHDL for embedded system design

12.1 INTRODUCTION

Embedded system refers to a computer which performs a specified task against the general purpose computers and includes both hardware as well as software. Broadly speaking, it is a digital integrated system which has a one or more special purpose processors embedded in them. These special purpose processors are generally referred to as processor cores. Processor cores run the special function that is required to be performed by the embedded system. In fact, with technological advances almost every electronics gadget is an example of embedded system at some level of complexity, for example, cell phones, music players, digital cameras, pop vending machines, electronic toy cars, automobile security systems, home automation systems and so on.

Embedded system is generally a part of a bigger system which has a specific task to perform and for carrying out its functions it makes use of a processing element—that is, its Central Processing Unit (CPU). CPU can be a microprocessor or a microcontroller depending on the level of the complexity of the system. Low-end consumer products may use a microprocessor while complex applications always use microcontroller as their processing base. VHDL as a design language is now being increasingly used for embedded system design because of the flexibility achieved by using FPGAs instead of the processor cores.

Characteristics of Embedded Systems

- Embedded systems are programmable hardware controlled with software to perform specific functions. The software used is called firmware.
- Embedded system is generally not a standalone system but a part of a bigger system which may have mechanical parts and the processing part is done by CPU of the embedded system.
- Low cost for application-specific systems is one of the major concerns in the embedded systems along with performance against general computing machines where performance is not bound by cost.

- Mostly, the real-time processing is done in the embedded systems which requires a high degree of reliability as the entire system fails even if a small error occurs in the responding mechanism to real situations.

12.2 ARCHITECTURE OF EMBEDDED SYSTEM

Though the embedded system architecture depends on the specific task it is going to perform, in general, the embedded system has the following components:

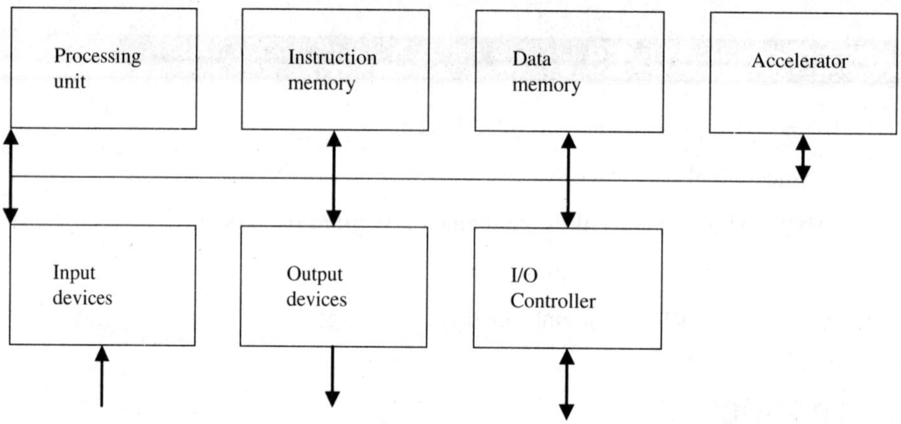

Figure 12.1 Architecture of Embedded System

Embedded system architecture revolves around CPU which can be microprocessor or microcontroller. The choice depends on the processing functions that are to be performed. If simple arithmetic, jump and logical operations are required, the microprocessor is used as the CPU; but if complex computations and control actions are required, then microcontrollers are used. With microprocessor memory and I/O controllers are to be included separately whereas in microcontrollers' memory, CPU and I/O controllers are all included in a single chip. Another alternative to very sophisticated, complex problems is using CPU as an FPGA component which gives the advantage of being application specific, but at the same time has the flexibility of being programmable. This facilitates customizing of I/O controllers as per user requirements. Next to CPU is Memory—RAM and ROM. In general, the embedded systems are Harvard architecture where separate memory space is used for data and instruction, the reason being the specific application for the embedded system which fixes the instruction memory.

I/O devices are the usual PC interfaced devices such as keyboards, mouse, keypads, microphones, sensors, A/D convertors, knobs, and so on, and output devices include display, LEDs, and actuators for physical systems.

I/O controllers process the data from input devices in a readable form for embedded system and at output end convert data from embedded processor to required output form for output device. The I/O controllers depending on the application can be in the simplest form an input register to read and store input values from input device and an output register to get the output values and give them to the output device. For complex applications, the I/O controllers can be in the form of programmable autonomous units where the function is not just reading and storing the values but to convert them in the required forms at the input as well as in the output.

Accelerator is an embedded system application-specific component which helps in faster processing and gives an increased throughput of the embedded system; hence the name accelerator.

All these components are interconnected by bus. Again, depending on complexity of application a single bus may be used or a multiple buses may be used.

12.3 DESIGN FLOW FOR EMBEDDED SYSTEM DESIGN

Embedded system is a combination of hardware and software to drive the hardware.

For an embedded system design using VHDL or, for that matter, using any HDL hardware and software, a codesign approach is followed as it helps to achieve a good design by balancing hardware and software concurrently. Also safe, faster and reliable design is obtained while meeting strict design constraints for power, speed and area optimizations. Firstly, the architecture of the system under design has to be specified wherein all blocks composing the complete system need to be identified along with their functional requirements. While partitioning the design into hardware and software, the degree of programmability has to be determined. The degree of programmability means determining who has the access to programming and what is the level of programming. Once the hardware and software partitioning of the design is finalized, concurrent coding of software part and hardware part is done, which is then verified at each step. This approach of codesign helps to speedup the design process manifold as hardware as well as software parts and can be designed, verified and modified independently. Once the correct functionality has been achieved for both, the next step is integrating the two system divisions into one functional system which is synthesized and simulated to check its functionality. After post-synthesis simulation, physical implementation of system is done by mapping onto an FPGA and finally, if design is correct it is manufactured as a ready-to-market product.

12.4 ISSUES IN EMBEDDED SYSTEM DESIGN

12.4.1 Design Optimizations

Area: Reducing the chip area is an important consideration in the embedded design to reduce the cost of fabrication. Area optimization is achieved in the early stages of design process and little is left to be optimized at the final physical design stage. At the first step of design—architecture exploration and partitioning, the maximum can be done to minimize the area. Placement of blocks in the partitioned design should be such that a minimum area is only used. Alongside, an estimate should be made about the pin count required. Estimating the pin count at an early stage helps as alternative architectures with reduced pin count can be considered at the initial stage only. The next important stage for area optimization is synthesis. Often area optimization is specified as a constraint in the synthesis tool wherein the tool can be directed to minimize the area in the synthesized design.

Finally, at the physical design stage area optimization can be achieved in floorplanning, routing and placement steps. But the changes in the design at this stage involves a lot of rework; and hence it leaves scope of minimal changes to design.

Timing: Timing optimization minimizes the time required per operation and hence serves as an important performance measure. It can be achieved by parallelism constrained by data dependencies in the system. Speed optimization is also a part of most synthesis tools but area and power optimization are conversely related to timing optimization. So, tradeoff is made between area and speed optimization. At the initial stages of architecture exploration and partitioning, timing optimization

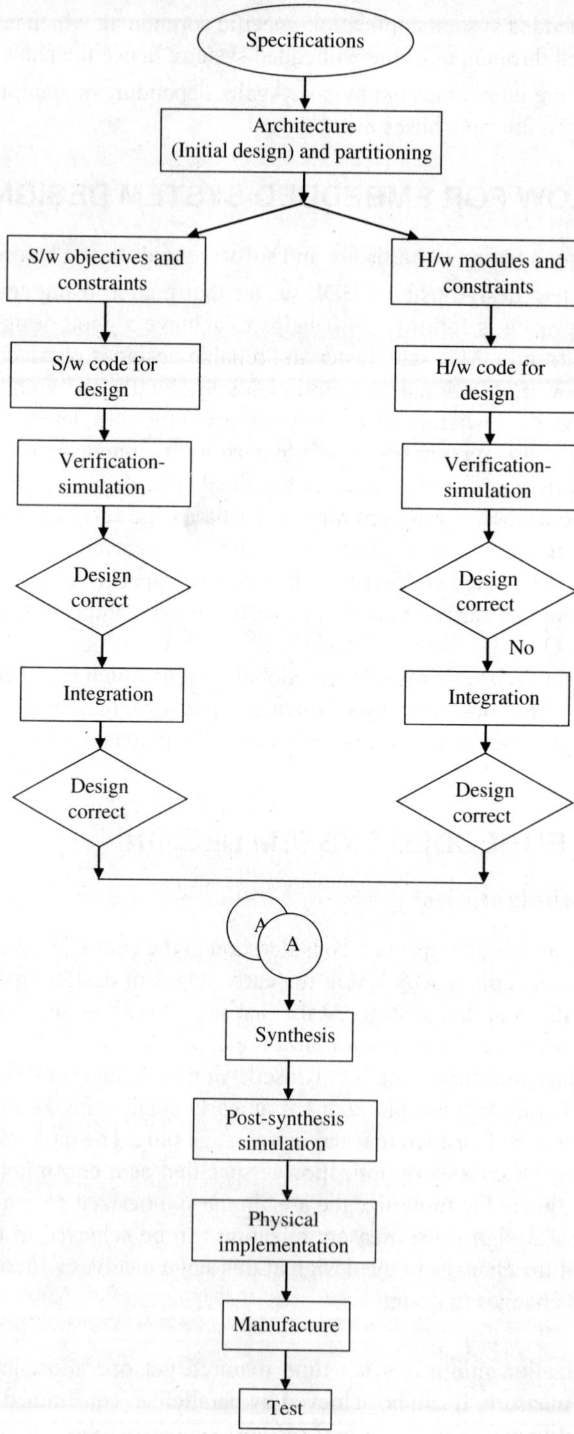

Figure 12.2 Design Flow for Embedded System Design

requires clock frequency as clock period constrains the propagation delay of the combinational circuit. At the synthesis stage timing optimization is influenced by choice of components and is done by tool by choosing timing optimization over area and tool synthesizes circuit to meet the timing constraints specified in a UCF file. Static timing analysis is also used to analyze synthesized design after place and route stage wherein timing estimates are used for technology library that are used to compute wire load models. Finally, in the physical design the placement of components determines the final time required per operation.

Power: Power optimization is a desirable one for the embedded systems, in particular, for battery-operated devices not only to give longer life to device on single battery but also to minimize cost as more power means more heat dissipation and more of heat sink methods required to absorb the dissipated power. Most of the area optimization methodologies also help to achieve power optimization as lesser area means lesser components, and hence less power. Apart from these different approaches are being used for power saving. One such approach is clock gating where the clock is switched off to components as flip-flops and registers whose value is not getting changed. This helps to save the dynamic power consumption of components whose values are not getting changed. A similar approach is to remove power from components which are in an idle state so that power is supplied to components which are operational; this saves the static power consumption.

Fault Diagnosis: Fault diagnosis and rectification of faults is an integral part of any embedded system design. The first step to fault diagnosis is to verify if the system has any fault or not. Generally, a set of test vectors with the known expected values of output(s) is used for fault diagnosis in the system. If the output given by the system is in agreement with the expected values then it means that the system has no fault; but if there is some mismatch between the actual and the expected outputs then that means fault or faults exist in the system design. If carefully chosen, test vectors help to locate the faults too. Else, fault models are used to locate faults. This approach requires a study of commonly occurring faults based on which fault model is made for each type of fault. This model then injects faults at different locations—one at a time—to get the incorrect results. The location and fault model for which the incorrect output results locates the fault. Once located, the design can be modified to remove the faults. Once the faults are located the next stage is to remove those from the system design before the final manufacturing.

Scan Design: Scan design techniques are used to test sequential circuits which are often not covered by conventional fault location techniques, particularly if sequential circuits are buried within the circuit. In scan design techniques, the sequential circuits are converted as a chain of shift registers with a combinational circuit connecting them. Test vectors are given as input to the shift register chain whose output is then checked for expected output. A test mode input is provided under which the sequential circuits are converted to the chain of shift registers. So, scan design techniques makes sequential circuits controllable as well as observable but adds system overheads.

Boundary scan extends the idea of scan design across chips to PCBs. It helps to detect different kinds of faults as broken connections, short points, open points, and so on with the help of test vectors. It actually creates chip internal test points which facilitate continuity testing. JTAG is the IEEE 1149.1 standard for the boundary scan. The advantages of JTAG boundary scan are increased observability and control of sequential components with a minimal on chip resources. Also it is not limited by chip functionality or complexity but depends on the board structure. It provides for in-chip testing of components as well as interconnections. It automates the testing process with the help of EDA tools with a little scope of errors. The boundary scan requires each component to have

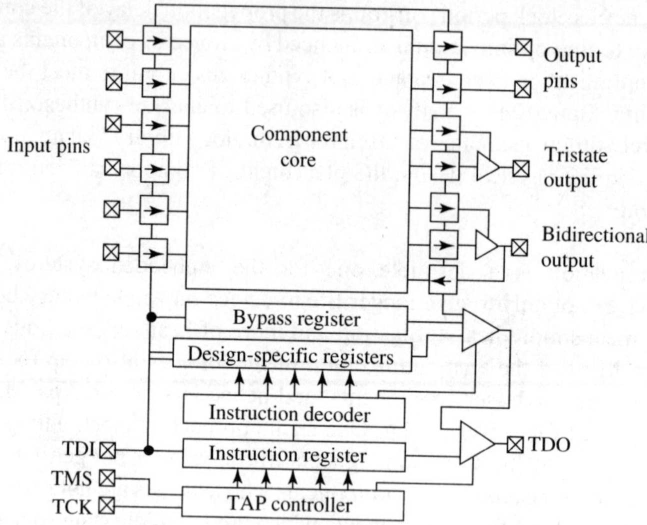

Figure 12.3 JTAG Boundary Scan Logic

Test Access Port (TAP). A chain of boundary scan cells is inserted between the component core and external pins.

TAP controller is the core which controls the test logic. A number of registers are used to facilitate testing. Boundary scan and data registers operate under the control of an instruction register. A chain of boundary scan cells is inserted between the component core and the external pins. Input and output pins each require only one cell whereas pins with high impedance state require two cells, one additional for controlling output enable function. Data is scanned from TDI to TDO with the help of TAP controller. TAP controller is synchronized with the clock applied at TCK and state selection is done by the TMS input.

Though boundary scan was a significant leap in testing a major disadvantage was increased overhead and time consumed in shifting of test mode and normal operational mode. In test mode, full speed was not used which made the testing process still slower. These problems served as a motivation for BIST (built-in self-test). BIST incorporates the test vectors generating circuits in the system itself. This helps to exploit the full system speed for the testing mode. Also multicycle test sequences can be generated to expose different types of faults. BIST does have more area and resource overheads but they are compensated by better fault coverage in less time when compared to other techniques. Moreover, BIST is there along with the system throughout; so, at any stage it can be used to diagnose faults if the system is not functioning properly.

12.5 LIMITATIONS OF VHDL IN THE EMBEDDED SYSTEM DESIGN

Designing an embedded system with VHDL is better than the most existing languages for the design process. It offers flexibility with FPGA to reprogram. Simulation results help to verify the functionality before the actual design in the hardware. It allows the use of IP cores which eases the process of design with inbuilt components and achieves optimization along with the design process. VHDL is a discrete event-driven language which effectively captures the system hierarchy of an embedded system and ignores the idle portions of design. But it has certain limitations as a design language as follows:

- VHDL allows a single driver only for a signal whereas multiple drivers exist for buses. In VHDL resolution functions are user defined.
- Not all constructs are synthesizable; care has to be taken for constructs which are synthesizable for final implementation. The change in the design code at the later stage of synthesis requires modification from the first step.

SUMMARY

- Embedded system is a system with specific application and having an embedded computer.
- It differs from a general purpose computer as it is meant for a specific task only.
- Embedded system essentially consists of a processing unit.
- Accelerators are co-processors as math processor to speed up the execution time in the embedded system.
- Processing unit of an embedded system can be a microprocessor or a microcontroller depending on the complexity of application for which it is designed.
- Embedded system design has both hardware and software part and concurrent design of both is called hardware–software codesign approach.
- Design optimizations are important at the initial stages rather than at the final physical design stage.
- Area and time optimizations are inversely related. One is achieved at the cost of another while power optimization goes hand in hand with area optimization.
- Codesign helps to speed up the design process as well as to achieve the optimization in terms of speed, power and area.

QUESTIONS

Objective Questions

1. Embedded system architecture is based on ....

   a. Von-Neumann b. Harvard c. Hybrid d. none of these

2. Embedded system contains a computer ....

   a. no b. yes

3. JTAG refers to ....

   a. joint testing action group b. joint terminal access group c. joint test for area gain

4. Embedded system essentially consists of ....

   a. processor core b. accelerator c. clock circuit d. synchronizing circuit

5. Boundary scan and BIST are different techniques for same purpose. (True/False)
6. Give full forms of TAP, BIST, ATE.

Descriptive Questions

1. What is embedded system?
2. List the characteristics of embedded system.
3. Discuss the architecture of embedded system.
4. What is meant by hardware–software codesign? What are the advantages of using this approach for the design of an embedded system?
5. Discuss design methodology for embedded system design.
6. Write a note on fault diagnosis and detection techniques.

13 MICROCOMPUTER DESIGN

CHAPTER OBJECTIVES

- Introduction to microcomputer
- Basic components of a microcomputer
- Implementation of microcomputer using VHDL

13.1 INTRODUCTION

Microcomputer is a computer that is available on small scale and it depends on a microprocessor. Typically, a microcomputer consists of a central processing unit, memory, and an input–output subsystem with buses for data transfer.

13.1.1 Basic Components of a Microcomputer

In its simplest form, a microcomputer architecture is as shown in Figure 13.1.

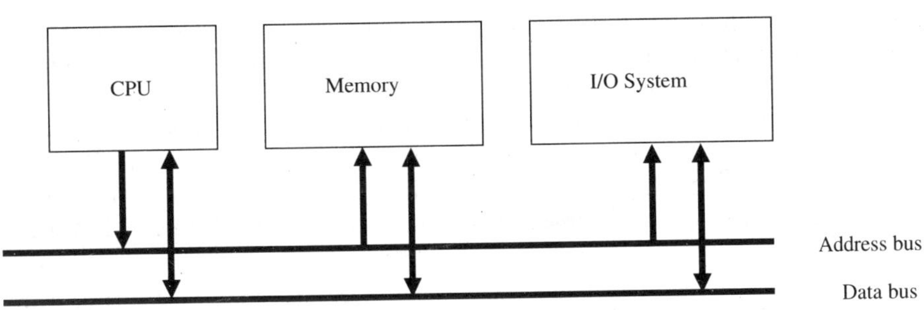

Figure 13.1 Microcomputer Architecture

Central processing unit (CPU) is central to microcomputer processing, consisting of a control unit (CU) and arithmetic logic unit (ALU). Memory consists of random access memory (RAM) and read only memory (ROM).

Central Processing Unit: Central processing unit (CPU) is the brain of microcomputer. It is responsible for all the mathematical and logical operations, temporary storage to store the results and for

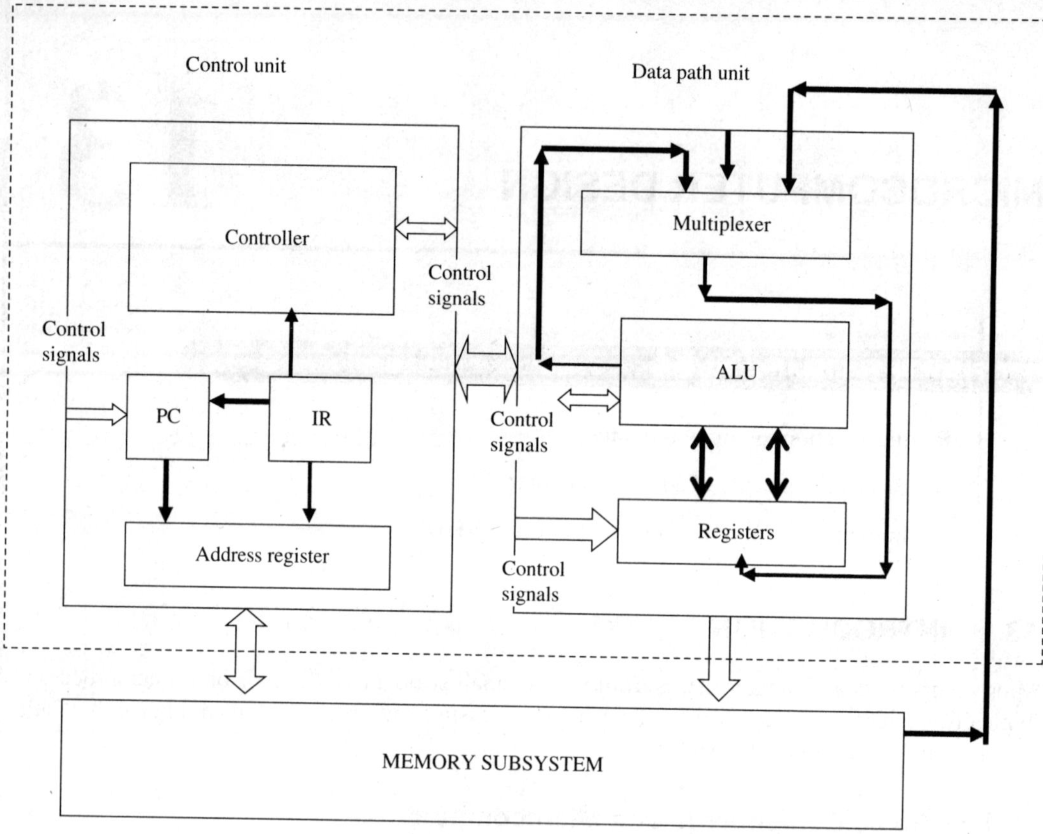

Figure 13.2 Block Diagram of a CPU

synchronizing and scheduling operations. It consists of a control unit that is responsible for synchronizing and scheduling functions whereas data path unit is responsible for mathematical and logical functions. The detailed block diagram of CPU is shown in Figure 13.2. It consists of a control unit and a datapath unit.

The control unit of CPU consists of a controller to synchronize the functioning of each subsystem, a program counter (PC) to store the address of the next instruction, instruction register (IR) to fetch the instruction and decode its opcode, and Multiplexer which takes the inputs from PC, IR and ALU and sends the data from the selected input to memory. The dashed area in the Figure 13.2 shows the CPU of a simple microcomputer.

Data path unit consists of a small 2:1 multiplexer which selects its inputs from the data from memory or the data from ALU and gives the output from the selected input line to the registers for storage. ALU performs all mathematical and logical operations and receive input data from registers or immediate data or from memory; and output too can be written to register, memory or the output peripheral. Registers form the temporary storage area for operations performed by ALU.

13.2 IMPLEMENTATION OF MICROCOMPUTER USING VHDL

To implement a microcomputer using VHDL, the hierarchical approach is followed wherein the whole architecture is divided into smaller subsystems:

- CPU
- Memory
- I/O
- Buses

Then each subsystem is subdivided into smaller units. Address bus 16-bit wide is used and a data bus of 16 bits is used. Following this approach, each subsystem will be considered one at a time and coded in VHDL.

Datapath unit:

ALU (Arithmetic Logic Unit)

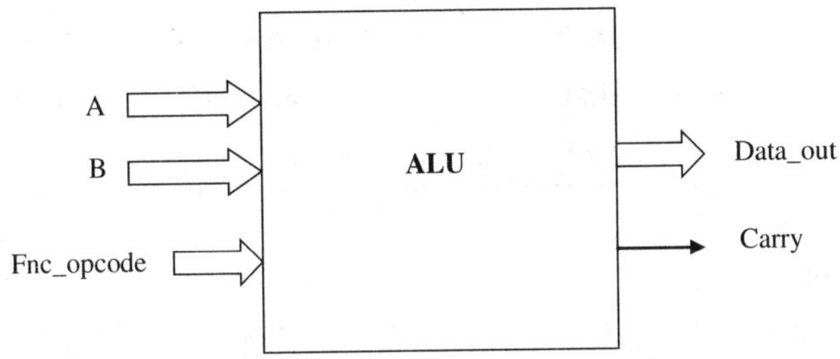

Figure 13.3 Block Diagram of ALU

ALU performs arithmetic and logic functions. It has two inputs of 16-bit each, which come from accumulator register 'A' and general purpose register 'B'—both coming from the register bank. So the operands for a binary function are in 'A' and 'B'. Another input function opcode is the selector, the operation to be performed and origin of inputs. Inputs or operands come in register 'A' and 'B' but data can come to these registers either from the input peripheral as external data or can be moved to register from memory location. The multiplexer in data path unit does this selection according to signal ext_d received by the IR.

Function opcode is 5-bit signal. Eight functions are defined to be performed by ALU as listed in the following table:

The operations to be performed are indicated by lower 3 bits of function opcode (fnc-opcode). While upper bits define the addressing mode they indicate if data is from external source or from memory or from register.

If

00 - no second operand
01 - from register
10 - from memory
11 - external data

So, function opcode of 110000 indicates a ADD operation with first operand in Accumulator and second operand is external immediate data.

Decoding of complete function opcode is done by the controller which sends signal to IR to set the function opcode. IR stores the complete instruction set.

Operation	Opcode
ADD	0000
SUB	0001
MUL	0010
MOV A, M	0011
MOV A, #	0100
NOT	0101
AND	0110
OR	0111
JMP	1000
HALT	1001
MOV A, R	1010

Instruction is in 16-bit format as:

A1	A2	O1	O2	O3	O4					R1s	R2s	R3s	R1	R2	R3

Single-word instructions

M15/d15	..	..	..	..	..	..	..	..	..	..	..	..	..	..	M0/d0

Second word for double-word instructions

```
LIBRARY IEEE;
USE IEEE.STD_LOGIC_1164.all;
USE IEEE.STD_LOGIC_ARITH.all;
USE IEEE.STD_LOGIC_UNSIGNED.all;

ENTITY alu IS
PORT(A,B: INOUT STD_LOGIC_VECTOR(15 DOWNTO 0)
      Fnc_opcode: IN STD_LOGIC_VECTOR(3 DOWNTO 0;
       Carry: INOUT STD_LOGIC;
      alu_out: OUT STD_LOGIC_VECTOR(15 DOWNTO 0));
END alu;

ARCHITECTURE behave OF alu IS -- Behavioural architecture model of the entity
  SIGNAL temp_reg:STD_LOGIC_VECTOR(15 DOWNTO 0):= "000000000000000";
  SIGNAL data_32: STD_LOGIC_VECTOR(31 DOWNTO 0):=
"0000000000000000000000000000000";
```

```
BEGIN
  P1: PROCESS(A,B, fnc_opcode)
     BEGIN
          CASE fnc_opcode IS
             WHEN "0000" => temp_reg<= A + B;
             WHEN "0001" => temp_reg<= A – B;
             WHEN "0010" => data_32<= A * B;
                              temp_reg<= data_32(15 DOWNTO 0);
             WHEN "0101" => temp_reg<= NOT (A);
             WHEN "0110" => temp_reg<= A AND B;
             WHEN "0111" => temp_reg<= A OR B;
          END CASE;
           END PROCESS P1;
alu_out<= temp_reg;
B<= data_32(31 downto 16);
END behave;
```

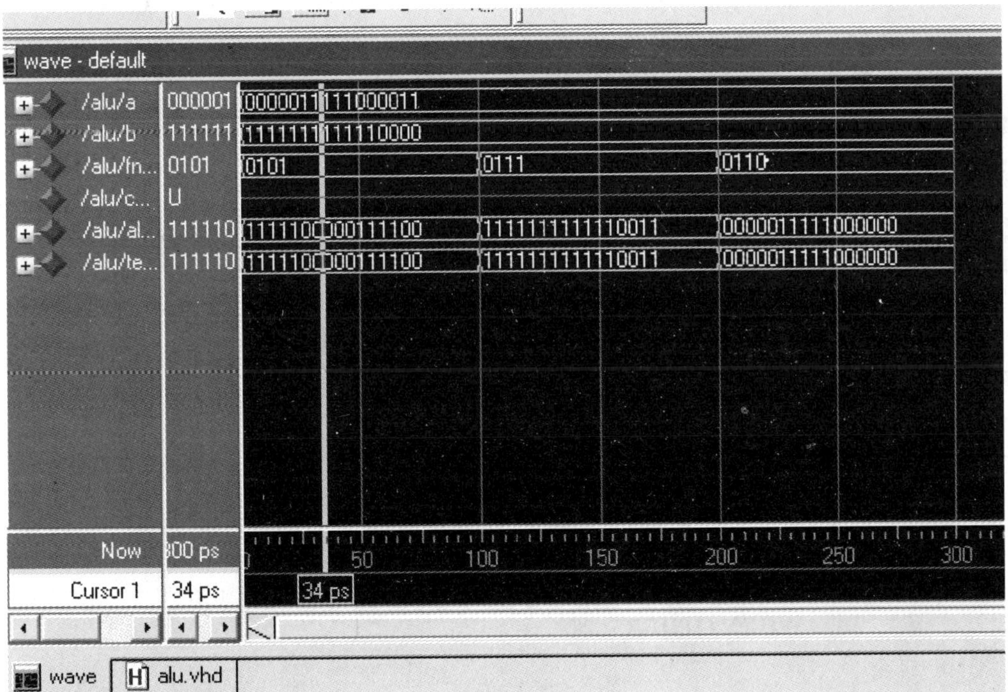

Figure 13.4 Simulation Waveforms for ALU

Multiplexer: Datapath unit uses 4:1 multiplexer to choose the input that is to be written to registers. Input can be data fetched from a memory location or it can be an external input or it can be an immediate data from Alu output. So, MUX selects the input and the selected input is sent to the output to be stored in working registers A or B.

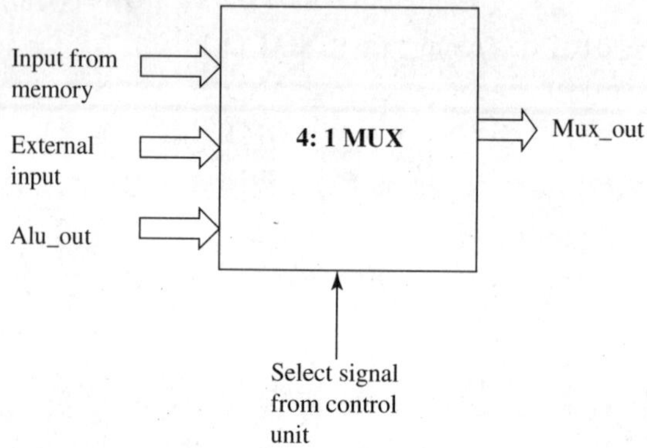

Figure 13.5 Block Diagram of Multiplexer (4:1)

```
LIBRARY IEEE;

USE IEEE.STD_LOGIC_1164.all;

ENTITY mux IS
PORT(in1,in2, alu_out: IN STD_LOGIC_VECTOR(15 DOWNTO 0);
        sel:IN STD_LOGIC_VECTOR(3 DOWNTO 0);
        mux_out: OUT STD_LOGIC_VECTOR(15 DOWNTO 0));
END mux;

ARCHITECTURE mux1 OF mux IS
BEGIN

 mux_out <= in1 WHEN sel= "00" ELSE
            in2 WHEN sel= "01" ELSE
            alu_out WHEN sel= "01" ELSE
                 WHEN OTHERS;
END mux1;
```

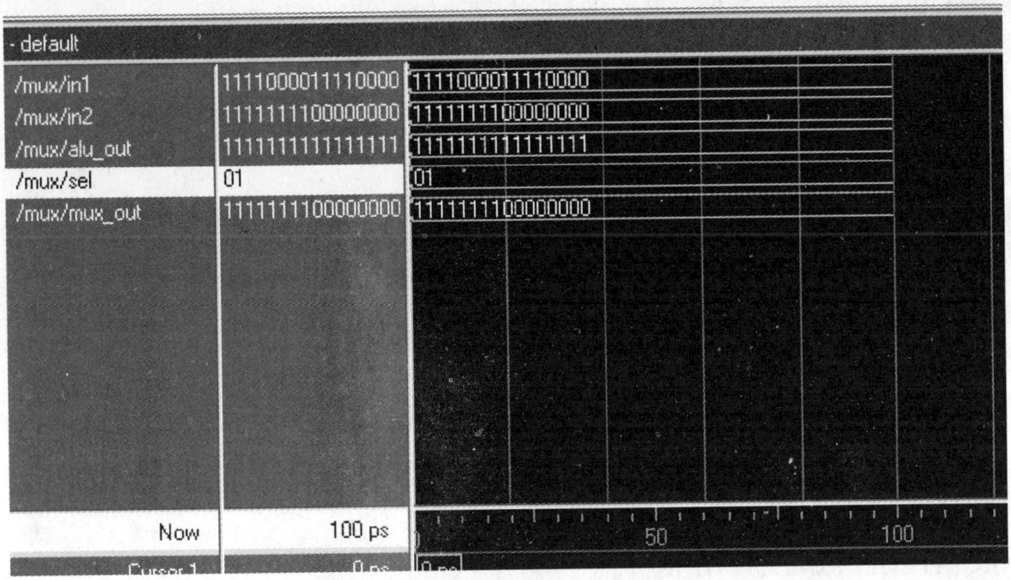

Figure 13.6 Simulation Waveforms for Mux

Register Bank: Register bank consists of eight registers—A, B, R1, R2,R3,R4,R5 and R6—each of 16-bit width.

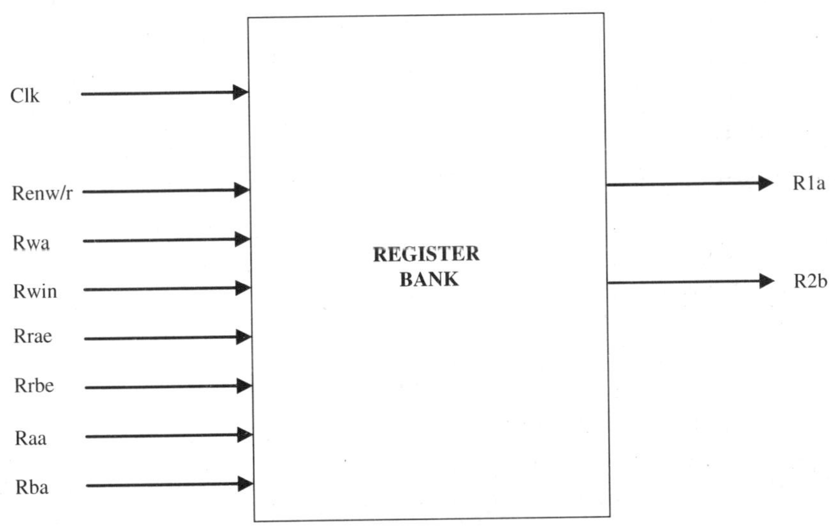

Figure 13.7 Block Diagram of Register Bank

Register bank is governed by clock (clk) and reset signal (rst).

Register bank has eight registers with each having address as follows:

```
LIBRARY IEEE;
USE IEEE.STD_LOGIC_1164.all;
USE IEEE.STD_LOGIC_ARITH.all;
USE IEEE.STD_LOGIC_UNSIGNED.all;

ENTITY reg_bank IS
PORT(clk, Renw,Renr Rrae, Rrbe: IN STD_LOGIC;
        Rwa,Raa, Rba:IN STD_LOGIC_VECTOR(0 TO 2);
        Rwin: IN STD_LOGIC_VECTOR(15 DOWNTO 0);
        R1a, R2b: OUT STD_LOGIC_VECTOR(15 DOWNTO 0));
END reg_bank;

ARCHITECTURE behave OF reg_bank IS
TYPE reg IS ARRAY (0 TO 2) OF STD_LOGIC_VECTOR(15 DOWNTO 0);
SIGNAL r_temp: reg;
BEGIN

Write: PROCESS( clk, Renw, Rwa)
        BEGIN
                IF clk'EVENT AND clk='1' THEN
                    IF Renw = '1' THEN
                        R_temp(conv_integer(Rwa)) <= Rwin;
                    END IF;
                END IF;
        END PROCESS write;
END behave;

Read1 : PROCESS( clk, Renr, Rrae,Raa)

        BEGIN
                IF clk'EVENT AND clk='1' THEN
                    IF Renr = '1' THEN
                        IF Rrae= '1' THEN
                            R1a<= R_temp(conv_integer(Raa));
                        END IF;
                    END IF;
```

```
            END IF;
        END PROCESS read1;

Read2 : PROCESS( clk,Renr, Rrbe, Rba)

        BEGIN

            IF clk'EVENT AND clk='1' THEN
                IF Renr = '1' THEN
                    IF Rrbe= '1' THEN
                        R2b<= R_temp(conv_integer(Rba));
                    END IF ;
                END IF;
            END IF;

        END PROCESS read2;
END behave;
```

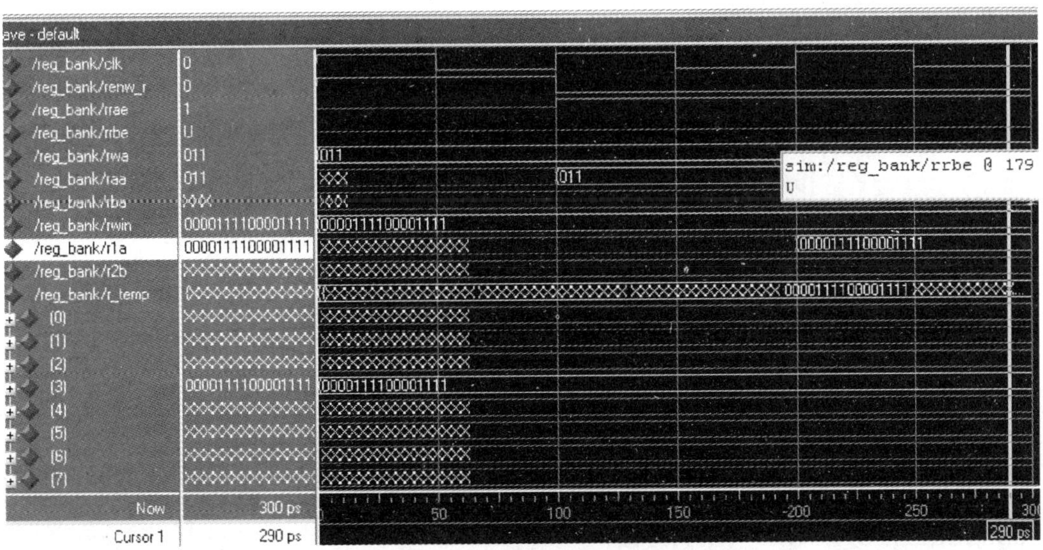

Figure 13.8 Simulation Waveforms for Register Bank

Datapath Unit:

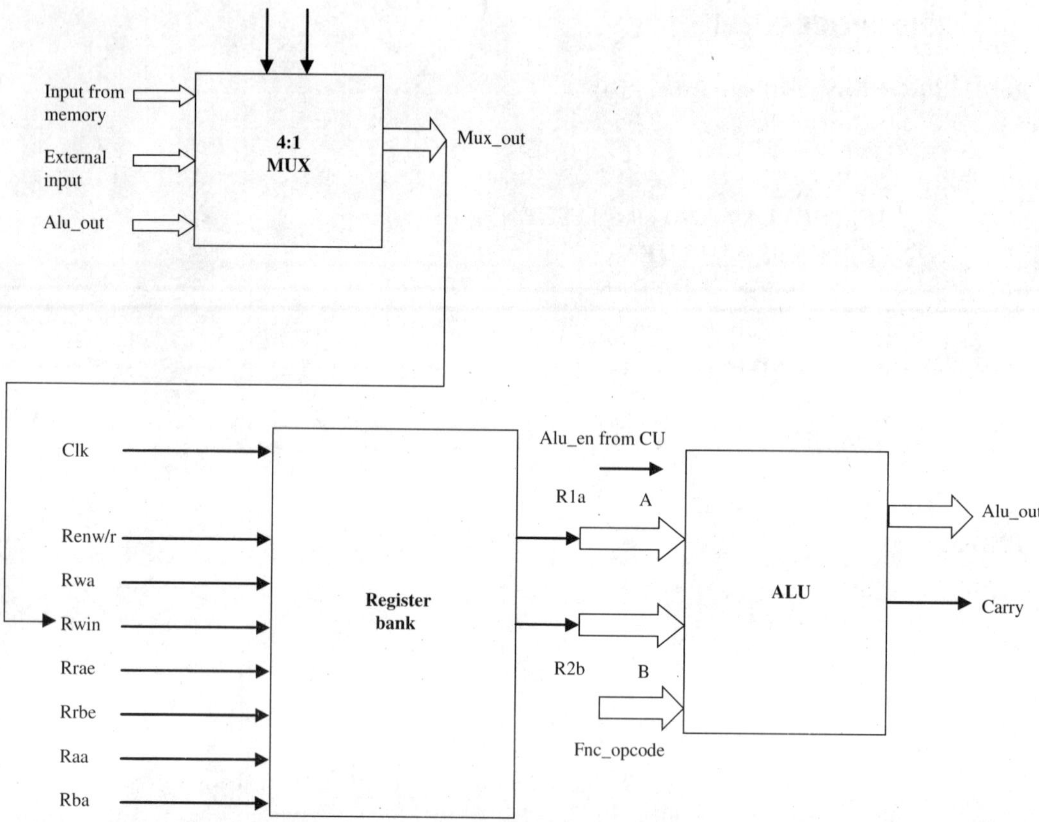

Figure 13.9 Datapath Unit

All the control signals come from the control unit.

Control Unit: Program Counter (PC)

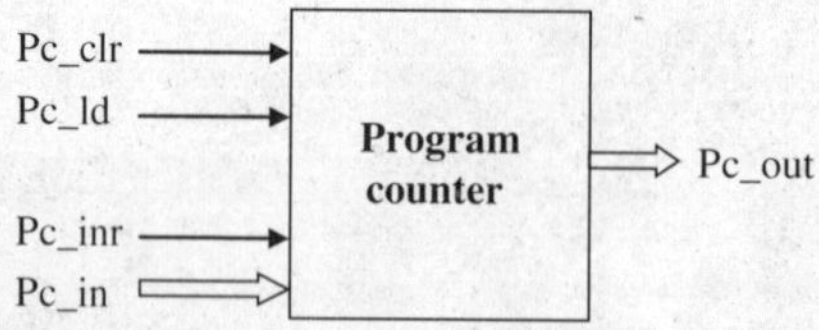

Figure 13.10 Block Diagram of Program Counter

```
LIBRARY IEEE;
USE IEEE.STD_LOGIC_1164.all;
USE IEEE.STD_LOGIC_ARITH.all;
USE IEEE.STD_LOGIC_UNSIGNED.all;
```

```
ENTITY Pc IS
PORT(Pc_clr,Pc_ld, Pc_inr: IN STD_LOGIC;
        Pc_in: IN STD_LOGIC_VECTOR(15 DOWNTO 0);
        Pc_out: OUT STD_LOGIC_VECTOR(15 DOWNTO 0));
END Pc;
ARCHITECTURE behave OF Pc IS
SIGNAL Pc_temp: STD_LOGIC_VECTOR(15 DOWNTO 0);
BEGIN
Pc1: PROCESS( Pc_clr,Pc_ld,Pc_inr)
     BEGIN
          IF Pc_clr= '1' THEN
             Pc_temp "0000000000000000";
          ELSIF Pc_ld = '0' THEN
           IF Pc_inr = '1' THEN
               Pc_temp <= Pc_temp + 1;
           END IF;
          ELSIF Pc_ld = '1'THEN
              Pc_temp <= Pc_in;
          END IF;
     END PROCESS Pc1;
Pc_out <= Pc_temp;
END behave;
```

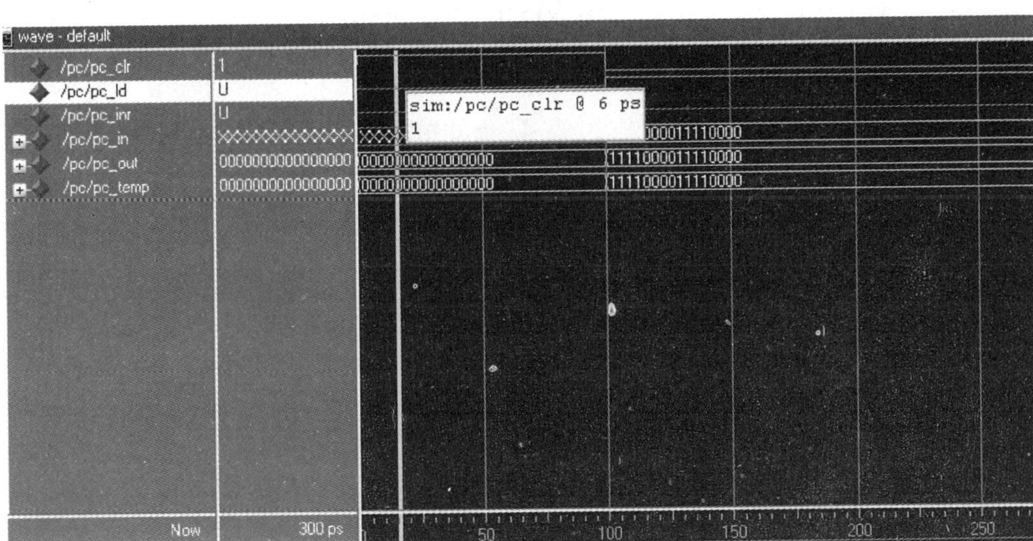

Figure 13.11 Simulation Waveforms for Program Counter

Instruction Register: Instruction register (IR) is responsible for decoding the instruction and stores the current instruction that is to be executed.

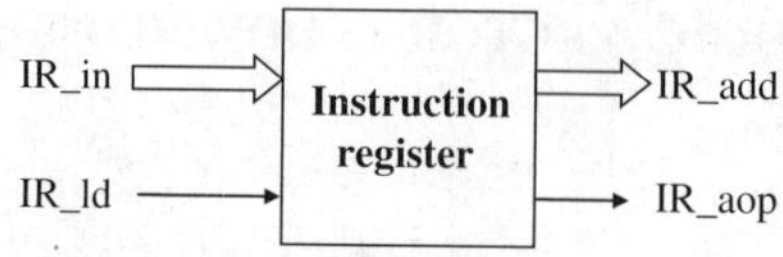

Figure 13.12 Block Diagram of Instruction Register

```
LIBRARY IEEE;
USE IEEE.STD_LOGIC_1164.all;
USE IEEE.STD_LOGIC_ARITH.all;
USE IEEE.STD_LOGIC_UNSIGNED.all;


ENTITY Ir IS
PORT(Ir_ld: IN STD_LOGIC;
        Ir_in: IN STD_LOGIC_VECTOR(15 DOWNTO 0);
        Ir_op: OUT STD_LOGIC_VECTOR(5 DOWNTO 0);
        Ir_add: OUT STD_LOGIC_VECTOR(15 DOWNTO 0));
END Ir;


ARCHITECTURE behave OF Ir IS
SIGNAL Ir_temp,Ir_temp1: STD_LOGIC_VECTOR(15 DOWNTO 0);
BEGIN
 Ir1: PROCESS( Ir_in, Ir_ld)
      BEGIN
          IF Ir_ld= '1' THEN
             Ir_temp <= Ir_in;
             Ir_op <= Ir_in(15 DOWNTO 10);
          END IF;
              IF Ir_op(5) = '0' THEN
                 Ir_ld <= '0'
              ELSE
                 Ir_temp1 <= Ir_in
              END IF;
   END PROCESS Ir1;
Ir_add <= Ir_temp1;
END behave;
```

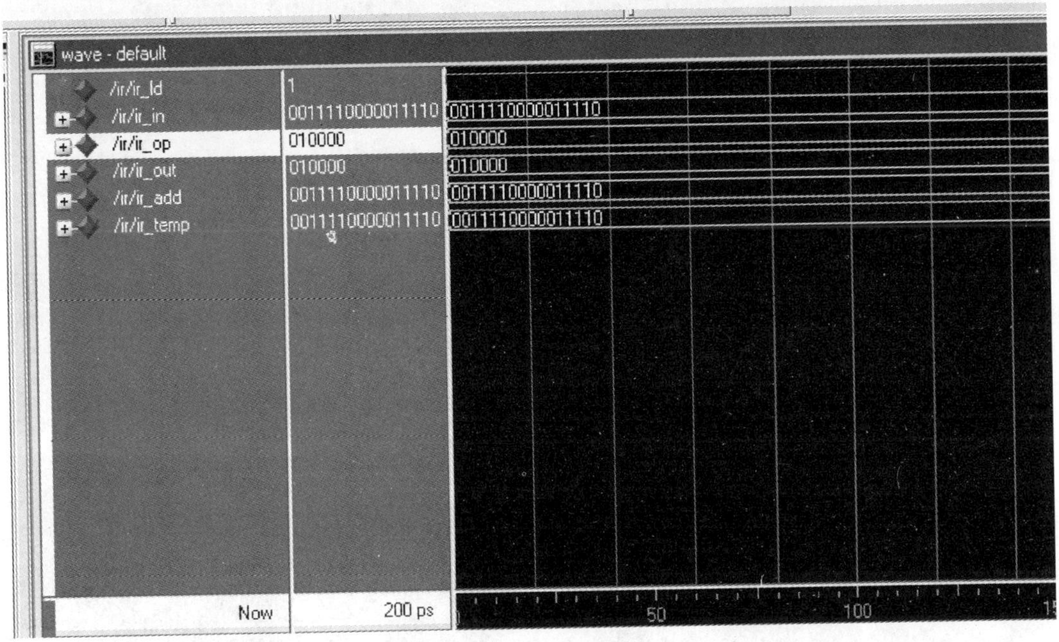

Figure 13.13 Simulation Waveforms for Instruction Register

Memory Address Register: It is a part of ROM only which contains mapping of assemble language to binary opcodes.

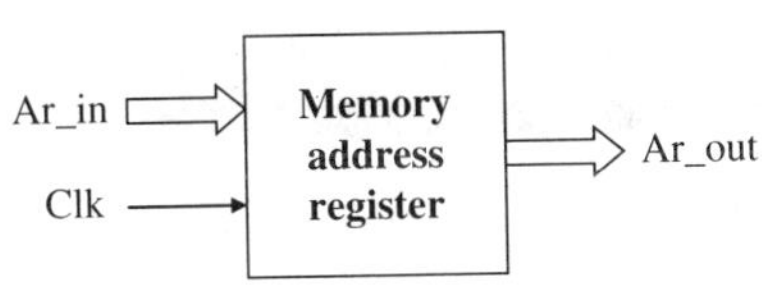

Figure 13.14 Block Diagram of Memory Address Register

```
LIBRARY IEEE;
USE IEEE.STD_LOGIC_1164.all;
USE IEEE.STD_LOGIC_UNSIGNED.all;
ENTITY Ar IS
PORT(clk: IN STD_LOGIC;
      Ar_in: IN STD_LOGIC_VECTOR(15 DOWNTO 0);
      Ar_out: OUT STD_LOGIC_VECTOR(15 DOWNTO 0));
END Ar;
ARCHITECTURE behave OF Ar IS
BEGIN
  Ar1: PROCESS( clk,Ar_in)
      BEGIN
            IF clk'EVENT AND clk= '1' THEN
```

```
            Ar_out<= Ar_in;
        END IF;
    END PROCESS Ar1;
END behave;
```

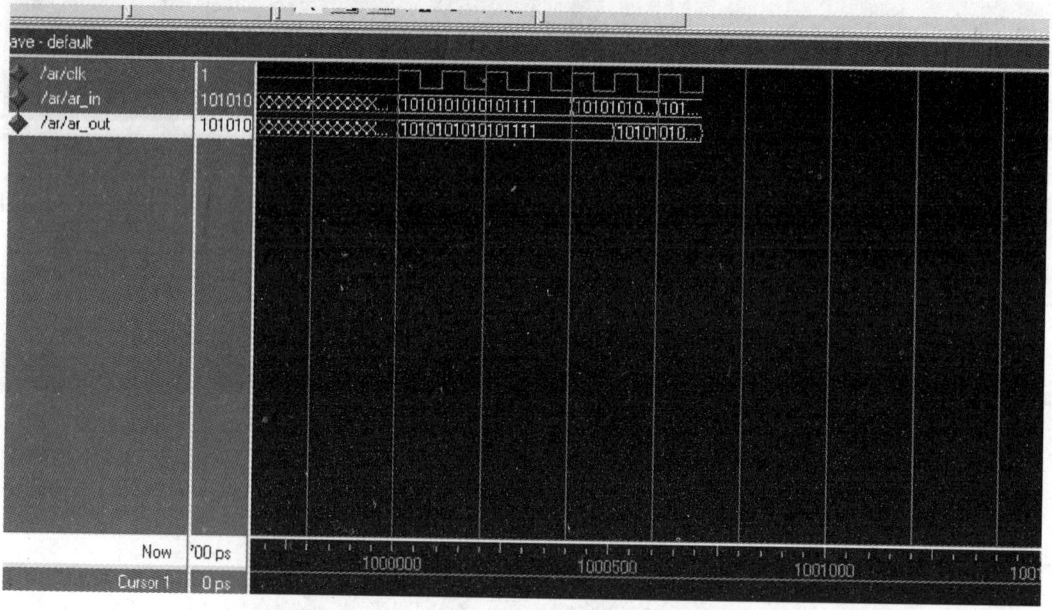

Figure 13.15 Simulation Waveforms for Memory Address Register

Memory:

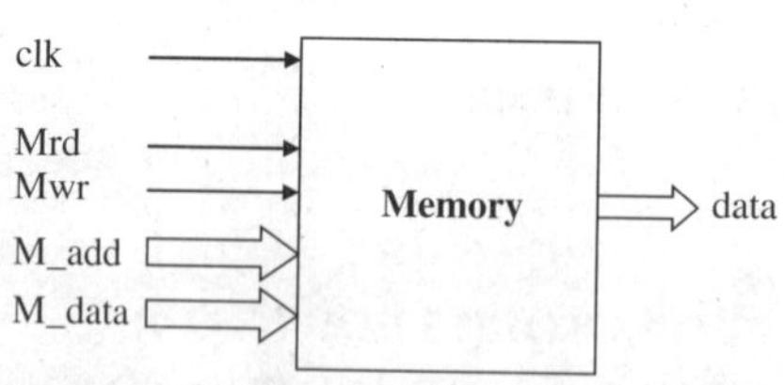

Figure 13.16 Block Diagram of Memory Unit

```
LIBRARY IEEE;
USE IEEE.STD_LOGIC_1164.all;
USE IEEE.STD_LOGIC_ARITH.all;
USE IEEE.STD_LOGIC_UNSIGNED.all;
```

```
ENTITY memory IS
PORT(clk, Mre,Mwe, ir_en: IN STD_LOGIC;
        addr,M_data : IN STD_LOGIC_VECTOR(15 DOWNTO 0);
        data: OUT STD_LOGIC_VECTOR(15 DOWNTO 0);
       data_ir: OUT STD_LOGIC_VECTOR(31 DOWNTO 0));
END memory;

ARCHITECTURE behave OF memory IS
TYPE ram_type IS ARRAY (0 TO 255) OF STD_LOGIC_VECTOR(15 DOWNTO 0);
SIGNAL tmp_ram: ram_type;

BEGIN
  write: PROCESS( clk, Mwe,M_data, addr)
        BEGIN

              IF clk'EVENT AND clk = '1' THEN
                IF M_wr = '1'THEN
                 temp_ram(Conv_integer(addr)) <= M_data;
                END IF;
              END IF;

        END PROCESS write;
  Read: PROCESS( clk, Mre, addr)
        BEGIN

              IF clk'EVENT AND clk = '1' THEN
                IF M_rd = '1'THEN
                  IF ir_en= '1' THEN
                     data_ir(15 DOWNTO 0) <= temp_ram(conv_integer(addr));
                     data_ir(31 DOWNTO 16) <= temp_ram(conv_integer(addr+1));
                  ELSE
                     data <= temp_ram(Conv_integer(addr));
                  END IF;
                END IF;
              END IF;
        END PROCESS read;
 END behave;
```

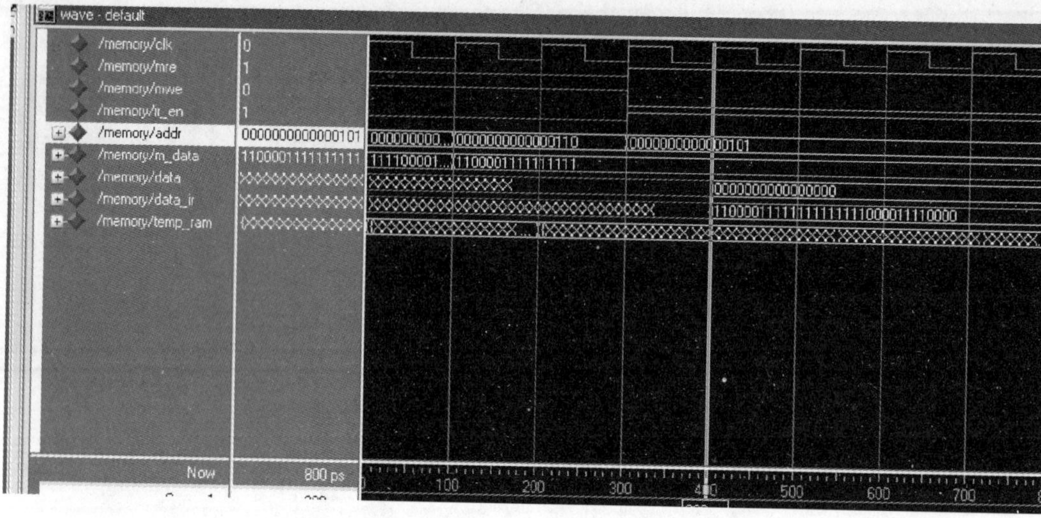

Figure 13.17 Simulation Waveforms for Memory

Controller: Controller is the main controlling unit which is modelled as an FSM. The state diagram for controller FSM is as follows:

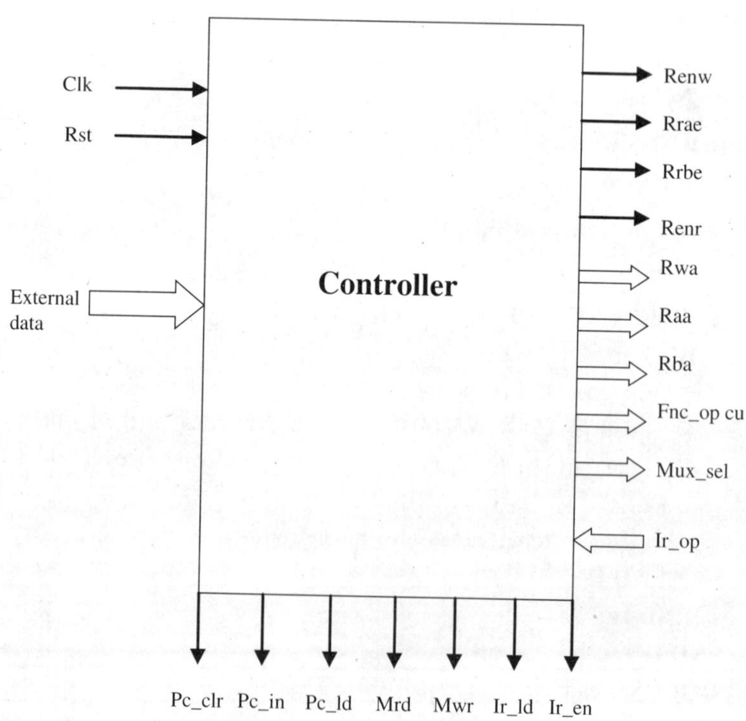

Figure 13.18 Block Diagram of Controller

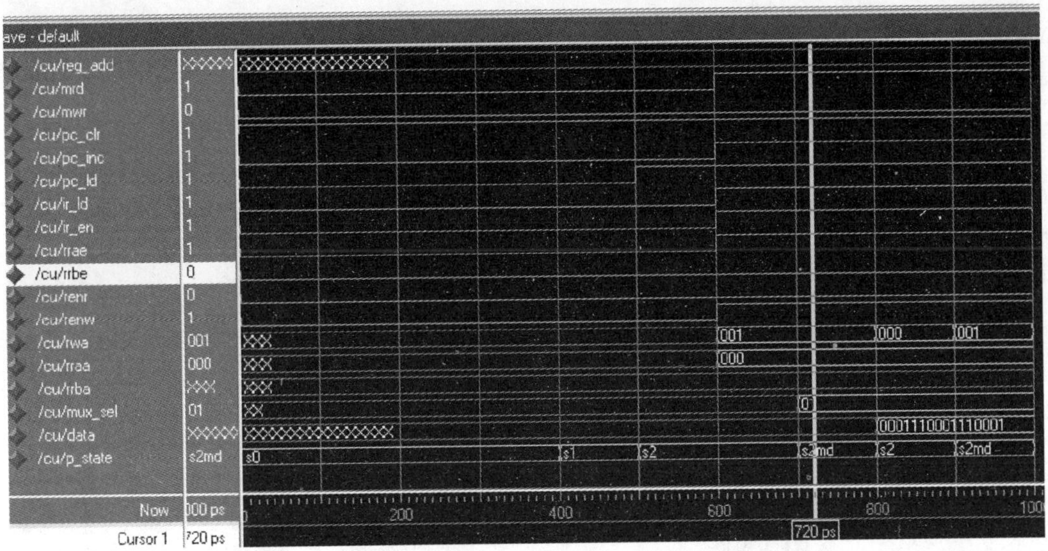

Figure 13.19 State Diagram for Controller

```
LIBRARY IEEE;
USE IEEE.STD_LOGIC_1164.all;
USE IEEE.STD_LOGIC_ARITH.all;
USE IEEE.STD_LOGIC_UNSIGNED.all;
ENTITY cu IS
PORT(clk, rst: IN STD_LOGIC;
        Ext_data : IN STD_LOGIC_VECTOR(15 DOWNTO 0);
        Ir_op: IN STD_LOGIC_VECTOR(5 DOWNTO 0);
        Reg_add: IN STD_LOGIC_VECTOR(15 DOWNTO 0);
        Mrd,Mwr, Pc_clr,Pc_inc, Pc_ld, Ir_ld, Ir_en, Rrae,Rrbe,Renr, Renw: OUT STD_LOGIC;
        Rwa, Rraa, Rrba: OUT STD_LOGIC_VECTOR( 2 DOWNTO 0);
        mux_sel: OUT STD_LOGIC_VECTOR(1 DOWNTO 0);
        data: OUT STD_LOGIC_VECTOR(15 DOWNTO 0));
END cu;
ARCHITECTURE fsm OF cu IS
TYPE state IS ( S0, S1, S2,S2a, S2s, S2m, S2mm , S2md, S2n, S2an, S2or, S2j, S2h, S2mr, S3);
SIGNAL P_state: state;
BEGIN
  PROCESS( clk, rst, P_state,Ir_op)
      BEGIN
          IF clk'EVENT and clk= '1' THEN
          CASE P_state IS
          CASE P_state IS
              WHEN S0 =>
            IF rst= '1' THEN
           Pc_clr<= '1';
```

```
Pc_inr<= '0';
Pc_ld<= '0';
Mr<= '0';
Mw<= '0';
Ir_ld<= '0';
Ir_en<= '0';
Renw<= '0';
Renr<= '0';
Rrbe<= '0';
Rrae<= '0';
P_state<= S0;
ELSE
P_state <= S1;
END IF;
     WHEN S1 =>
                    Pc_ld <= '1';
                    P_state <= S2;
     WHEN S2 => Pc_inc<= '1';
                    Mrd<= '1';
                    Ir_en<= '1';
                    Ir_ld<= '1';
                    IF Ir_op(5 DOWNTO 4)= "10" THEN
                       mux_ sel<= "00";
                    ELSIF Ir_op(5 DOWNTO 4)= "11" THEN
                       Mux_sel<= "01";
                    ELSIF Ir_op(5 DOWNTO 4)= "01" THEN
                       Renw<= '0';
                    END IF;
     CASE ir_op(3 DOWNTO 0) IS
          WHEN "0000"=> P_state<= S2a;
          WHEN "0001"=> P_state<= S2s;
          WHEN "0010"=> P_state<= S2m;
          WHEN "0011"=> P_state<= S2mm;
          WHEN "0100"=> P_state<= S2md;
          WHEN "0101"=> P_state<= S2n;
          WHEN "0110"=> P_state<= S2an;
          WHEN "0111"=> P_state<= S2or;
          WHEN "1000"=> P_state<= S2j;
          WHEN "1001"=> P_state<= S2h;
          WHEN "1010"=> P_state<= S2mr;
     END CASE;
```

```
   IF P_state = S2n OR P_state = S2md OR P_state = S2mm OR P_state = S2j OR
P_state = S2mr THEN
                Renw<= '0';
                Rrae<= '1';
                Rraa<= "000";
                Rrbe<= '0';
    ELSE
                Rrae<= '1';
                Rraa<= "000";
    END IF;
    WHEN S2a => P_state<= S2;
   Rbae<= '1';
   Rba<= "001";
    WHEN S2s => P_state<= S2;
   Rbae<= '1';
   Rba<= "001";
    WHEN S2m => P_state<= S3;
   Rbae<= '1';
   Rba<= "001";
    WHEN S2mm => P_state<= S2;
    WHEN S2md => P_state<= S2;
     WHEN S2n => P_state<= S3;
    WHEN S2an => P_state<= S3;
   Rbae<= '1';
   Rba<= "001";
    WHEN S2or => P_state<= S3;
   Rbae<= '1';
   Rba<= "001";
    WHEN S2j => Pinc<= '0';
               P_state<= S1;
     WHEN S2h => P_state<= Sh;
                 IF rst = '1' THEN
                  P_state<= S0;
                 END IF;
     WHEN S2mr=> P_state<= S3;
     WHEN S3 => P_state<= S0;
     WHEN OTHERS =>
    END CASE;
    END IF;
 END PROCESS;
 END fsm;
```

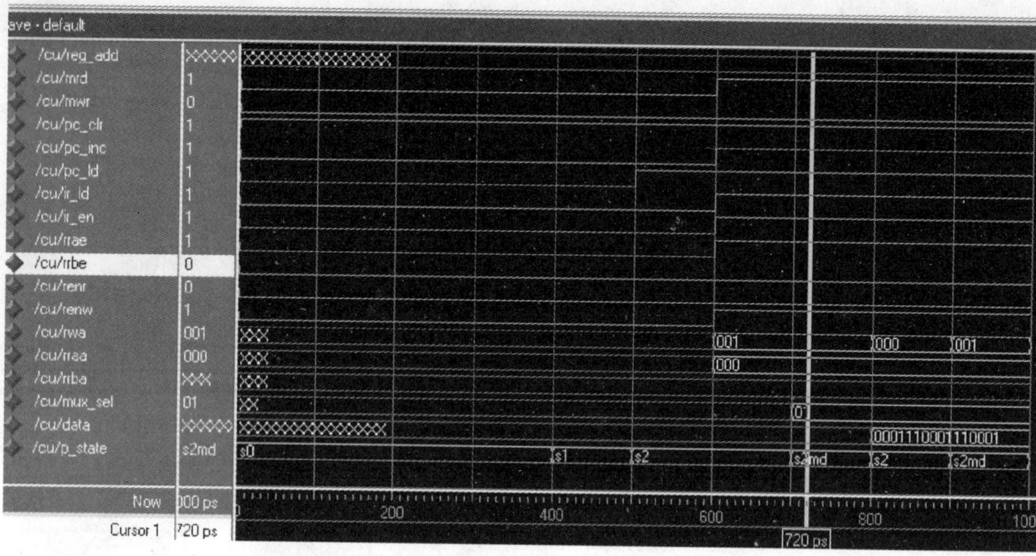

Figure 13.20 Simulation Waveforms for Control Unit

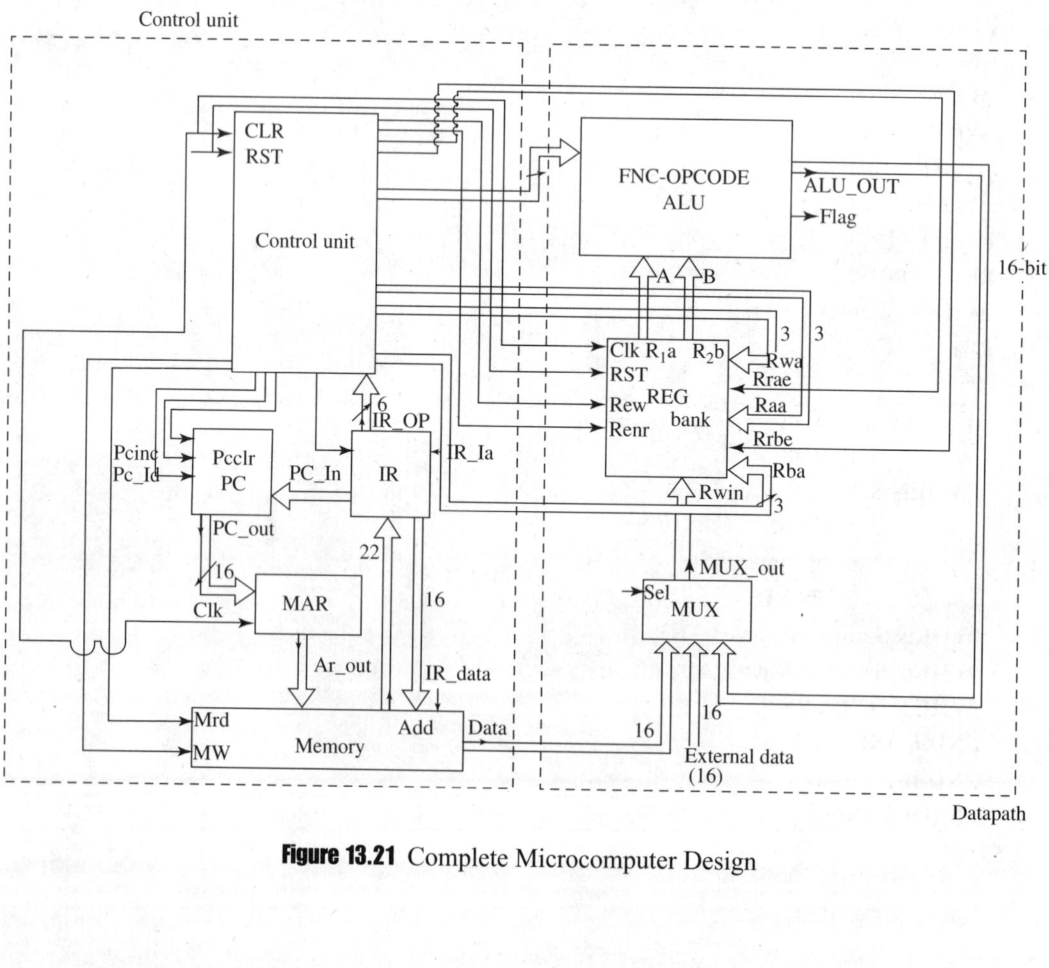

Figure 13.21 Complete Microcomputer Design

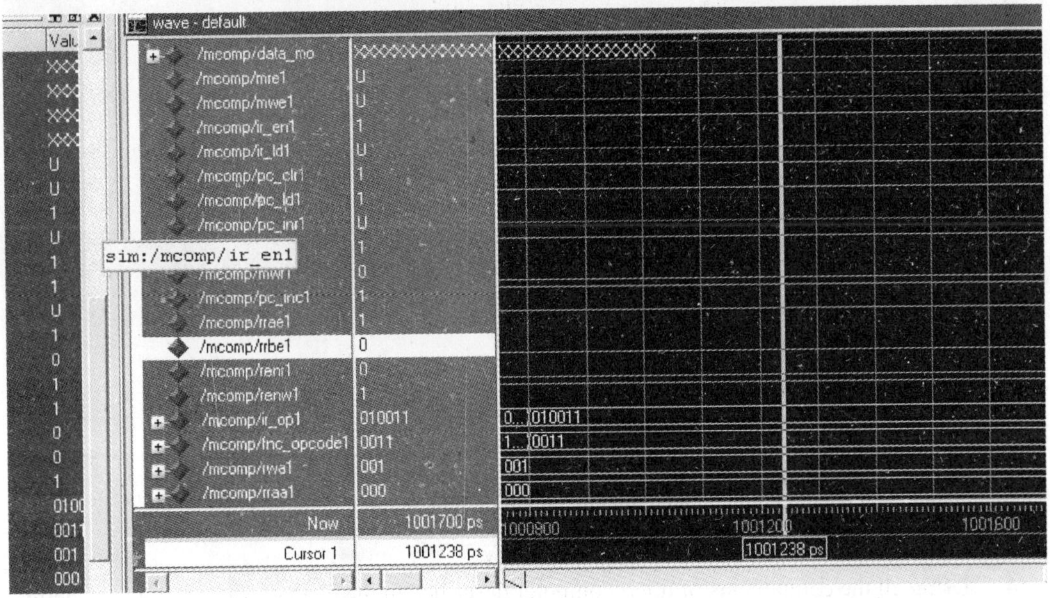

Figure 13.22 Simulation for the Functioning of Microcomputer

Note: Xilinx 9.1 is used for VHDL and Modelsim 6.2 is used for simulation of above given VHDL codes.

SUMMARY

- Microcomputer is a computer on small scale that depends on a microprocessor.
- Microcomputer essentially consists of a processing unit, memory and I/O system.
- Processing unit is the core of microcomputer whose internal structure may vary for any microcomputer, but generally has an ALU, PC, IR and register storage.
- Signals in microcomputer can be control signals, address lines and data lines.
- Microcomputer has its assembly language instruction set described by mnemonics.

QUESTIONS

Objective Questions

1. Microcomputer is a/an .... on a smaller scale.

   a. machine b. computer c. processor d. embedded system

2. The .... contains the address of the next instruction that is to be executed.

   a. program counter b. instruction register c. memory d. control unit

3. Instruction set is stored in ....

   a. instruction register b. memory address register c. RAM d. ROM

4. A 64-word memory will have .... address lines.
   a. 6 b. 8 c. 16 d. 32
5. Instruction opcode consists of ....
   a. function opcode, operand
   b. opcode, address of next instruction
   c. data, data
   d. anything stored in ROM
6. Basic components of microcomputer are ....
   a. Processor, I/O system, Memory
   b. I/O system, Timing ckt, Processor,
   c. ROM, ALU , I/O
   d. Processor and Memory

Descriptive Questions

1. Explain the basic components of a microcomputer.
2. Why is the processor called the core of microcomputer?
3. Microcomputer is an embedded system. Justify.
4. Using all the components of microcomputer described in the chapter use structural modelling to implement the complete microcomputer.
5. Design a 4-bit ALU in VHDL.
6. What is the function of memory address register?
7. Describe the detailed architecture of a 16-bit microcomputer.

14 PROGRAMMABLE LOGIC DEVICES

CHAPTER OBJECTIVES

- What are PLDs
- Need for PLDs
- Types of PLDs
- Introduction to FPGAs and CPLDs
- Differences between FPGAs and CPLDs
- Implementing digital design using FPGAs/CPLDs

14.1 INTRODUCTION

Programmable logic devices (PLD) refers to integrated circuits which can be programmed to implement any digital system. It gives an immediate realization of the logic circuit and can be reprogrammed allowing the same IC to be used again, and PLD can be upgraded in the field without any need for removing it from the chipboard. PLDs were introduced in 1970s with the idea to design a single purpose chip which can be programmed to meet particular hardware requirements. Earlier systems used SSI and MSI ICs as TTL 7400 series which was not usable for the design of large circuits. It came as a big leap to replace SSI and MSI circuits that were used for combinational design. With the advent of PLDs digital design has grown substantially from simple gates to high-density designs.

A PLD consists of logic circuit elements and interconnection switches. These switches are programmable connections which allow the PLD to be programmed as per the user needs.

14.2 TYPES OF PLDs

PLDs can be broadly classified in the following categories:

SPLDs are simple PLDs which can handle up to 10–20 logic equations only and are meant for simple designs. A typical SPLD is a two-level array of logic gates AND–OR with programmable switches and gives the output in an SOP (sum of products) form. SPLDs are of four types:

- PLAs
- PALs

- Registered PLAs
- GALs

CPLDs and FPGAs are dealt in detail after the types of SPLDs.

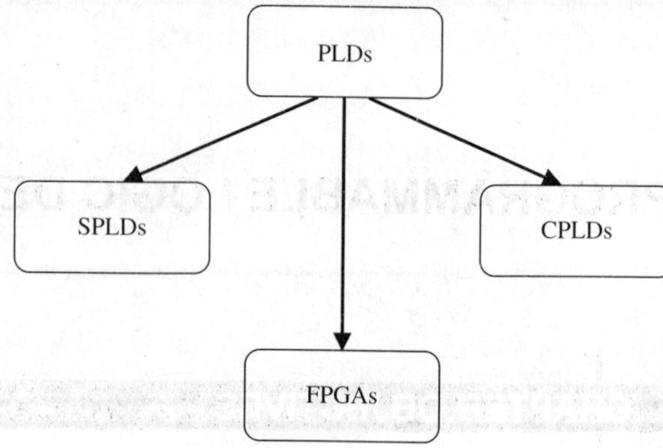

Figure 14.1 Types of PLDs

14.2.1 PLAs (Programmable Logic Arrays)

The first user-programmable chip introduced was PROM (programmable read-only memory) to implement logic functions which used address lines as logic circuit inputs and data lines as outputs. But it was an inefficient way to implement logic circuits as logic circuit outputs contain only few product terms whereas PROM uses a full decoder for its address lines with most of them left unused. This led to the introduction of first PLD — that is PLA.

PLA (programmable logic array) is an SPLD. It was introduced by monolithic memories in the mid 1970s. It has two levels of logic implementation wired AND-plane and wired OR-plane. Both AND and OR planes are programmable. The basic structure of a PLA is as shown in Figure 14.2.

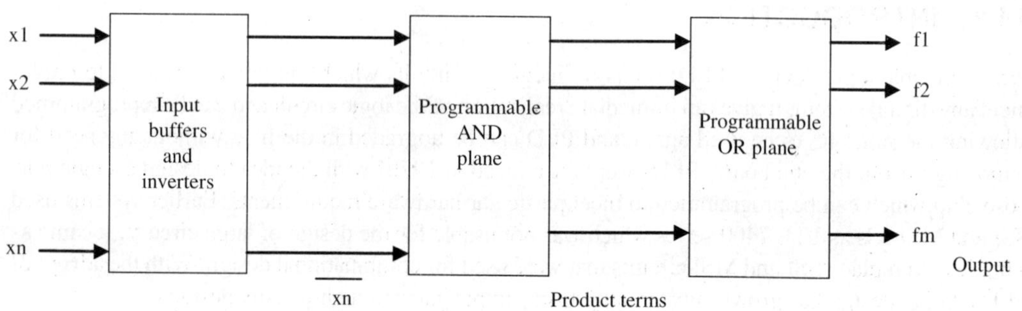

Figure 14.2 Basic Structure of PLA

The output of PLA is in SOP form. The gate-level structure of PLA is as shown as follows:

PLA can be directly used to implement a logic function up to few input variables only. A single PLA can be used for implementation of multilogic equations depending on the PLA used. Though the two levels of programmability offered flexibility to design any logic circuit it also led to low-speed performance of PLAs. Moreover, making both AND–OR planes configurable made the manufacturing cost quite high. This motivated the designers to come forward with PLD with single-level device programmability and thus PALs (programmable logic arrays) were introduced.

Use of PLAs for Logic Implementation: Consider a function F(A,B,C,D) given by Σ(2 4,6, 8, 9, 10, 12,15).

The implementation of the above logic function is shown in Figure 14.4 using 4X3X4 PLA.

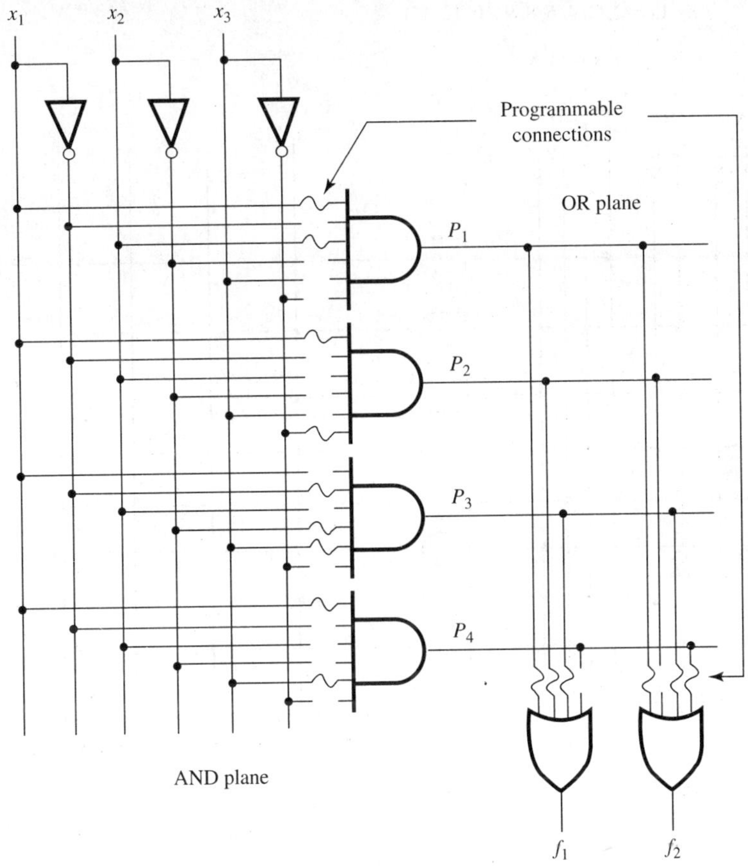

Figure 14.3 Internal Structure of PLA

14.2.2 PALs

PAL is a PLD with only AND plane programmable and OR plane fixed. They too realize the output in the SOP form. Generally, the outputs of fixed OR gates are connected to flip-flops so that sequential circuits can be realized. PALs had low cost and high pin to pin speed performance when compared to PLAs and hence, quite popular for digital design. The basic structure of a typical PAL is shown in Figure 14.5

Use of PALs for Logic Implemcntation: Consider the logic function F(A,B,C)= Σ(2,4,6,7).
The implementation of the above logic function is shown in Figure 14.6 using 4 X 2 PAL.

14.2.3 Registered PALs

PALs using flip-flops at the output for implementation of sequential circuits are called registered PALs. The commonly used PAL is PAL 16R8 which means PAL with 16 inputs, 2 outputs and 6 Inout Pins. R stands for registered. If outputs of OR gates are not connected to flip-flop then PAL can implement only combinational circuits. Such PALs have a notation as 16L8.

$F_1 = \Sigma(2, 4, 6, 8, 9, 10, 12, 15)$

using $4 \times 8 \times 4$ PLA

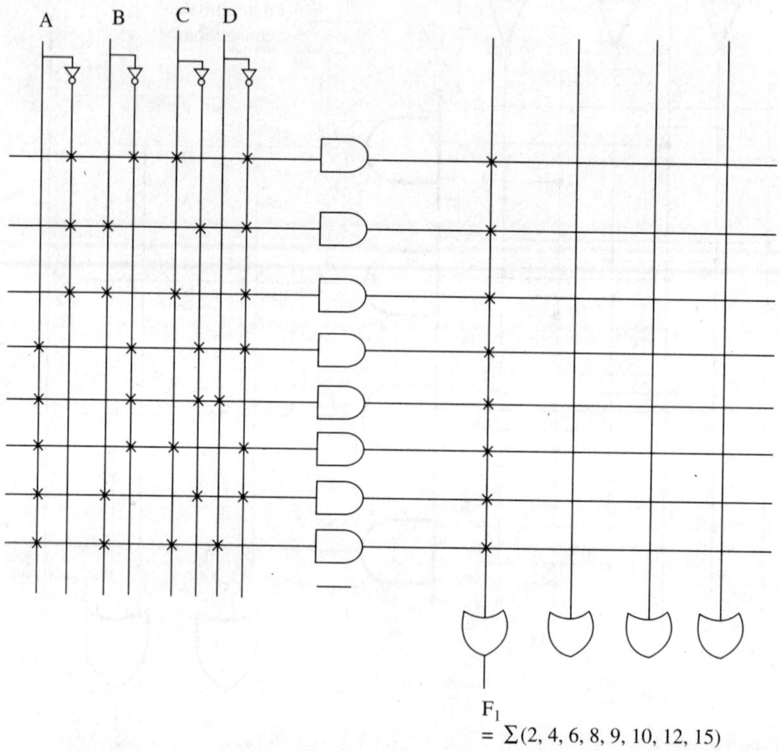

Figure 14.4 Implementation of Logic Function Using PLA

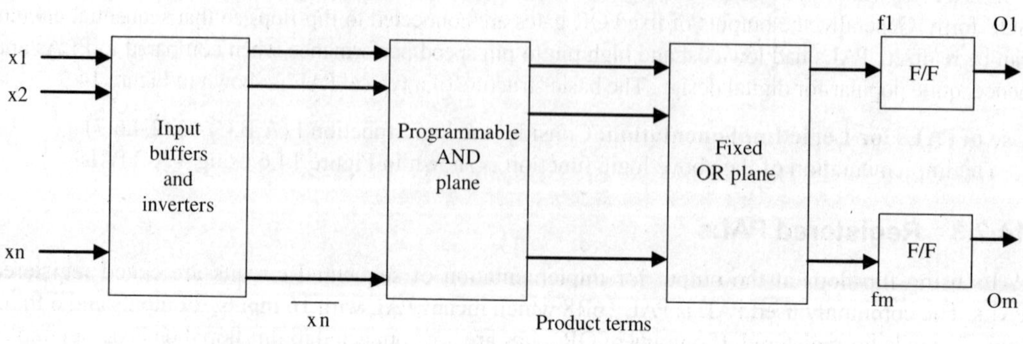

Figure 14.5 Basic Structure of a PAL/Registered PAL

using PAL (4×2)

Implement $\Sigma(2, 4, 6, 7) = \Sigma(A,B,C)$

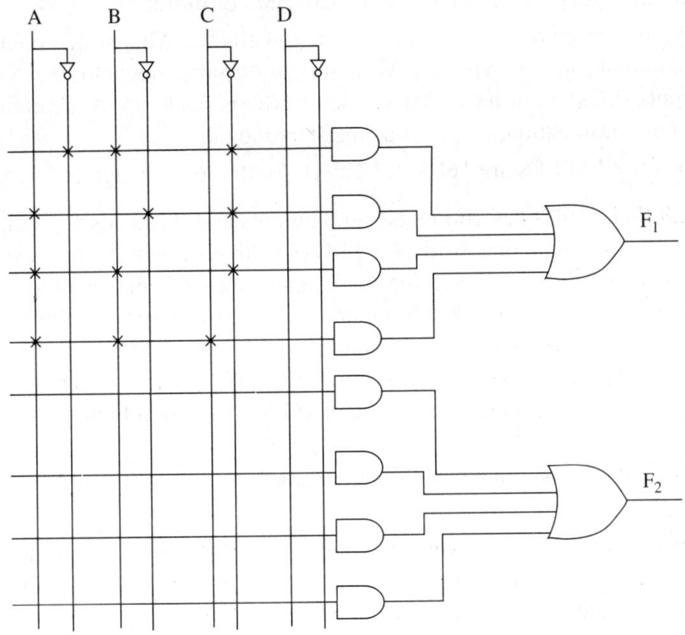

Figure 14.6 Implementation of Logic Function Using PAL

14.2.4 GALs (Generic Array Logic)

GALs were introduced in 1980s as an improvement in SPLDs by Lattice. GAL devices instead of using flip-flops connected at the outputs used macro-cell structure. Each macrocell consisted of flip-flop, gates and multiplexers. Further, these internal components of macrocell were programmable which added to the versatility of GALs. E^2PROM/FLASH technology was used for device programmability which was a significant improvement in device reconfigurability. GAL architecture as shown in Figure 14.7 consists of a reprogrammable array of AND gates, fixed array of OR gates and programmable output logic. The reprogrammable array is a grid of conductors forming rows and columns with an electrically erasable CMOS (E2CMOS) cell at each crosspoint. Each row

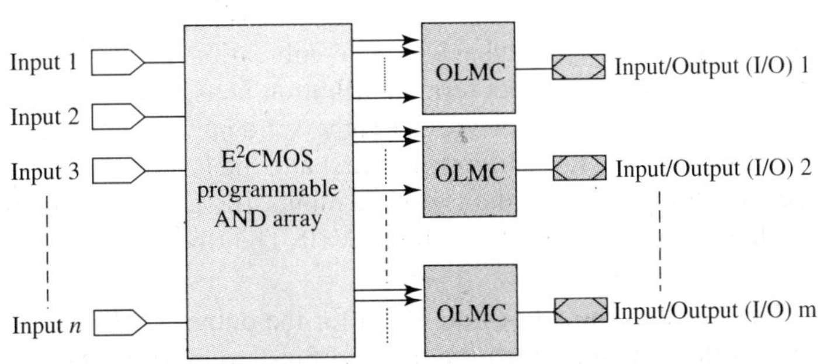

Figure 14.7 Architecture of GAL

is connected to the input of an AND gate and each column is connected to a variable or its complement. By programming each E^2CMOS cell to be either on or off, if a cell on it effectively connects its corresponding row and column. The cell can be electrically erased and programmed. Typical E2CMOS cell can retain its programmed state for 20 years or more.

The AND array outputs go to the output logic macro cells (OLMC) which contain the OR array and programmable output logic. A typical GAL has eight or more inputs to its AND array and eight or more input/outputs (I/Os) from its OLMCs. The OLMC is made up of logic circuits that can be programmed either as combinational logic or as registered logic.

Two most commonly used SPLDs are 16L8 and 22V10. Both are discussed briefly.

Example of a 16L8 PAL: 16L8 has 16 inputs and 8 outputs and 'L' stands for logic array. Of the 16 inputs available 8 dedicated input pins while 8 are I/O pins, that is, they can be used as either input or output pins. In the ouput mode it works as a feedback input coming from the output of OR gate output. Programmable AND plane consists of 64 AND gates each with 16 inputs (or their complements) and the fixed OR array has 8 OR gates with seven inputs to each from AND plane. Three state-inverting buffers are used at 8 I/O pins to AND plane to control the bidirectional operation of the pin and also at the outputs to flip-flops. The device architecture of 16L8 is shown in Figure 14.8.

SPLD PAL 22V10

It is a PAL with 22 inputs and 10 outputs and V stands for versatile. The architecture for 22V10 is shown in Figure 14.9.

22V10 is a macrocell-based PLD in which the output of OR gates is given to macrocell. The macrocell architecture for 22V10 is shown in Figure 14.10. The device can be programmed for variable product terms as AND plane is programmable and at the level of the macrocell. It can work either as a simple PAL or as a registered PAL.

14.2.5 Macrocell Structure

Here C0 and C1 are configuration bits which select polarity of macrocell output as well as mode of output as registered or not. The following table represents the mode of macrocell with C0C1 bits.

C0	C1	Mode
0	0	Registered/active low output
0	1	Registered/active high
1	0	Combinational/active low
1	1	Combinational/active high

Also the macrocell output can be given back to the combinational circuit which adds to device flexibility and versatality. Variable product term distribution is used in 22V10, that is, some OR gates require more product terms when compared to others. So number of product terms assigned to each OR gate vary from 8 (min) to 16 (max). The first and the last OR gate have 8 inputs while the middle OR gate has 16 inputs while the rest have inputs 10, 12, 14 depending on their position. This feature helped to keep the cost low for 22V10. The timing parameters for data sheet are as follows:

t_{pd} (propagation delay) is the amount of time taken for the output to change once the input has changed. Set-up time is the time for which the input to flip-flop must be stable before the active edge of the clock occurs and hold time is the time for which the input must be stable after the rising

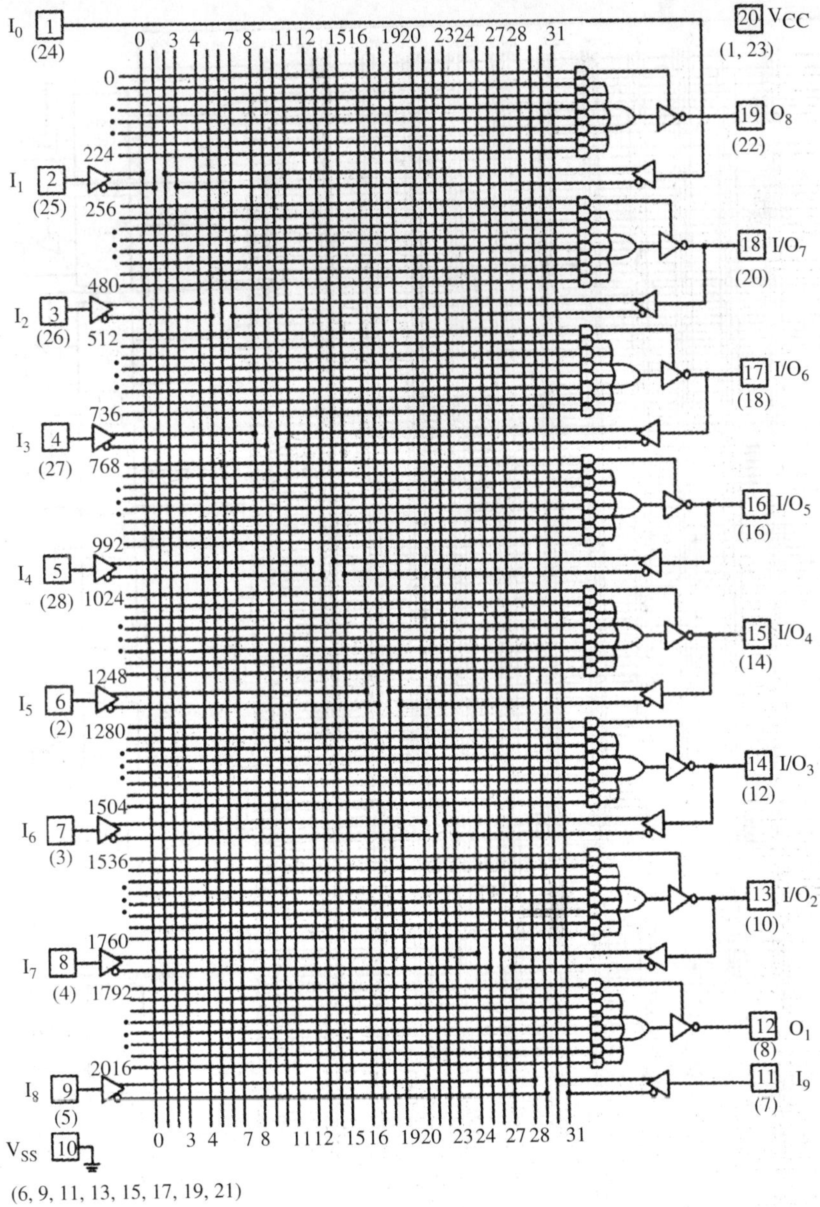

Figure 14.8 Internal Structure of 16L8 PAL

edge of the clock. The clock to output delay is the delay if the output of flip-flop is not directly taken to output pin, but instead is fed to a combinational circuit for further use. The system clock to system clock delay is the measure of minimum time required for the register to register an operation. It helps to determine the maximum frequency of operation for the registered outputs.

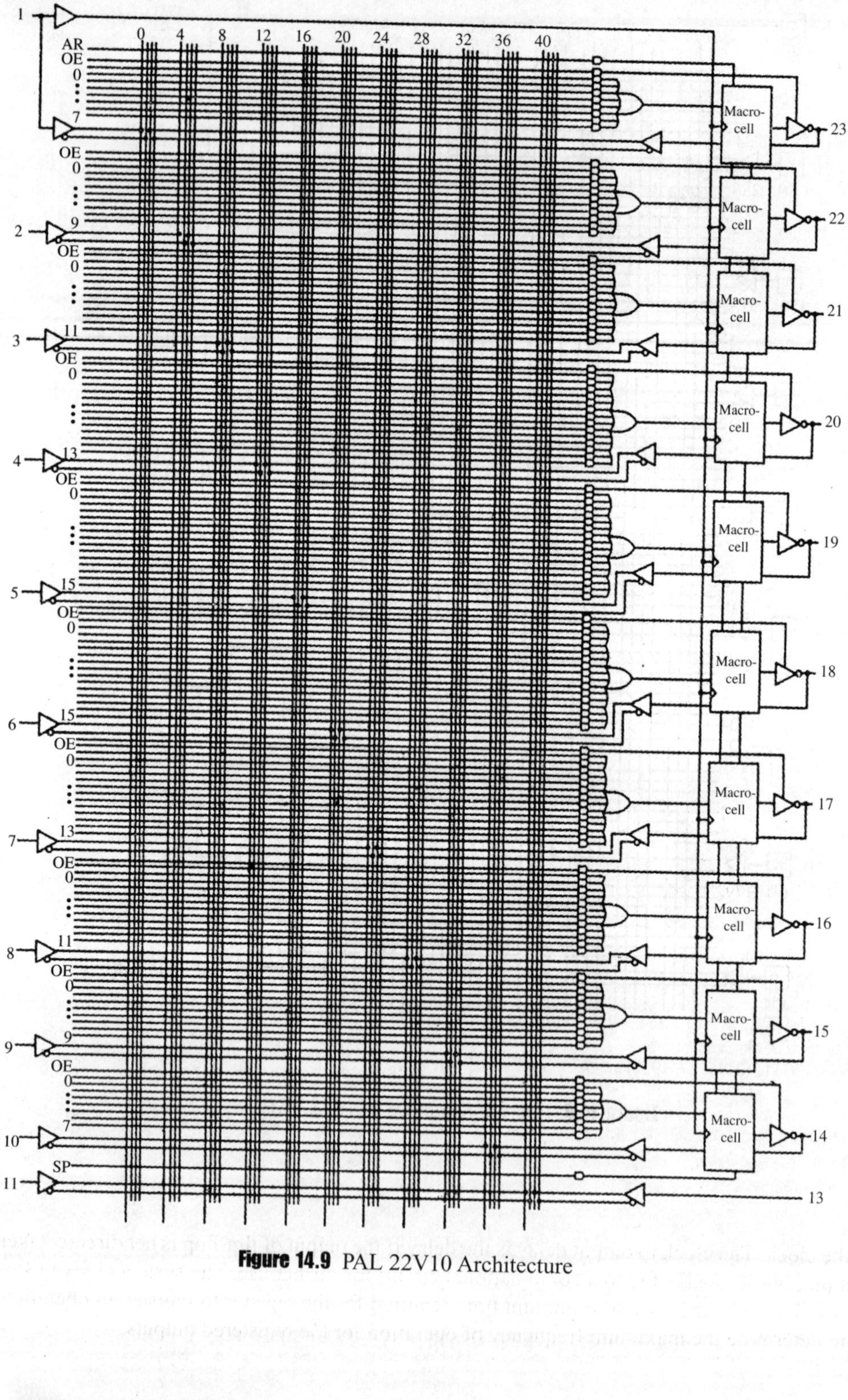

Figure 14.9 PAL 22V10 Architecture

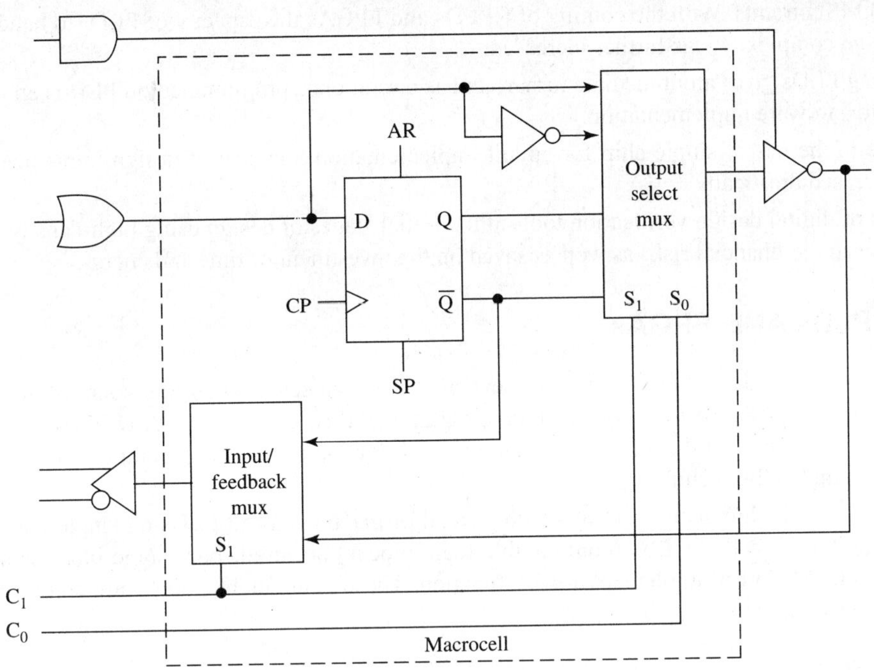

Figure 14.10 Macrocell Structure of PAL 22V10

Parameter	Description	Min	Max
tpd	Propagation delay		4 ns
ts	Set up	2.5 ns	
tH	Hold time	0	
tco	Clock-to-output delay		3.5 ns
tc02	Clock-to-output delay (through logic array)		7 ns
tscs	System clock to system clock delay		5.5 ns

14.3 ADVANTAGES OF USING PLDs

Digital design nowadays revolves around PLDs only. In fact, due to the manifold advantages of PLDs high-density PLDs (CPLDs and FPGAs) are used for modern digital design.

- The biggest advantage of a PLD lies in its programmability. A PLD is not a one-time device. It can be reprogrammed again and again to meet the hardware requirements of the user. This facilitated ease of changes in the design.
- To reprogram a PLD, it is not required to be removed from the printed circuit board but can be reprogrammed on the field. Due to this reason, PLDs are also known as FPLDs (field PLDs).
- Coming of PLDs was a new revolution in digital design as it could handle large, complex design. It could be implemented with single PLD in comparison to the previously used SSI

and MSI circuits. With the coming of CPLDs and FPGAs the capacity of PLDs to handle the design complexity was further increased.

- Using PLDs gives an immediate turnaround as circuit once programmed on PLD is an immediate hardware implementation.
- Due to the use of single chip for circuit implementation, the cost of design implementation decreased drastically.
- Use of digital design verification tools at the initial stages of design using high-density PLDs reduced the financial risks as well as saved on the investment of time and energy.

14.4 CPLDs AND FPGAs

CPLDs are complex PLDs which consist of several PLDs connected by an interconnection matrix which is programmable. So it offers two-level programmability, one at each PLD block in CPLD and the other at the level of the interconnection matrix. It facilitates designing complete large digital systems on a single CPLD chip.

FPGA is also a high-density PLD in which instead of placing several PLDs on single chip as in CPLDs, logic blocks called CLBs (configurable logic blocks) are used. Each logic block is a programmable unit which can implement a logic function. These configutable blocks are connected by programmable switches.

14.4.1 CPLDs

CPLD stands for complex PLD structure in which SPLDs serve as basic building blocks connected by programmable interconnects. The purpose behind CPLDs was to facilitate high-density complex digital design. CPLD chips were introduced by ALTERA.

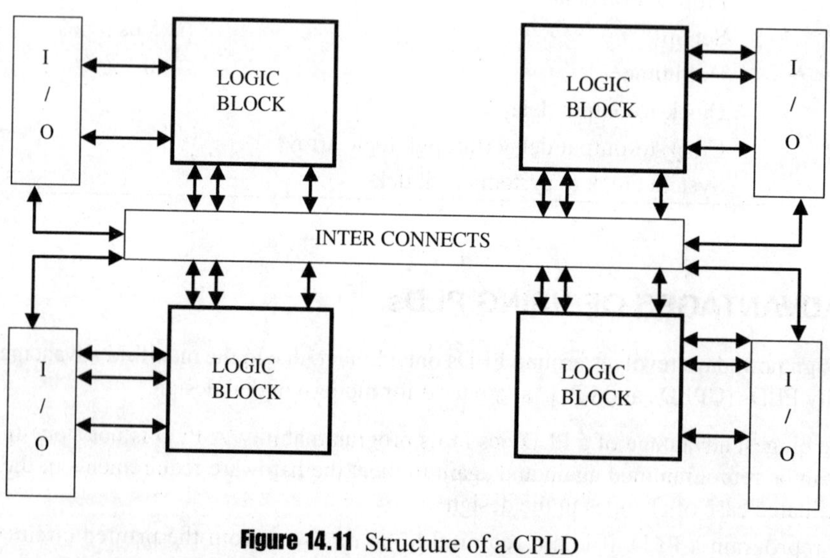

Figure 14.11 Structure of a CPLD

CPLD structure contains multiple logic blocks interconnected through programmable switches. CPLDs lead to better performance, more utilizsation of chip area and reduced cost. Logic blocks can be PAL or PLA but preferably a PAL only.

The general structure of a logic block is

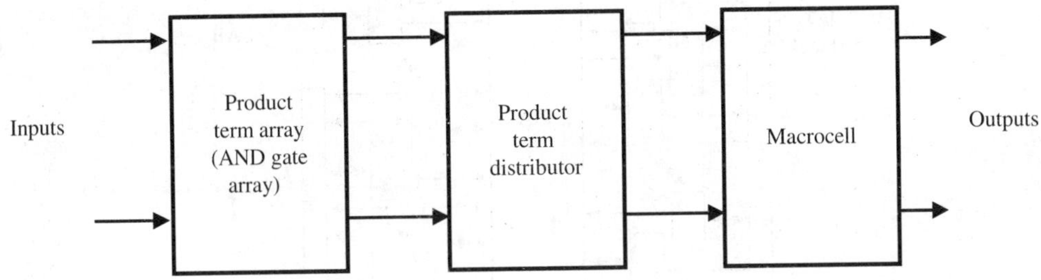

Figure 14.12 Internal Structure of a Logic Block

Macrocells used are more flexible than the ones used in SPLDs. They can be configured not only for polarity but also for the type of flip-flop that is to be used (D-type or T-type). Macrocells used in CPLDS can be of three types:

- I/O macrocell
- Input macrocell
- Buried macrocell

I/O macrocell is associated with I/O. It consists of flip-flop, multiplexer and tri-state buffers. The output of a combinational circuit is fed to flip-flop in the macrocell, if required. The configuration bits are set to choose the mode of output (registered or not registered), polarity of output and type of flip-flop to be used. Thereafter, the output from the product distribution scheme through macrocell is propagated to the I/O. Input macrocell is comparatively simpler and is connected at input to logic blocks. Buried macrocell does not direct output to I/O cell; instead, it provides a feedback path to programmable interconnect so that the feedback is facilitated in a sequential circuit design.

Programmable interconnects control the device configuration. It can be an array form or multiplexer based. In an array or matrix-based interconnects, there is a matrix of horizontal lines and vertical lines. Horizontal lines are inputs to AND gate array and vertical lines are input to the logic block. Each horizontal line can be connected to only some of the vertical lines that it crosses but not all. In case of multiplexer-based interconnects, one multiplexer is used for each input to a logic block. Inputs that are to be given to logic block are connected at the inputs of multiplexer and select lines of multiplexer are programmable. Increased routability can be achieved at the cost of chip area as it would require a wider multiplexer with a higher number of inputs.

Example: Altera MAX 7000 CPLD chip

The architecture of MAX 7000 consists of logic array blocks (LAB) and programmable interconnects called programmable interconnect array (PIA). It connects any LAB to any other LAB through their input and outputs. EPROM/EEPROM programming technology for PIAs is used for MAX7000.

A LAB typically consists of two sets of eight macrocells. Each macrocell further contains a set of programmable product terms that gives input to an OR-gate and a flip-flop. The number of inputs to the OR-gate in a macrocell is variable. The OR gate output is connected to XOR gate. The flip-flops can be configured as D type, JK, T, SR, or can be transparent. An OR-gate can have input of up to five product terms within the macrocell; and in addition, can have up to 15 extra product terms from macrocells in the same LAB. Product term flexibility adds to chip efficiency in terms of chip area because

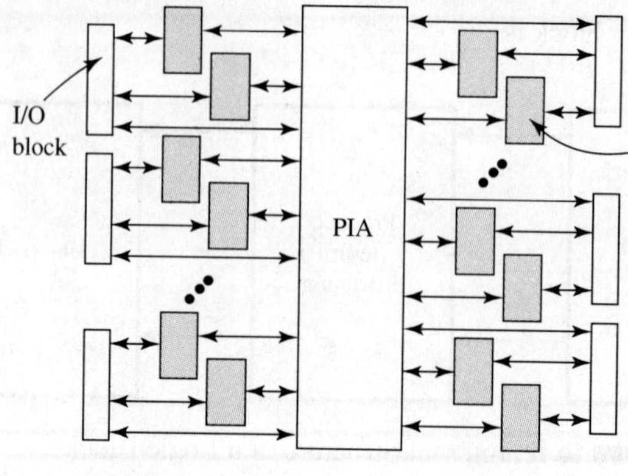

Figure 14.13 Altera MAX7000 CPLD Chip

typical logic functions do not need more than five product terms and the architecture supports wider functions when they are needed.

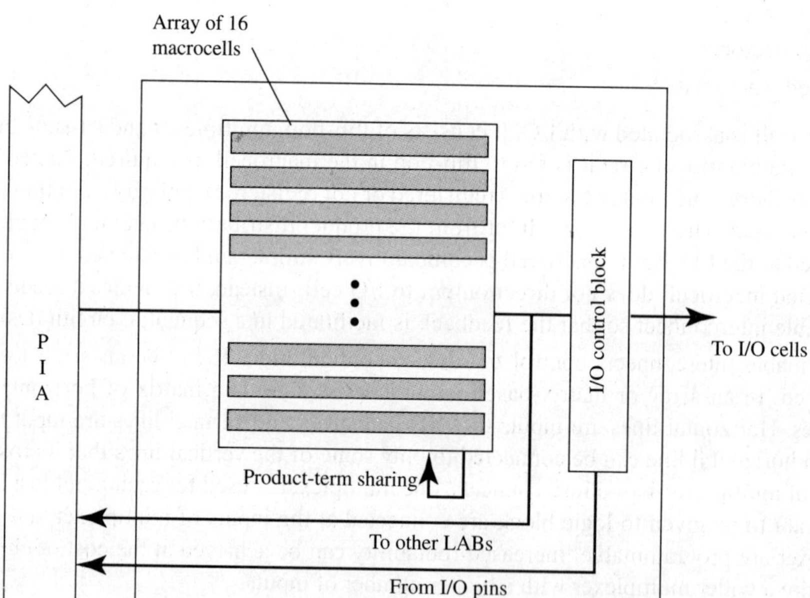

Figure 14.14 LAB Structure for CPLD Altera MAX7000

The internal structure of macrocell in CPLD Altera MAX7000 is shown in the Figure 14.15. It consists of a product select matrix which selects the product terms which are to be given as input to OR gate. The inputs to product select matrix are the outputs of programmable AND array and also input from other LABs in the chip. A global clock signal is used for synchronizing the functionality of the chip. The output of OR/EXOR gate is given to flip-flop. Type of flip-flop can be set by configuration bits. The output of the flip-flop is not directly given at I/O cell but through a multiplexer and PIA. This helps to use the output as input if needed in the circuit.

Other CPLD chip series available from Altera are 5000 series and 9000 series. 5000 series is almost obsolete now. Whereas 9000 has some advanced features than 7000. Else AMD CPLDs are Mach family and Xilinx popular CPLDs are XC7000 and XC9500 series.

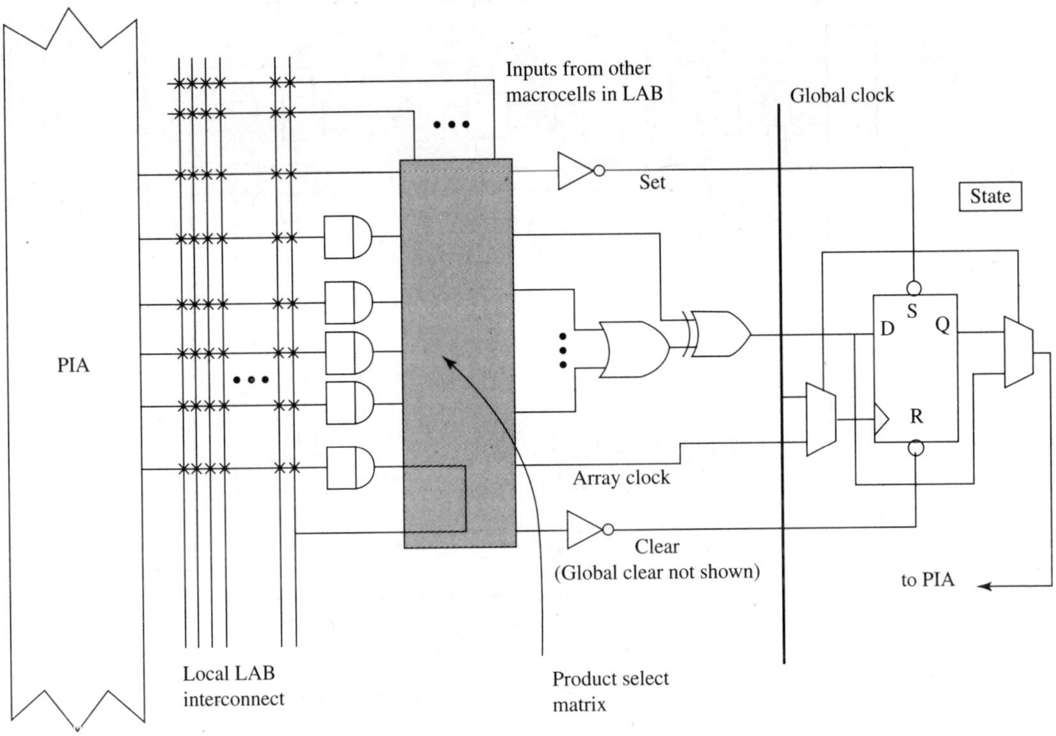

Figure 14.15 Macrocell Structure of CPLD Altera MAX7000

14.4.2 FPGAs (Field Programmable Gate Array)

FPGAs are high-density PLDs meant for complex system design. The capabilities of FPGA to design complex circuits in size and functionality far outweigh than CPLDs. The gate equivalent of FPGAs has about 10,000 to hundreds of thousand gates. FPGAs instead of using SPLDs as basic building blocks use CLBs (Configurable Logic Blocks)—that is, they do not contain AND or OR gate planes.

The general architecture of FPGA is as shown in Figure 14.16.

CLBs and I/O blocks are connected routing channels. Routing channels consist of horizontal and vertical wires connected using programmable switches. CLB typically has a small number of inputs and one output. The density of logic block used in an FPGA depends on the length and number of wire segments used for routing. The number of segments used for interconnection typically is a tradeoff between the density of logic blocks used and the amount of area used up for routing.

Generally, CLB is implemented as a look-up table (LUT) which contains storage cells for implementing small logic functions. Each storage cell can store a single bit value and are volatile storage. So, LUT acts as a one-bit memory. Logic block can be implemented using transistor pairs, multiplexers, combinational circuits and Wide fan-in AND-OR structures depending on the level of density. Each

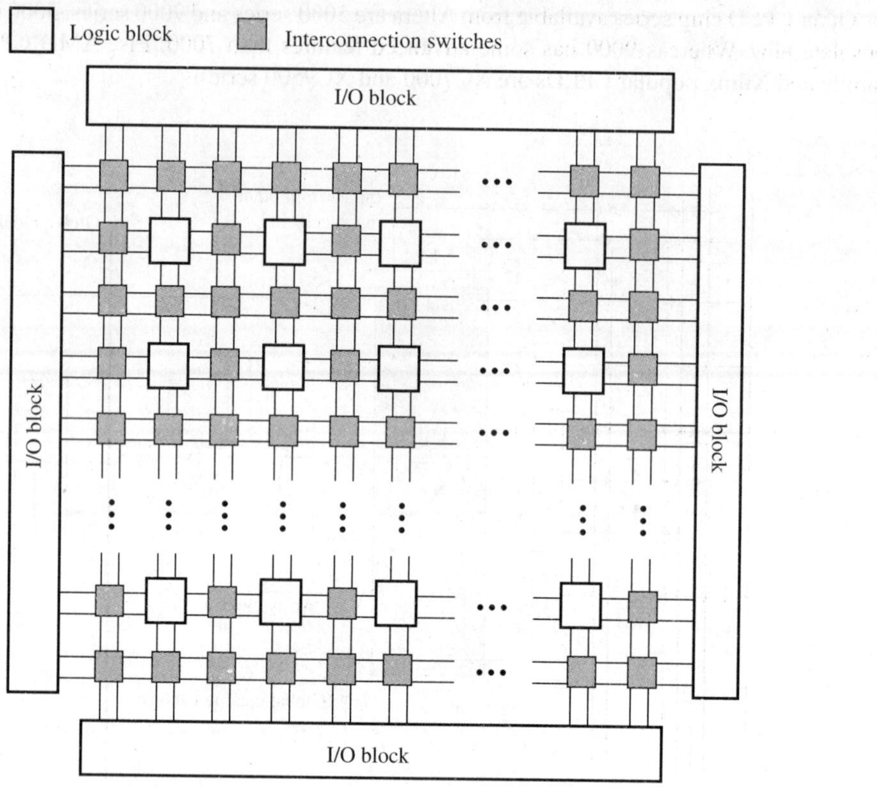

Figure 14.16 General Structure of an FPGA

storage cell can store a single-bit value and are volatile storage. So, LUT act as a one-bit memory. LUT is implemented using SRAM. A k-input logic function is implemented using 2^k X 1 size SRAM. The number of different possible functions for k input LUT is 2^2^k. This allows an implementation of a large number of logic functions.

Diagram for LUT: Consider a 2-input circuit with a truth table of XOR gate: A and B are inputs and Y is the output. LUT for this is

A	B	Y
0	0	0
0	1	1
1	0	1
1	1	0

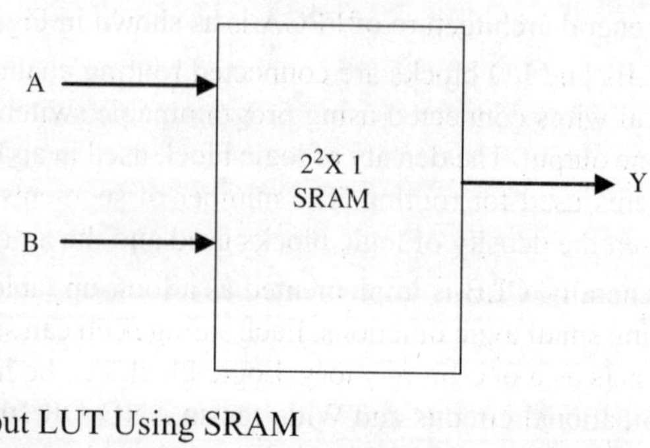

Figure 14.17 2-input LUT Using SRAM

FPGA devices can be based on flash, SRAM, EEPROM or antifuse connectivity. SRAM technology is most popular. SRAM uses pass transistors for establishing connections between horizontal and vertical wires. Each SRAM cell consists of five transistors due to large space requirements. Not every interconnection between horizontal and vertical lines is made using SRAM as it would add to increased cost and slower performance. Instead, SRAm cells are strategically placed so to make an optimum compromise between routability, density and performance. SRAM cell is reprogrammable and are volatile, that is, its loses its contents when the power is switched off. SRAM-based FPGAs have logic blocks with fewer inputs and outputs which end up in using the full capacity of logic cells.

Another popular technology is one-time programmable antifuse technology. Antifuse elements are one-time programmable connections formed in the via or the space between the layers of metal in the transistor used for antifuse link. Antifuse elements are densely placed and to avoid any short circuit each antifuse element is isolated using pass transistors. Antifuse technology makes a better use of chip than SRAM cells. Due to this reason, the antifuse-based FPGAs have more inputs and outputs and routing is no limitation. Though they support ISP (in-system programmability) SRAM is preferred because of added costs in reconfiguring the antifuse-based FPGAs.

FPGA Example Xilinx XC4000:

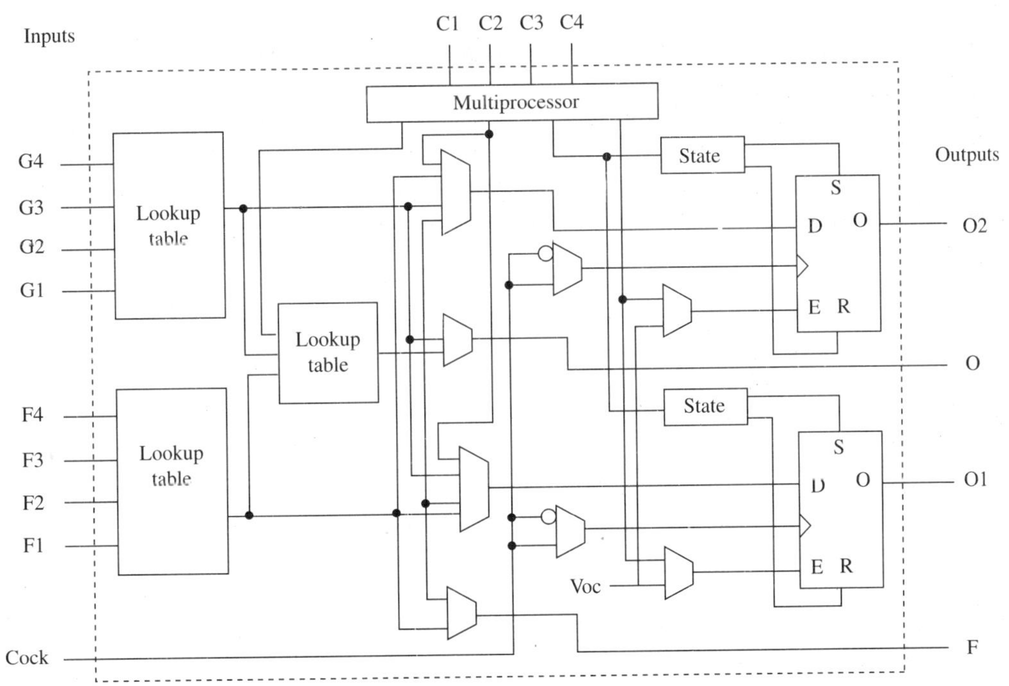

Figure 14.18 Architecture of FPGA Xilinx XC4000

Xilinx XC4000 is one of the most popular FPGAs used. Though Xilinx also offers XC3000 and XC5000, Xc4000 is discussed here as XC3000 is almost obsolete and Xc is the improved form of XC4000 only.

XC4000 FPGA consists of a 2-d array of CLBs interconnected through vertical and horizontal routing channels. It is a SRAM-based chip with SRAM acting as an LUT. It contains two 4-input LUT, each of which can implement 2^{16} logic functions with four inputs. The output of LUTs is fed to flip-flops optionally to implement sequential circuits or to have the registered output. A 3-input LUT is also there which allows an implementation of logic functions with 5/6 variables. Since LUT are implemented using SRAM cells they can be combined to form large memory blocks.

14.4.3 Differences Between FPGAs and CPLDs

Though FPGAs and CPLDs are both high-density PLDs, they differ in architecture, programming technology, circuit capacity, density and implementation.

Table 14.1 Differences Between FPGAs and CPLDs

FPGA	CPLD
FPGA consists of configurable logic blocks, I/O blocks interconnected by routing channels.	CPLD typically is a collection of PAL like structure interconnected by programmable switch matrix.
They are much more versatile than CPLDs with much higher logic capacity and hence used for complex-level designs.	Though it can implement complex design but is limited only to moderate level of design complexity.
They do not contain AND-OR planes.	AND–OR planes form the backbone of CPLD structure.
FPGAs generally use SRAM or antifuse technology for programmable switches.	They rely on E2PROM/Flash technology for programming interconnects.
FPGAs uses narrow logic resources and higher ratio of flip-flops to logic resources than CPLDs.	CPLDs use logic resources (gates/combinational circuits) with wider inputs.
*Logic capacity of FPGAs is much more higher than CPLDs (from 1000 to few thousands (40,000)).	Logic capacity is lower than FPGAs (from few hundreds to 12,000).
Logic density of FPGAs is higher than CPLDs.	Lower logic density.

*Logic capacity is a measure of the amount of digital logic that can be mapped into a single FPD. This is usually measured in the units of equivalent number of gates in a traditional gate array. That is, the capacity of an FPD is measured by the size of the gate array that is comparable to and generally be measured as "number of 2-input NAND gates".

14.4.4 Implementation Using FPGAs and CPLDs

For implementing a logic design with the FPGA or CPLD, CAD tools play a very important role starting right from design entry level to synthesis of circuit and its placement on chip.

1. First step is to enter description of the logic circuit using a hardware description language (HDL), VHDL or Verilog being the most popular ones. An alternative approach is to use a schematic capture tool as ORCAD.
2. Using a simulation tool as a Modelsim to convert the schematic of HDL file to RTL view.

3. Using a synthesis tool as a synopsis to transform the HDL or schematic into a *netlist*. The netlist is just a description of the various logic gates in our design and how they are interconnected.
4. Using implementation tools map the logic gates and interconnections into the FPGA/CPLD. At this stage, design with FPGAs requires additional steps. More sophisticated CAD tools are

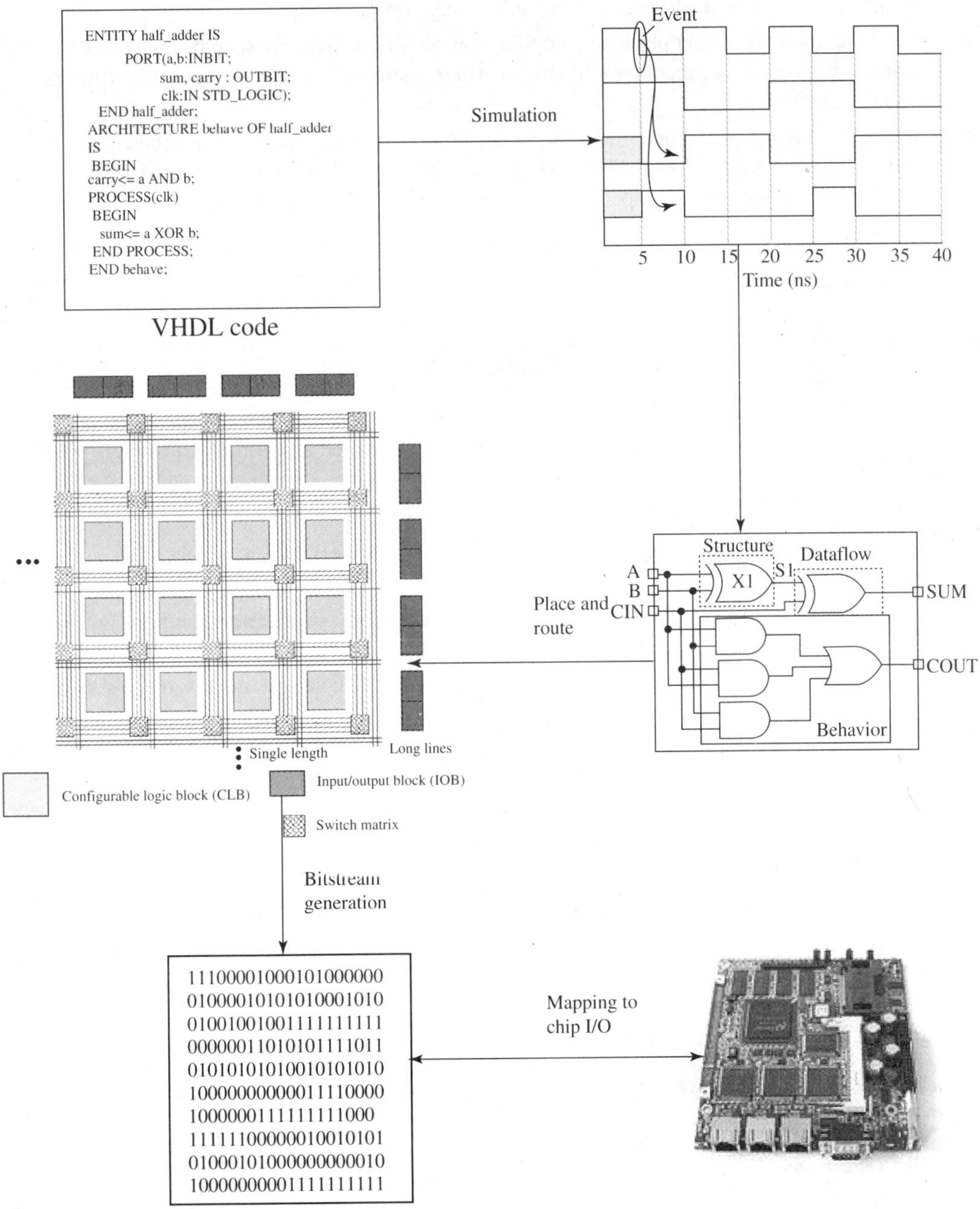

Figure 14.19 Complete Process for Implementation Using FPGAs/CPLDs

required for FPGA design. The configurable logic blocks in the FPGA can be further decomposed into look-up tables that perform logic operations. The CLBs and LUTs are interwoven with various routing resources. The mapping tool collects your netlist gates into groups that fit into the LUTs and then the place and route tool assigns the gate collections to specific CLBs while opening or closing the switches in the routing matrices to connect the gates together. The optimization for speed and area is done at this stage only.

5. Once the implementation phase is complete, a program extracts the state of the switches in the routing matrices and generates a bitstream where the ones and zeroes correspond to open or closed switches.
6. The bitstream is downloaded into a physical FPGA chip. The electronic switches in the FPGA open or close in response to the binary bits in the bitstream. Upon completion of the downloading, the FPGA performs the operations specified by HDL code or schematic.

Figure 14.19 shows the sequence of steps to be followed for the implementation of design using PLDs.

SUMMARY

- PLDs offer reconfigurable architecture for logic function implementation.
- Depending on the level of complexity a PLD can be classified as a SPLD, CPLD or FPGA.
- SPLDs are simple array of gates with programmable inputs.
- SPLD can be PALs, PLAs and GALs.
- PAL (programmable array logic) has both AND gate plane and OR gate plane which gives output in SOP form and both the planes are programmable.
- Making both AND–OR planes programmable leaves many connections unused and making them programmable adds significantly to the cost.
- PLAs have only AND plane programmable with fixed OR plane and outputs are generally connected to flip-flops to implement sequential circuits.
- GALs consist of programmable AND array whose outputs are connected to output logic macro cells (OLMC) which contain the OR array and programmable output logic.
- CPLDs are used at a higher level of design implementation of bigger circuits. It connects multiple SPLD blocks with programmable switches.
- CPLDs generally use EPROM/EEPROM technology for programmability of interconnects.
- FPGAs are used for implementing very large circuits. They have much higher logic capacity than CPLDs.
- FPGA consists of CLBs (configurable logic blocks), I/O blocks and programmable routing channels to connect them.
- FPGAs generally use SRAM or antifuse technology for programming interconnects.

QUESTIONS

Objective Questions

1. PLDs are .... devices.

   a. reconfigurable b. fixed function c. complex d. SoC

2. PALs have .... efficiency than PLAs.
   a. better b. higher c. lower d. comparable
3. FPGAs have .... logic capacity than CPLDs.
   a. higher b. lower c. lesser d. comparable
4. For complex system designs .... are preferred over CPLDs.
   a. GALs b. PLAs c. FPGAs d. none of these
5. FPGAs use .... in their internal structure.
   a. SPLDs b. LUTs c. AND plane d. EXOR gate plane
6. HDLs are important in design with PLDs as ....
   a. backend tool
   b. physical verification
   c. frontend tool
   d. schematic tool
7. PALs and PLAs differ in ....
   a. programmable OR plane
   b. programming technology
   c. logic capacity
   d. programmable AND plane

Descriptive Questions

1. Compare ROMs, PALs and PLAs for logic implementation.
2. What is the difference between FPGAs and CPLDs?
3. Discuss the architecture of an FPGA.
4. What is programmable logic? What are its advantages?
5. Discuss the advantages of FPGAs over custom/fixed function ICs.
6. Why are SRAM-based FPGAs more popular than antifuse-based FPGAs?
7. Briefly discuss 22V10 PLD.
8. Design BCD to Excess-3 converter using
   a. PAL b. PLA
9. Design a 2-input LUT each for AND, OR and EXOR gates.
10. Implement $F(A,B,C) = \Sigma(2,4,6,7)$ using PLA.
11. Discuss the complete process of implementing a circuit using PLDs.
12. What are macrocells? What is their use?
13. Discuss the different types of macrocell used in CPLDs.
14. Write a note on different programming technologies for PLDs.
15. Implement $F1=\Sigma(2,3,7,9,12,14,15)$
    $F2=\Sigma(2,4,5,10,11,14,15)$
    $F3=\Sigma(2,3,5,7)$
    using a 4 X 4 PAL.
16. Implement the above using 4 X 6 X 4 PLA.

CASE STUDY—GENETIC ALGORITHM PROCESSOR

15

CHAPTER OBJECTIVES

- Concept of hardware GA
- Advantages of GAP over software GA
- Architecture of genetic algorithm processor
- Example of VLSI bipartitioning

15.1 INTRODUCTION TO GENETIC ALGORITHMS

Genetic Algorithms (GA) are evolutionary computational models based on Charles Darwin's theory of natural evolution based on the concept of the survival of the fittest. The original GA and its many variants, collectively known as genetic algorithms, are computational procedures that mimic the natural process of evolution. The theories of evolution and natural selection were first proposed by Darwin to explain his observations of plants and animals in the natural world. He observed that, as variations are introduced into a population with each new generation, the less-fit individuals tend to die off in the competition for food, and this survival of the fittest principle leads to improvements in the species. The concept of natural selection was used to explain how species have been able to adapt to the changing environments and how, consequently, species that are very similar in adaptivity may have evolved.

All information required for the creation of appearance and behavioural features of a living organism is contained in its chromosomes. Reproduction generally involves two parents, and the chromosomes of the offsprings are generated from portions of the chromosomes taken from their parents. In this way, the offsprings inherit a combination of characteristics from their parents. GAs attempt to use a similar method of inheritance to solve various problems such as those involving adaptive systems. The objective of the GA is then to find an optimal solution to a problem. Since GAs are heuristic procedures, modelled as function optimizers, they are not guaranteed to find the optimum, but are able to find very good solutions for a wide range of problems.

GAs work by evolving a population of individuals over a number of generations. A fitness value is assigned to each individual in the population, where the fitness computation depends on the application. For each generation, individuals are selected from the population for reproduction and they are crossed to generate new individuals, and the new individuals, in turn, are mutated with some low mutation probability. The new individuals may completely replace the old individuals in the population with distinct generations evolved. Alternatively, the new individuals may be combined with the old individuals in the population. In this case, the population's size can be maintained at a particular constant, for example, by selecting the best individuals from the population. Both these approaches have yielded good results. The choice of which approach is used may depend on the application. Since the selection is biased towards more highly fit individuals, the average fitness of the population tends to improve from one generation to the next. The fitness of the best individual is also expected to improve over time, and the best individual may be chosen as a solution after several generations.

All GAs work on a population or a collection of several alternative solutions to the given problem. Each individual in the population is called a string or a chromosome, in analogy to chromosomes in the natural systems. Often individuals are coded as binary strings, and the individual characters or symbols in the strings are referred to as genes. In each iteration of the GA, a new generation is evolved from the existing population in an attempt to obtain better solutions. The population size determines the amount of information stored by the GA. The GA population is evolved over a number of generations.

15.2 HARDWARE GA

Though GAs can be effectively implemented in a software, even it is supplied as a toolbox for software technical languages like MATLAB but the high run time and consumption of large memory resources limit the use of GAs. The software implementation further adds to high resource consumption and limits the extensibility of GA as a general-purpose optimization tool. The hardware implementation of GA aims to use modified compact GA which can act as a general-purpose GA processor (GAP) with flexibility to change application-specific operations such as fitness evaluation. Further, the coming of reconfigurable devices on FPGAs—the utility and significance of hardware GAP increases all the more. The past several years have witnessed a sharp increase in work with reconfigurable hardware systems. Reconfigurability is essential in a general-purpose GA engine because certain GA modules require changeability (fitness function, chromosome encoding). Thus a hardware-based GA is both feasible as well as desirable.

15.3 ARCHITECTURE OF GAP

In this case study we attempt to design a reconfigurable GAP having the following architecture.

The basic architecture of a GAP is shown in Figure 15.1. GAP essentially consists of a processing unit which has modules for performing basic genetic operations. The input to GAP is encoded chromosome. The encoding mechanism can be real numbers or simple binary or backed BCD depending on the problem. A chromosome is actually any random solution to the problem. This chromosome is then used for generating random population of a predefined size. The population of chromosomes represents a set of possible solutions for a given problem. The initial population is stored in the main memory. How good a solution is (chromosome) is determined by the fitness criteria for the chosen problem. Generally, the better a solution higher is its fitness value. This is done by the evaluation module which computes fitness for each member of the generated population. The computed fitness values for each member is stored in the fitness memory corresponding to the address of its chromosome in the main memory.

The initial values for the size of population, number of runs of GA, probability of mutation are stored in the register bank. The proposed GAP allows three basic genetic operations: selection, crossover and mutation.

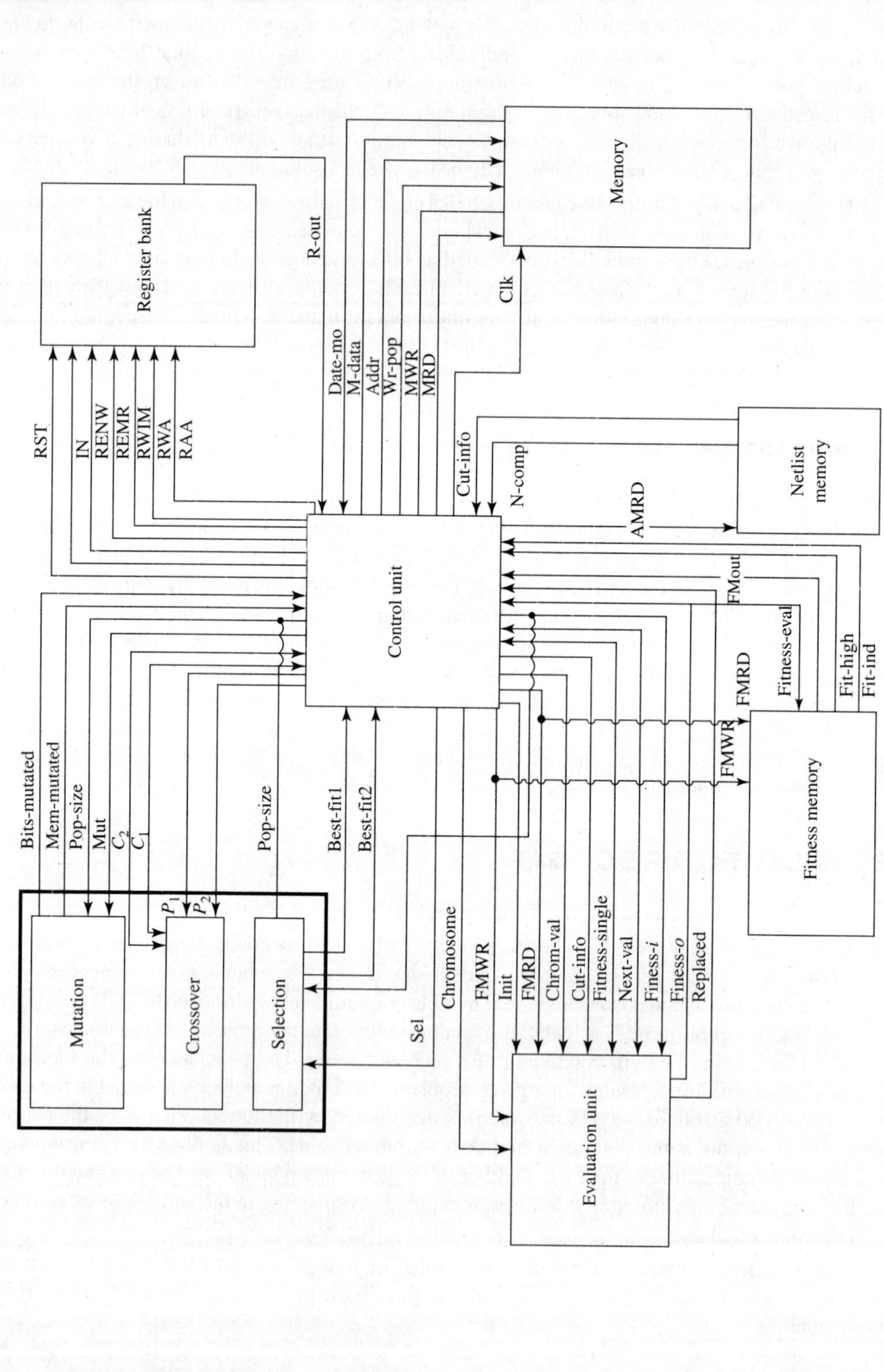

Figure 15.1 Architecture of GAP

Selection is based on the roulette wheel selection which is a proportionate selection scheme in which slots of a roulette wheel are sized according to the fitness of each individual in the population as shown in Figure 15.2. Spinning the roulette wheel and noting the position of the marker select an individual. The probability of selecting an individual is therefore proportional to its fitness. The selected parents then undergo a crossover to generate expectedly better-fit offsprings.

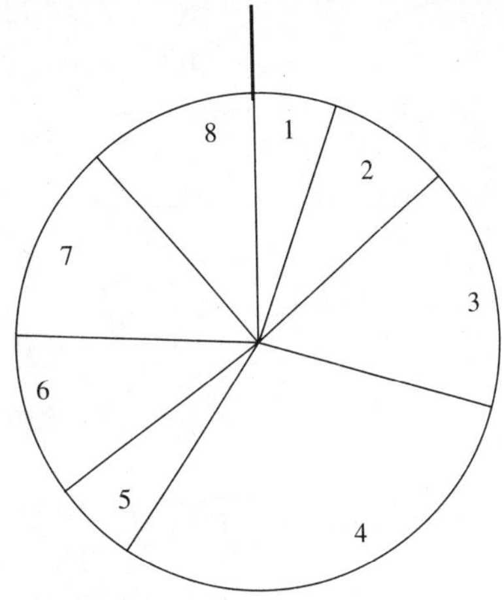

Figure 15.2 Roulette Wheel Selection

The selected parents are then given as input for a crossover to yield offsprings of fittest individuals in the population. The crossover can be a one-point crossover or a uniform crossover. Selected parents after the crossover yield offsprings which are evaluated. The offsprings are evaluated for fitness and replace the worst-fit individuals in the population. Mutation operation is used to introduce diversity in the population. It is very useful to spread the search area which helps to cover the search space for finding global optimas to the problem instead of finding the best solution in the limited search space which may be a local mimimum. The GA runs over and over till the stopping criteria is met. The stopping criteria to stop the GA is either completing the number of runs or reaching the optimal value within the tolerance limits, whichever earlier.

15.4 IMPLEMENTING THE MODULES FOR GAP

In this section, implementation of GA-based processor has been discussed. Though GAP is reconfigurable, some sections of GAP are application specific which can be well explained only with reference to its application area. The proposed GAP is implemented and discussed with reference to the example of bipartitioning a circuit. Though the idea to use GAP in the partitioning process is meant for VLSI circuits, here we discuss a simple example of bipartitioning a simple circuit with eight components alone.

For a partitioning example a chromosome of length eight bits is chosen and a binary encoding is used for bipartitioning.

15.4.1 Processing Unit

Processing unit is responsible for performing genetic operations. Processing unit as shown in Figure 15.3 has three modules. It performs three operations:

- Selection
- Crossover
- Mutation

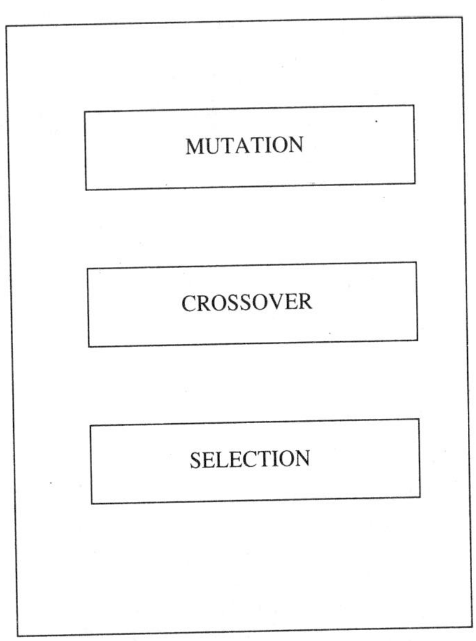

Figure 15.3 Processing Unit

Selection: As discussed earlier, selection is used to choose best-fit individuals for reproduction in an effort to generate more fit individuals in the population. Figure 15.4 shows the selection unit.

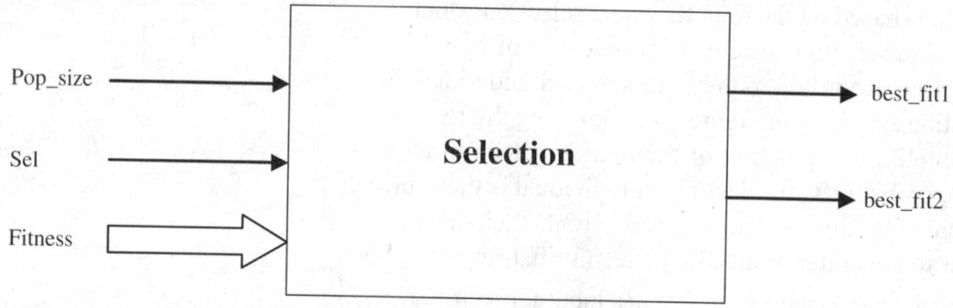

Figure 15.4 Selection Module

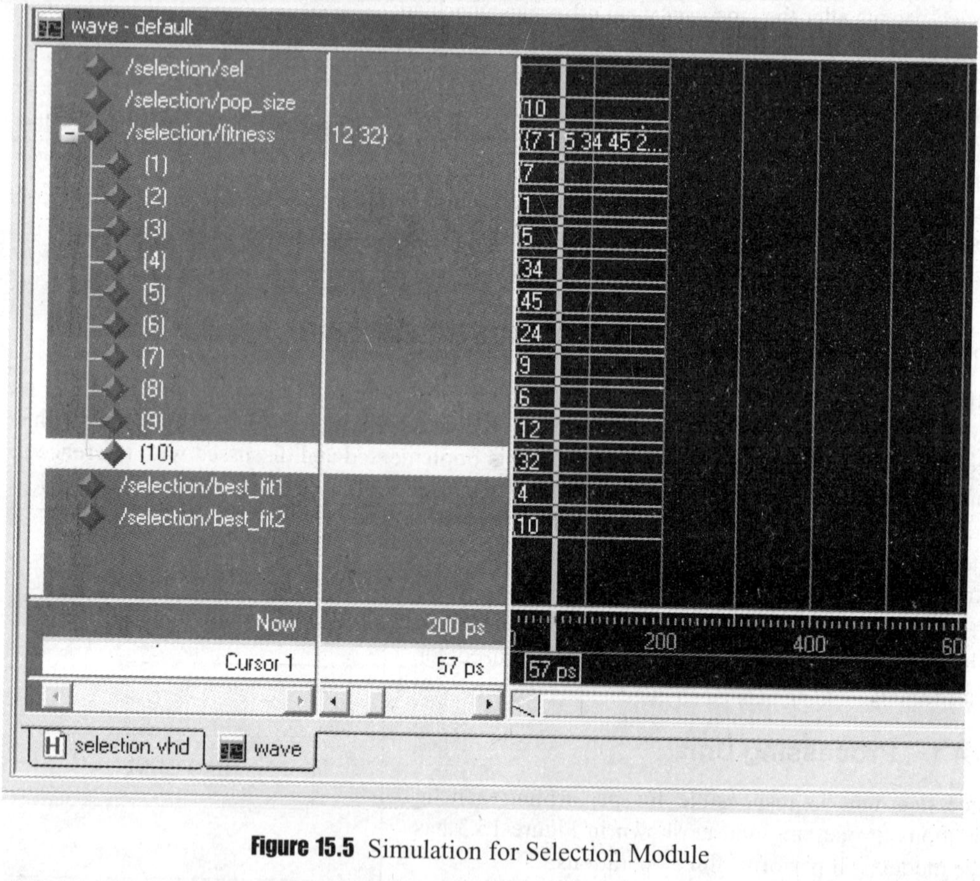

Figure 15.5 Simulation for Selection Module

```
ENTITY selection IS
   PORT( sel:IN std_logic;
      pop_size:IN integer range 1 to 15:= 10;
         fitness:IN fitness_arr ;
         best_fit1,best_fit2:OUT integer range 1 to 15);
END selection;
```

```
ARCHIECTURE Behavioural OF selection IS
BEGIN
p11:PROCESS(sel)
VARIABLE selectpt,sel_roul:INTEGER range 0 to 1000:=0;
VARIABLE s1:integer range 0 to 15:= 1;
VARIABLE s2:integer range 0 to 1000;
VARIABLE s3:integer range 0 to 1000:= 1;
VARIABLE int_val, sum_fit:INTEGER range 0 to 1000;
BEGIN
IF sel='1' THEN
FOR i in 1 to 10 LOOP
int_val:= real(fitness(i));
sum_fit:= int_val+ sum_fit;
END LOOP;
random(seed1,selectpt);
sel_roul:= selectpt* sum_fit;
int_val:=0.0;
FOR i IN 0 TO pop_size LOOP
int_val:= real (fitness(i)) + int_val;
IF int_val >= sel_roul THEN
best_fit1 <= i;
EXIT;
ELSE
NEXT;
END IF;
END LOOP;
END IF;
END process p11;
p22:process(sel)
variable selectpt,sel_roul:integer range 0 to 1000:=0;
variable s1:integer range 0 to 15:= 1;
variable s2:integer range 0 to 1000;
variable s3:integer range 0 to 1000:= 105;
variable int_val, sum_fit:integer range 0 to 1000;
begin
    if sel<='1' then
for i in 1 to 10 loop
int_val:= fitness(i);
sum_fit:=  int_val+ sum_fit;
end loop;
```

```
s2:= sum_fit;
randg(s1,s2,s3,selectpt);
sel_roul:= selectpt;
int_val:=0;
for i in 1 to 10 loop
int_val:= (fitness(i)) + int_val;
if int_val>= sel_roul then
best_fit2 <= i;
exit;
else
next;
end if;
end loop;
end if;
end process p22;
end Behavioral;
```

Crossover: For this example, a one-point crossover operation has been chosen.

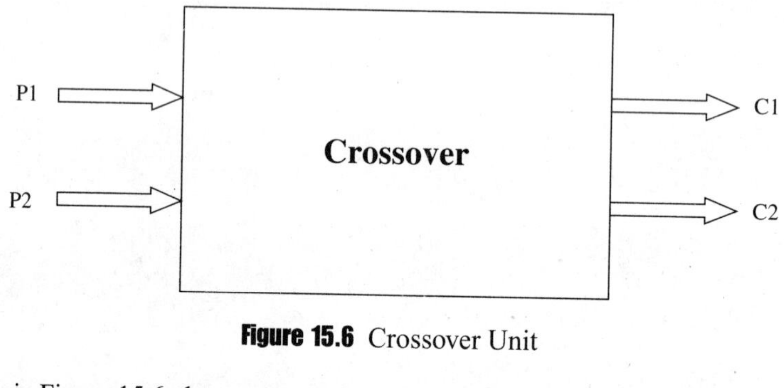

Figure 15.6 Crossover Unit

As shown in Figure 15.6, the crossover unit accepts two inputs which are two members of the population selected by the selection unit in an effort to replicate better-fit individuals.

```
entity crossover is
Generic (crosspt:INTEGER:=4);
      Port ( p1,p2 : IN STD_LOGIC_VECTOR(1 to 8);
                c1,c2: inOUT STD_LOGIC_VECTOR(1 to 8));
end crossover;
architecture Behavioural of crossover is
begin
PROCESS(p1,p2)
```

```
BEGIN
for i IN crosspt +1 to 8 loop
c1(i)<= p2(i);
c2(i)<=p1(i);
end loop;
for i IN 1 to crosspt loop
c1(i)<= p1(i);
c2(i)<= p2(i);
end loop;
end process;
end Behavioural;
```

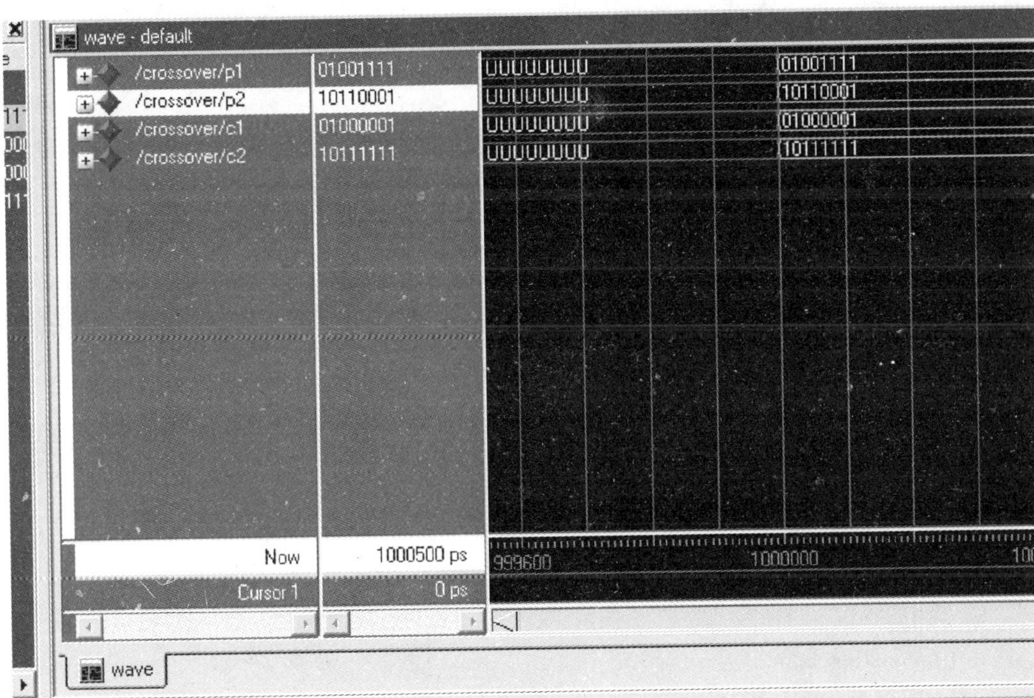

Figure 15.7 Simulation for Crossover

Mutation: Mutation operator is used to introduce diversity in the population and helps to cover the solution space. Figure 15.8 shows the mutation unit which starts the mutation operation when the signal 'mut' is high set by the control unit. It requires the size of the initial population to be input and generates two outputs mem_mutated which is an array of integrated values containing the index value of population member whose bit is to be mutated and bits_mutated contains the position of bits to be mutated in the corresponding member.

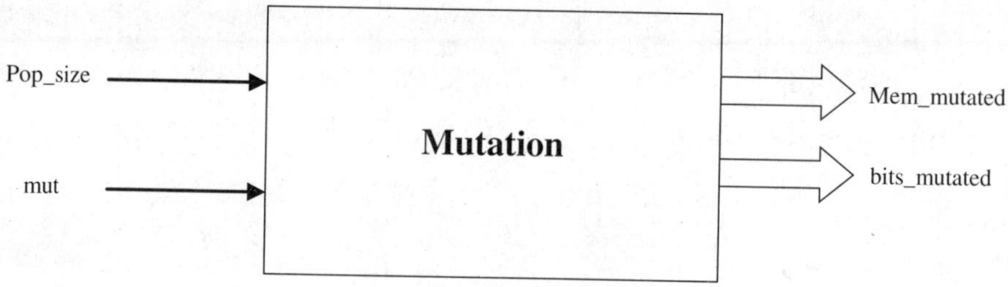

Figure 15.8 Mutation Unit

```
Library IEEE;
use IEEE.STD_LOGIC_1164.ALL;
use IEEE.STD_LOGIC_ARITH.ALL;
use IEEE.STD_LOGIC_UNSIGNED.ALL;
use work.rng.all;
entity mutation is
port(mut:in std_logic;pop_size:in integer range 0 to 15:=10;
      mem_mutated,bits_mutated:out arr);
end mutation;
architecture Behavioral of mutation is
begin
process(mut)
Variable s1,s4:integer:=1 ;
Variable s2:integer range 0 to 15;
Variable s3:integer range 0 to 1000:=5;
variable s5:integer range 0 to 1000:= 8;
variable ind_no, bit_mut:integer;
variable mut_ind, bit_pos:arr;
begin
   s2:= pop_size;
if mut='1' then
for i in 1 to mut_bits loop
randg(s1,s2,s3,ind_no);
mut_ind(i):=ind_no;
s3:= s3+ 2;
randg(s1,s5,s4,bit_mut);
          bit_pos(i):= bit_mut;
```

```
        s4:=s4 + 1;
end loop;
mem_mutated<= mut_ind;
bits_mutated<= bit_pos;
end if;
end process;
end Behavioral;
```

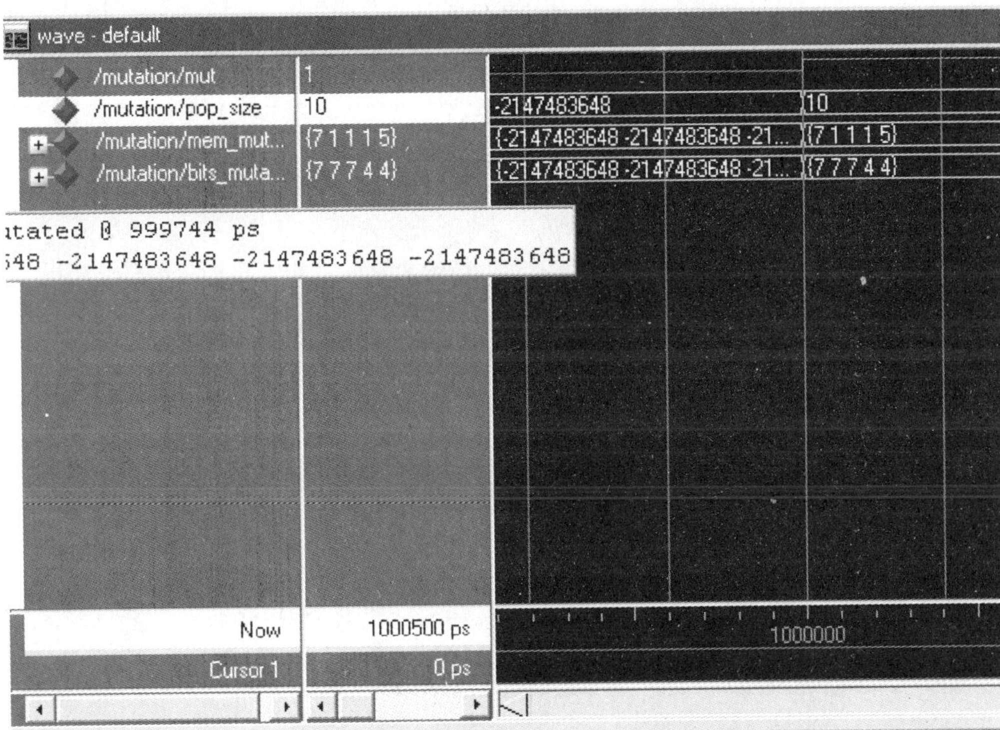

Figure 15.9 Simulation Waveform for Mutation Operator

Evaluation Unit: Evaluation unit is a computational unit which computes fitness for each member of the population—be it part of an initial population or a newly generated offspring. It uses the fitness criteria for determining the fitness of the chromosome. It accepts chromosome to be evaluated from control unit which reads either from memory or gives offsprings generated as input for fitness evaluation to evaluation unit.

For example, a partitioning fitness criteria is **Fitness=cut_value,** where the cut_value denotes the number of interconnections between the circuit components cut due to given partitioning.

To determine the cut_value for the given partition solution (chromosome) the evaluation unit needs information for interconnections between circuit components which is provided in the form of cut_info_n signal.

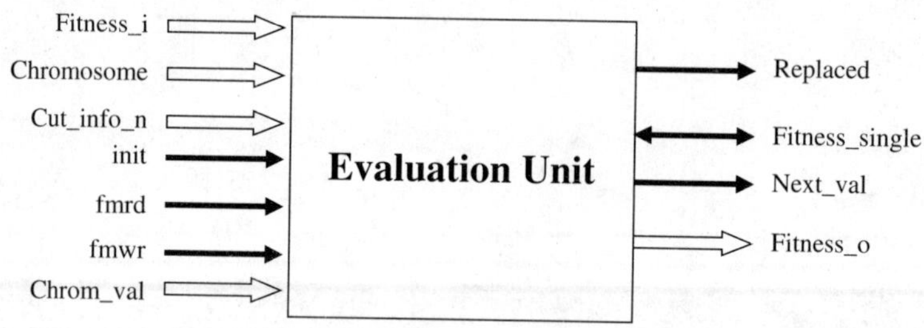

Figure 15.10 Evaluation Unit

```
library IEEE;
use IEEE.STD_LOGIC_1164.ALL;
use IEEE.STD_LOGIC_ARITH.ALL;
use IEEE.STD_LOGIC_UNSIGNED.ALL;
use work.rng.all;
entity eval is
port(chromosome:in std_logic_vector( 1 to 8);
     init,fmrd,fmwr:in std_logic;
     chrom_val: in chrom_matrix;
     cut_info_n:in adj_matrix;
     fitness_single:inout integer range 0 to 1000;
     next_val:inout std_logic;
     fitness_i:in fitness_arr;
     fitness_o:out fitness_arr;
     replaced:out std_logic);
end eval;
architecture Behavioral of eval is
signal temp_chrom: std_logic_vector( 1 to 8);
signal fitness_lowest:integer range 0 to 150;
begin
process(init,next_val, chromosome, cut_info_n,fmwr,fmrd)
variable count,count1, index:integer range 0 to 20:=0;
variable cut_val:fitness_arr;
```

```
variable cut_val_single,loop_fit:integer range 0 to 150 :=0;
variable temp_chrom_var:std_logic_vector( 1 to 8);
begin
if init='1' and next_val ='1' then
        count:=count+1;
        for i in 1 to pop_size loop
temp_chrom_var := chrom_val(i);
                temp_chrom <= temp_chrom_var;
            fitness_o(i)<= evalfnc(temp_chrom,cut_info_n);
        end loop;
             if count > pop_size then
                    next_val<= '0';
             else
                    next_val<='1';
             end if;
elsif init = '0' then
             fitness_single <= evalfnc(chromosome,cut_info_n);
end if;
if fmrd='1' then
             for i in 1 to 10 loop
                    for j in 1 to 10 loop
                           if fitness_i(j) < fitness_i(i) then
                              fitness_lowest<= fitness_i(j);
                    index:= j;
                    end if;
                end oop;
          end loop;
end if;
    if fmwr= '1' and fitness_lowest > fitness_single then
            fitness_o(index)<= fitness_single;
            replaced<='1';
    else
            replaced <= '0';
end if;
end process;
end Behavioral;;
```

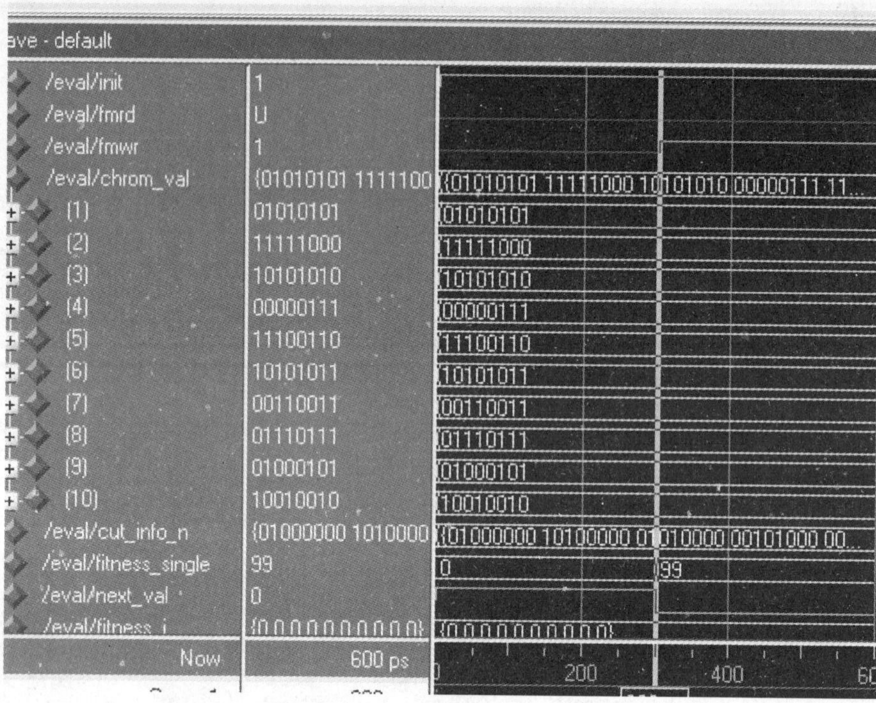

Figure 15.11 Simulation Waveforms for Evaluation Unit

Fitness Memory: The fitness memory shown in Figure 15.12, stores the fitness values for all the members of the initial population. With GA in process the fitness values tend to improve and lower fitness values replace the higher values constituting a family of solutions which are expected to give a good solution if not optimal.

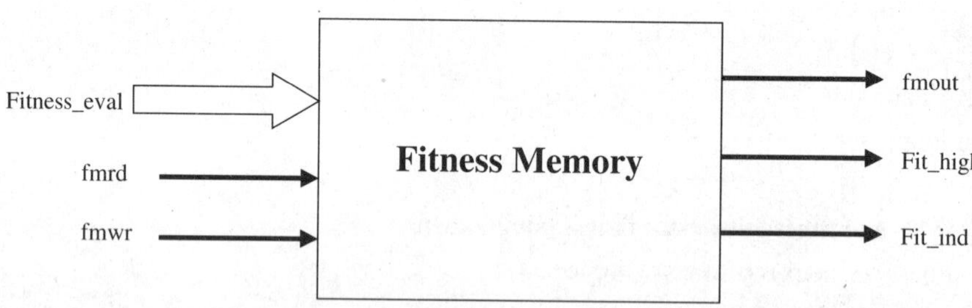

Figure 15.12 Fitness Memory

```
library IEEE;
use IEEE.STD_LOGIC_1164.ALL;
use IEEE.STD_LOGIC_ARITH.ALL;
use IEEE.STD_LOGIC_UNSIGNED.ALL;
use work.rng.all;
```

```
entity fitness_mem is
port(fitness_eval: IN fitness_arr;
      fmwr,fmrd:IN std_logic;
      fmout:OUT integer range 0 to 150;
      fit_high:out integer range 0 to 150; fit_ind: out integer range 1 to 15);
 end fitness_mem;
architecture Behavioral of fitness_mem is
signal fitness_lowest:integer;
signal temp_fit: fitness_arr;
begin
P_wr:Process(fmwr, fitness_eval)
 variable fmsize:integer range 0 to 15;
begin
fmsize := fitness_eval'high;
if fmwr='1' then
       for i in 1 to fmsize loop
temp_fit(i) <= fitness_eval(i);
end loop;
end if;
end process;
P_rd: process(fmrd)
variable fit_val, fit:integer range 0 to 150;
variable ind: integer range 1 to 15;
begin
if fmrd='1' then
for i in 1 to 10 loop
fit:= temp_fit(i);
for j in 1 to 10 loop
if temp_fit(j) > fit then
fit:= temp_fit(j);
ind:= j;
end if;
end loop;
end loop;
end if;
fmout<=fit_val;
fit_high<= fit;
fit_ind<= ind;
end process;
end Behavioral;
```

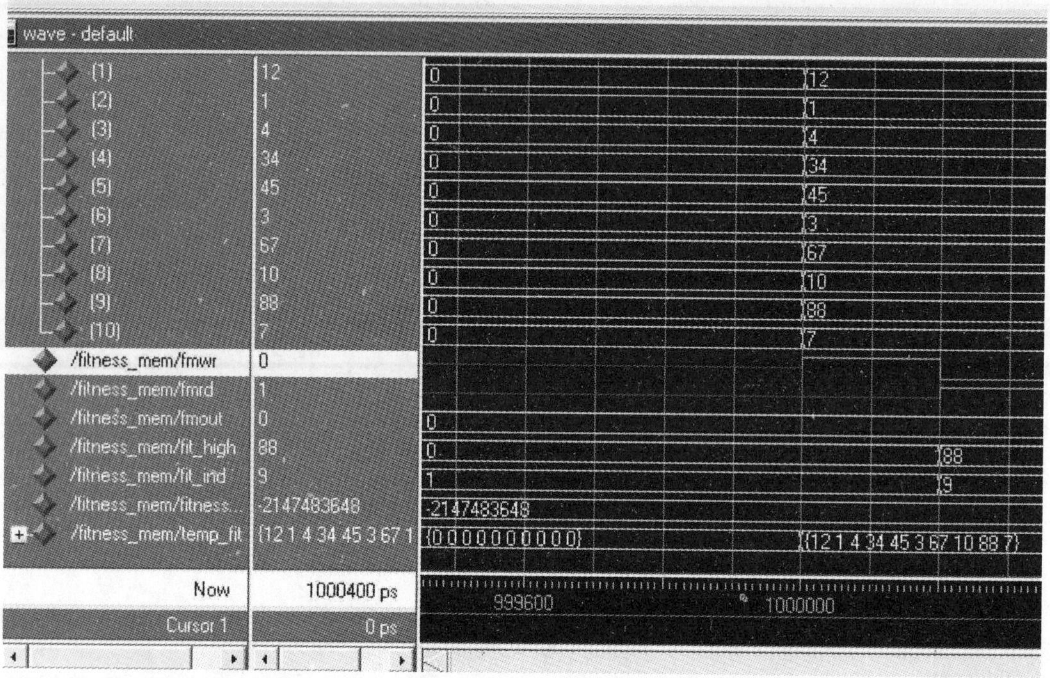

Figure 15.13 Simulation Waveforms for Fitness Memory Module

Netlist Memory: Netlist memory is a special purpose memory module which stores the information of the circuit that is to be partitioned. To partition a circuit, the information needed is as how many components are there? Which components are interconnected?

To get this information the netlist (text format) of circuit has to be read by the code and the required information to be extracted and stored in the netlist memory. The code for reading the netlist has been omitted as it differs depending on the format of netlist, the benchmarks used and the circuit analysis tool used. As discussing all these things are beyond the scope of VHDL they are not discussed here. An understanding is to be achieved by the reader that netlsit memory contains at least some basic information on the number of components and the interconnections between the components. The interconnection information is represented and stored in netlist memory as adjacency matrix for the components of the circuit, from the point of view to design a GAP. It is suffice to know that netlist contains the adjacency matrix for the interconnections between the circuit components.

The nelist memory module shown in the Figure 15.14 stores the complete information about the circuit that is to be partitioned in the form of adjacency matrix.

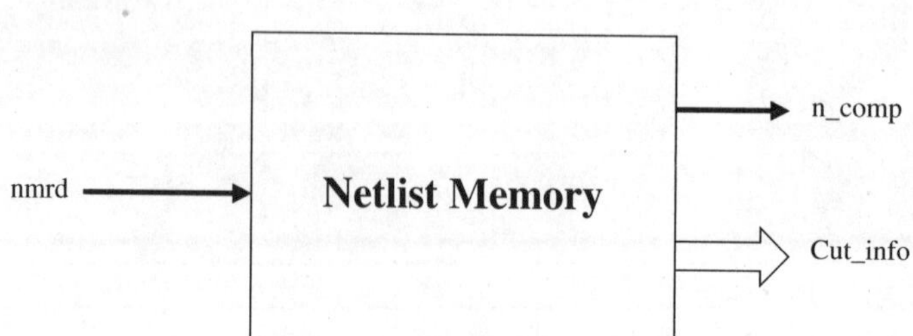

Figure 15.14 Netlist Memory

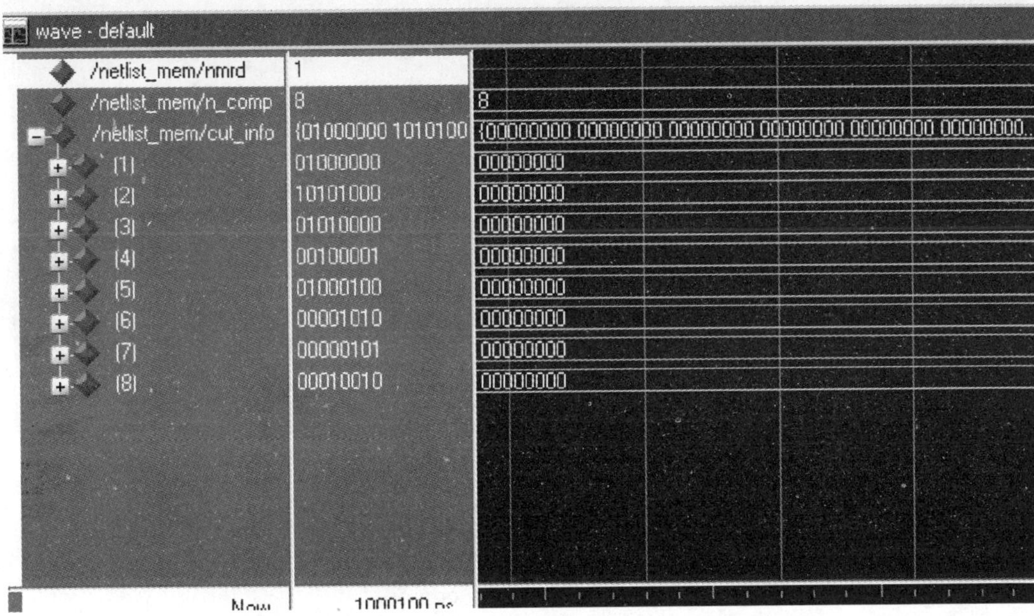

Figure 15.15 Simulation Waveforms for Netlist Memory

Register Bank: Register bank is a special purpose register which stores the user-defined input for GA. It stores the values for population size, that is, the number of individuals in the initial population, the mutation rate which is the value between 0 and 1, and the number of iterations of GA as shown by the structure of register bank in Figure 15.16. The mutation rate determines how many bits in the population have to be mutated in one go. Generally, the mutation rate is kept low and value less than 0.5 tends to give best results.

Figure 15.17 shows the control signals that are needed to write the values in the register bank or to read the values from the register bank.

Address	Parameter
00	Population size
01	Mutation rate
10	No. of GA runs
11	empty

Figure 15.16 Structure of Register Bank

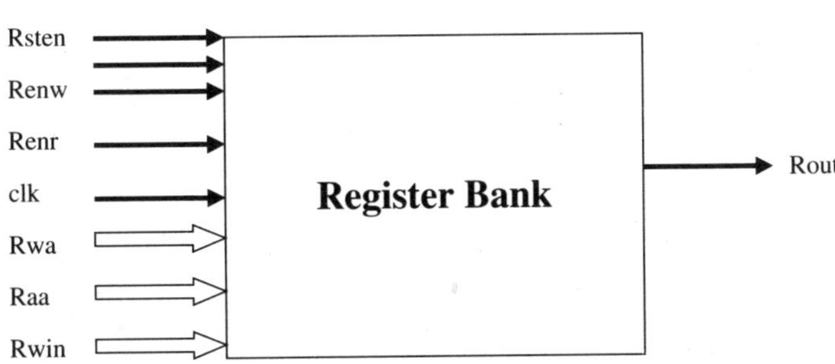

Figure 15.17 Register Bank

```
ENTITY reg_bank IS
PORT(clk,Renw,Renr,rst,en: IN STD_LOGIC;
      Rwa,Raa:IN STD_LOGIC_VECTOR(0 TO 1);
Rwin: IN real;
Rout:out real);
END reg_bank;
ARCHITECTURE behave OF reg_bank IS
TYPE reg IS ARRAY (0 TO 3) OF real;
SIGNAL r_temp: reg;
BEGIN
Write: PROCESS( clk, Renw, Rwa,en)
      BEGIN
            IF clk'EVENT AND clk='1' THEN
                                    if rst='1' then
                                    for i in 0 to 3 loop
                                    r_temp(i)<=0.0;
                                    end loop;
                                    end if;
                                    if rst='0' then
                                    if en='1'then
                                    IF renw = '1' THEN
            r_temp(conv_integer(rwa)) <= rwin;
          END IF;
        END IF;
      end if;
   end if;
   END PROCESS write;
Read1 : PROCESS( clk, Renr,Raa)
      BEGIN
         IF clk'EVENT AND clk='1' THEN
            IF Renr = '1' THEN
               Rout<= R_temp(conv_integer(Raa));
            END IF;
         END IF;
      END PROCESS read1;
end behave;
```

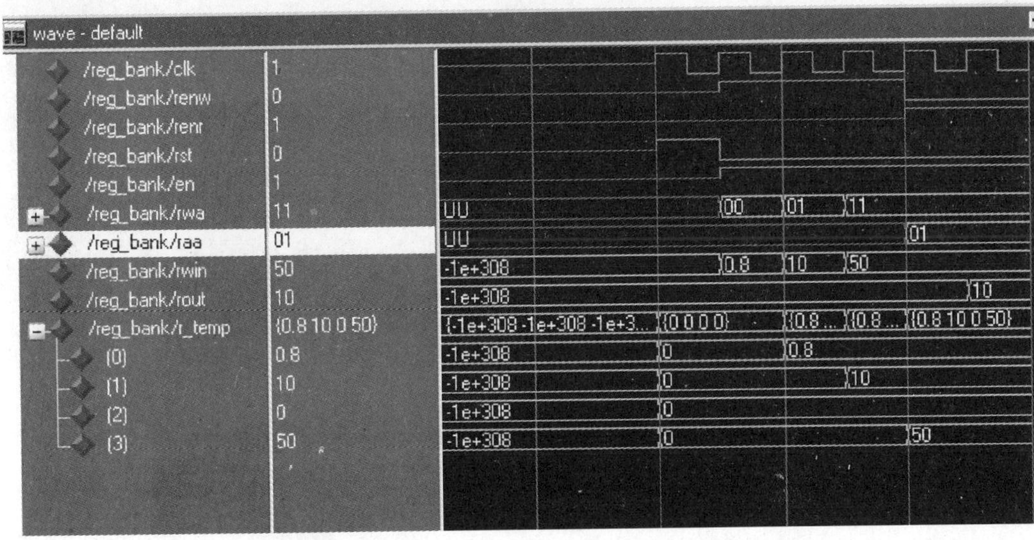

Figure 15.18 Simulation Waveforms of Register Bank

Memory: Memory is RAM-structured memory, as shown in Figure 15.19, which stores the randomly generated initial population. The size of memory is 256 × 8 , that is 256 memory locations each 8-bit wide.

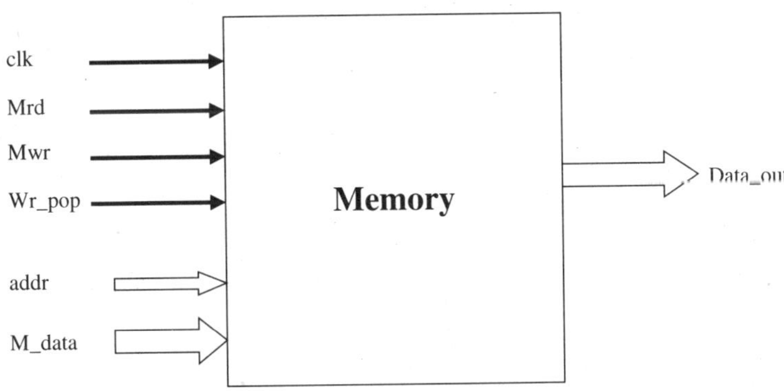

Figure 15.19 Memory Unit

```
ENTITY memory IS
PORT( clk,Mre,Mwe,wr_pop: IN STD_LOGIC;
addr:in std_logic_vector(2 downto 0); M_data : IN STD_LOGIC_VECTOR(1 TO 8);
data_mo: OUT STD_LOGIC_VECTOR(1 TO 8));
END memory;
ARCHITECTURE behave OF memory IS
TYPE ram_type IS ARRAY (0 TO 255) OF STD_LOGIC_VECTOR(8 DOWNTO 1);
SIGNAL temp_ram: ram_type;
BEGIN
```

```
write: PROCESS( clk,Mwe,M_data, addr)
variable s1:integer range 1 to 15:=1;
variable s2:integer range 1 to 300:=255;
variable gen_sig:std_logic_vector(8 downto 1);
variable s3:integer range 1 to 1000:=1;
variable pop:integer range 1 to 300; BEGIN
                IF clk'EVENT AND clk = '1' THEN
                    IF Mwe = '1'THEN
            if wr_pop= '1' then
                for i in 1 to pop_size loop
                    randg(s1,s2,s3,pop);
                    s3:=s3 + 2;
                    convert_bit(pop,gen_sig);
                            temp_ram(i) <= gen_sig;
                end loop;
            else
                    temp_ram(Conv_integer(addr))<= M_data ;
            end if;
                    END IF;
            END IF;
        END PROCESS write;
Read: PROCESS(clk, Mre, addr)
        BEGIN
            IF clk'EVENT AND clk = '1' THEN
                IF Mre = '1'THEN
                    data_mo<=  temp_ram(Conv_integer(addr));
                END IF;
            END IF;
        End PROCESS read;
END behave;
END PROCESS write;
Read: PROCESS( clk,Mre, addr)
        BEGIN
            IF clk'EVENT AND clk = '1' THEN
                IF Mre = '1'THEN
                    data_mo<=  temp_ram(Conv_integer(addr));
                END IF;
            END IF;
        END PROCESS read;
END behave;
```

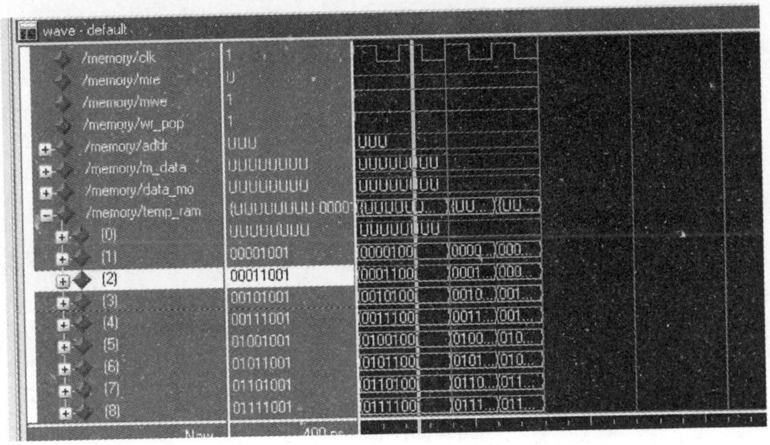

Figure 15.20 Simulation Waveforms for Memory

Control Unit: The control unit is the main synchronizing or the controlling unit. All other modules interact and exchange information through the control unit only.The basic block for the control unit is shown in Figure 15.21.

Figure 15.21 shows the complete signal diagram for the control unit. The control unit is modelled as FSM in VHDL. The Figure 15.22 shows the state diagram for the control unit FSM.

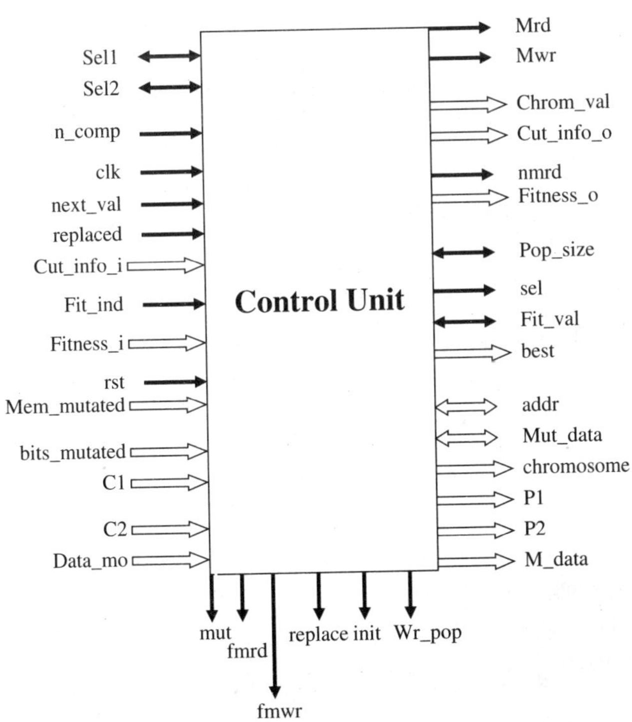

Figure 15.21 Control Unit

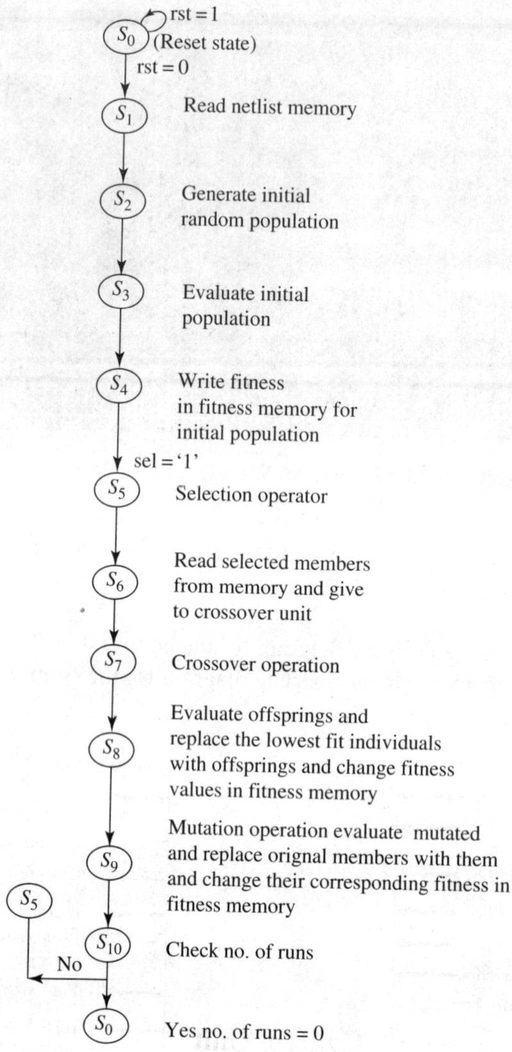

Figure 15.22 Control Unit FSM State Diagram

```
entity control_unit is
port(clk,next_val,replaced:in std_logic;
n_comp:in integer ;
 fit_ind:in integer range 1 to 15;
pop_size,sel1,sel2:inout integer range 1 to 15;
mem_mutated,bits_mutated:in arr;
fit_val: inout integer range 0 to 150;
rst:in std_logic;
```

```
fitness_o : out fitness_arr;
fitness_i:in fitness_arr;
 data_mo:in std_logic_vector(1 to 8);
addr:inout std_logic_vector(2 downto 0);
mwr,mrd,nmrd,fmwr,fmrd,init,replace,wr_pop,mut,sel:out Std_logic;
chromosome,p1,p2,M_data,best: out std_logic_vector(1 to 8);
c1,c2: in std_logic_vector(1 to 8);
chrom_val:out chrom_matrix;
cut_info_i:in adj_matrix;
cut_info_o:out adj_matrix;
 mut_data: inout std_logic_vector(1 to 8));
end control_unit;
architecture Behavioral of control_unit is
type state is (S0,s1,s2,s3,s4,s5,s6,s7,s8,s9,s10);
Signal P_state:state;
signal no_of_runs:integer range 0 to 100:=5;
begin
process(clk,p_state,rst)
variable i1, a1,a2, fit_i:integer range 1 to 15;
variable o1,o2, fit_o:std_logic_vector(8 downto 1);
begin
               IF clk'EVENT and clk= '1' THEN
               CASE P_state IS
                WHEN S0 =>
         IF rst= '1' THEN
      Mrd<= '0';
                Mwr<= '0';
                fmwr<= '0';
                fmrd<= '0';
                nmrd<= '0';
                init<= '0';
                replace<= '0';
      P_state<= s0;
                ELSE
              P_state <= S1;
   END IF;
  WHEN S1 => nmrd<='1';
      cut_info_o<= cut_info_i;
   P_state<= s2;
```

```
  WHEN s2=>
        mwr<= '1';
        wr_pop<= '1';
   P_state <= s3;
  WHEN  S3 => init<= '1';
mwr<='0';
wr_pop<='0';
addr<="000";
if next_val='1' then
mrd<='1';
addr<= addr + 1;
for i in 1 to 10 loop
chrom_val(i) <= data_mo;
end loop;
else
mrd<= '0';
end if;
   p_state <= s4;
  WHEN  S4 =>
   init<='0';
  fmwr<='1';
      P_state<= S5;
 WHEN  S5 =>           --selection
  sel<='1';
      P_state<= S6;
 WHEN  S6 =>
  mrd<= '1';
  a1:= sel1;
  a2:= sel2;
  convert_bit(a1,o1);
  addr<= o1( 3 downto 1);
  p1<= data_mo;
      convert_bit(a1,o2);
      addr<=o2( 3 downto 1);
  p2<= data_mo;
       mrd<='0';
  P_state<= S7;
WHEN  S7 => P_state<= S8;--crossover
 WHEN  S8 =>
 init<='0';
```

```
        chromosome<= c1;
replace<='1';
   chromosome<= c2;
   fmrd <= '1';
   fit_i:= fit_ind;
   fmwr <= '1';
   P_state<= S9;--evaluate offspring
 WHEN  S9=> --mutation
   replace<='0';
   for i in 1 to mem_mutated'high loop
      mrd<='1';
      i1:= i;
      convert_bit(i1,o1);
      addr<=o1(3 downto 1);
        for j in 1 to 8 loop
      mut_data(j)<=data_mo(j);
         if j= bits_mutated(i) then
         mut_data(j)<= not(data_mo(j));
      end if;
    end loop;
   fit_val <= evalfnc(mut_data,cut_info_i);
   fmwr<= '1';
   fitness_o(i)<= fit_val;
   fmrd<='0';
   mwr<='1';
   m_data<=mut_data;
   end loop;
   P_state<= S10;-- check no of runs
       if no_of_runs = 0 then
 P_state<= S0;
         else
       P_state<=s5;
    end if;
  WHEN OTHERS =>
 END CASE;
 END IF;
 fit_i:= fit_ind;
 convert_bit(fit_i,fit_o);
 addr<= fit_o(3 downto 1);
 END PROCESS;
```

```
Mrd<= '1';
best<= data_mo;
end Behavioral;
```

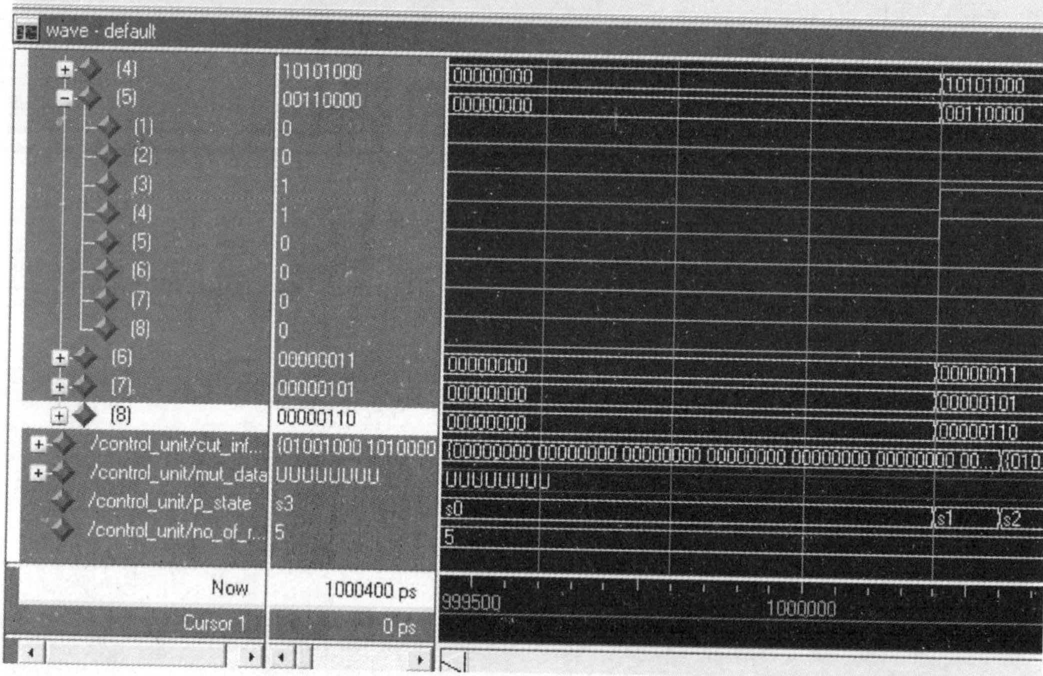

Figure 15.23 Simulation Waveform for Control Unit

A user defined package RNG.Vhd is used in the design which includes default values of GA parameters and subprograms for random number generation and converting integr to 8 bit std_logic_vector.

```
Library IEEE;
use IEEE.STD_LOGIC_1164.ALL;
use IEEE.STD_LOGIC_ARITH.ALL;
use IEEE.STD_LOGIC_UNSIGNED.ALL;
Type arr is array (1 to mut_bits) of integer;
type chrom_matrix is array (1 to pop_size) of std_logic_vector(1 to n_comp);
type fitness_arr is array (1 to pop_size) of integer range 0 to 1000;
--type fitness_arr1 is array( 1 to pop_size) of real;
type adj_matrix is array (1 to n_comp,1 to n_comp) of bit;
```

```
function evalfnc (signal chromosome: in std_logic_vector( 1 to 8); signal cut_info:adj_matrix)
return integer;
procedure randg (variable x,y,t:in integer range 0 to 1000; variable z: out integer);
procedure convert_bit(variable a:in integer;variable y:out std_logic_vector(8 downto 1));
end rng;
package body rng is
procedure randg (variable x,y,t:in integer range 0 to 1000; variable z: out integer) is
variable val, a:integer range 0 to 1000:= 0;
begin
if x>y then
a:= x-y;
else
a:= y-x;
end if;
if t > 3*a then
  val:= (t-a)/2;
  elsif t > a then
   val:= t-a;
    else
    val:= (x+y+t)/2;
    end if;
    z:= val;
    end randg;
    function evalfnc (signal chromosome: in std_logic_vector( 1 to 8); signal cut_info:adj_matrix)
return integer is
    variable fitness: integer range 0 to 500:= 0;
    variable cut_val: integer range 0 to 15:= 0;
    variable max_fit:integer range 0 to 360:=100;
    begin
     for i in 1 to n_comp loop
       for j in 1 to n_comp loop
        if cut_info(i,j)= '1' then
                      cut_val:= cut_val +1;
          end if;
     end loop;
    end loop;
     fitness := max_fit - cut_val;
     return fitness;
     end evalfnc;
```

```
procedure convert_bit(variable a:in integer ; variable y:out std_logic_vector(8 downto 1)) is
variable no:std_logic_vector(8 downto 1):="00000000";
variable num: integer range 0 to 256;
begin
   num:= a;
if num <= 255 and num>= 128 then
no(8):='1';
num:=num - 128;
end if;
if num < 128 and num >= 64 then
    no(7):='1';
    num:=num - 64;
end if;
    if num < 64 and num >= 32 then
    no(6):='1';
    num:=num -32;
end if;
if num < 32 and num >= 16 then
    no(5):='1';
    num:=num - 16;
      end if;
      if num < 16 and num >= 8 then
     no(4):='1';
     num:=num - 8;
        end if;
        if num < 8 and num >= 4 then
      no(3):='1';
       num:=num - 4;
           end if;
           if num < 4 and num >= 2 then
        no(2):='1';
       num:=num - 2;
            end if;
            if num < 2  then
        no(1):='1';
        end if;
y:= no;
end convert_bit;
end rng;
```

The top level module GAP.vhd is the genetic algorithm processor which is formed by all the above specified entities as components.

```
entity gap is
port(clk,rst:in std_logic ;
outp:out std_logic_vector( 1 to 8):= "00000001");
end gap;
architecture Behavioral of gap is
component control_unit is
port(clk,next_val,replaced:in std_logic;
n_comp:in integer;
fit_ind:in integer range 1 to 15;
pop_size,sel1,sel2:inout integer range 1 to 15;
mem_mutated,bits_mutated:in arr;
fit_val: inout integer range 0 to 150;
rst:in std_logic;
fitness_o : out fitness_arr;
fitness_i:in fitness_arr;
data_mo:in std_logic_vector(1 to 8);
addr:inout std_logic_vector(2 downto 0);
mwr,mrd,nmrd,fmwr,fmrd,init,replace,wr_pop,mut,sel:out Std_logic;
chromosome,p1,p2,M_data,best: out std_logic_vector(1 to 8);
c1,c2: in std_logic_vector(1 to 8);
chrom_val:out chrom_matrix;
cut_info_i:in adj_matrix;
cut_info_o:out adj_matrix;
mut_data: inout std_logic_vector(1 to 8));
end component;
component crossover is
Generic (crosspt:INTEGER:=4);
Port ( p1,p2 : IN  STD_LOGIC_VECTOR( 1 TO 8);
        c1,c2: inOUT STD_LOGIC_VECTOR(1 To 8));
end component;
component eval is
port(chromosome:in std_logic_vector( 1 to 8);
     init,fmrd,fmwr:in std_logic;
     chrom_val: in chrom_matrix;
     cut_info_n:in adj_matrix;
     fitness_single:inout integer range 0 to 1000;
     next_val:inout std_logic;
```

```
        fitness_i:in fitness_arr;
        fitness_o:out fitness_arr;
       replaced:out std_logic);
 end component;
 component fitness_mem is
 port(fitness_eval: IN fitness_arr;
       fmwr,fmrd:IN std_logic;
       fmout:OUT integer range 0 to 150;
       fit_high:out integer range 0 to 150; fit_ind:out integer range 1 to 15);
 end component;
 component mutation is
 port(mut:in std_logic;pop_size:in integer range 0 to 15:=10;
       mem_mutated,bits_mutated:out arr);
 end component;
 component netlist_mem is
 port( nmrd:in std_logic;
        n_comp:out integer range 0 to 15;
        cut_info:out adj_matrix);
 end component;
 component selection is
 port( sel:in std_logic;pop_size:in integer range 0 to 15:=10;
       fitness:in fitness_arr ;
        best_fit1,best_fit2:out integer range 1 to 15);
 end component;
 component memory IS
 PORT( clk,Mre,Mwe,wr_pop: IN STD_LOGIC;
        addr:in std_logic_vector(2 downto 0); M_data : IN STD_LOGIC_VECTOR(1 TO 8);
        data_mo: OUT STD_LOGIC_VECTOR(1 TO 8));
 END component;
 signal      init1,mut1,mwr1,mrd1,nmrd1,fmwr1,fmrd1,replace1,sel1,wr_pop1,      next_val1,
 replaced1:Std_logic;
 signal pop_size1,best_fit11,best_fit21,sel11,sel12, fit_ind1: integer range 1 to 15;
 signal n_comp1:integer;
 signal  fit_val1, fit_high1:integer range 0 to 150;
 --signal Rwa1,Raa1: STD_logic_vector(0 to 1);
 signal data_mem1, fitness2:fitness_arr;
 signal mem_mutated1,bits_mutated1:arr;
 signal fitness_i1,fitness_o1: fitness_arr;
 signal mem_add1,addr1: std_logic_vector(2 downto 0);
```

```
signal new_pop1: std_logic_vector(0 to 1);
signal p11,p21, c11,c21,data_mo1:std_logic_vector(1 to 8);
signal chromosome1, chromo_in1, m_data1, mut_data1:std_logic_vector( 1 to 8);
signal cut_info_n1,cut_info_o1:adj_matrix;
signal chrom_val1:chrom_matrix;
signal fmout1,Rout1,fitness_single11:integer range 0 to 1000;
signal fitness_single1:integer range 0 to 150;
begin
c1: crossover port map(p11,p21,c11,c21);
e1: eval port map(chromo_in1,init1,fmrd1,fmwr1,chrom_val1,cut_info_o1,fitness_single11,next_
val1,fitness_o1,fitness_i1, replaced1);
fm1:fitness_mem port map(fitness_o1,fmwr1,fmrd1,fmout1, fit_high1,fit_ind1);
m1:mutation port map(mut1,pop_size1,mem_mutated1,bits_mutated1);
n1: netlist_mem port map(nmrd1,n_comp1,cut_info_n1);
s1:selection port map (sel1,pop_size1,fitness_o1,best_fit11,best_fit21);
mem1: memory port map(clk,Mrd1,Mwr1,wr_pop1,addr1,mut_data1,data_mo1);
cu:control_unit port map(clk,next_val1,replaced1,n_comp1,fit_ind1,pop_size1,sel11,sel12,mem_
mutated1,bits_mutated1,fit_val1,rst,fitness_o1,fitness_i1,              data_mo1,addr1,mwr1,mrd
1,nmrd1,fmwr1,fmrd1,init1,replace1,wr_pop1,mut1,sel1,chromosome1,p11,p21,M_
data1,outp,c11,c21,chrom_val1,cut_info_n1,cut_info_o1, mut_data1);
end Behavioral;
```

The above designed processor was tested and synthesized for VLSI partionong problem with following default parameters:

Parameter	Value
Crossover point	4(mid point)
No. of runs	10
No. of components	8
No. of individuals	10
No. of bits mutated	5
Crossover type	One point crossover
Selection scheme	Roulette wheel selection

Note: All these values are read by CU from register problem but in given example default values are taken and register bank is omitted because of limitation of Xilinx XST to treat real values as not synthesizable.

The processor was implemented on CPLD XC9572- 15PC84 from 9500 Xilinx CPLD family with following synthesis results:

FINAL REPORT

```
Final Results
RTL Top Level Output File Name    : gap.ngr
Top Level Output File Name        : gap
Output Format                     : NGC
Optimization Goal                 : Area
Keep Hierarchy                    : YES
Target Technology                 : XC9500 CPLDs
Macro Preserve                    : YES
XOR Preserve                      : YES
wysiwyg                           : NO

Design Statistics
# IOs                             : 10

Cell Usage                        :
# BELS                            : 5426
#     AND2                        : 1684
#     AND3                        : 114
#     GND                         : 4
#     INV                         : 1581
#     OR2                         : 1362
#     OR3                         : 41
#     OR4                         : 1
#     VCC                         : 3
#     XOR2                        : 636
# FlipFlops/Latches               : 127
#     FD                          : 104
#     LD                          : 23
CPU : 73.61 / 75.59 s | Elapsed : 73.00 / 75.00 s
```

15.5 FUTURE SCOPE

GA processor has a bright scope for solving optimization problems. It has current topic of research interest with focus on evolutionary optimization techniques in near future. Further many researchers have already found hardware Gas to be faster and more efficient software GA.

Additional Case Studies

Digital Timer

A digital timer is a circuit which allows the user to set time and sets the alarm on when the time pulses are counted. Digital timer is a basic circuit which can be used as a component for a number of important complex digital systems such as safety alarm system, home security system, electronics safe system and so on.

The following is the basic block diagram for the proposed digital timer:

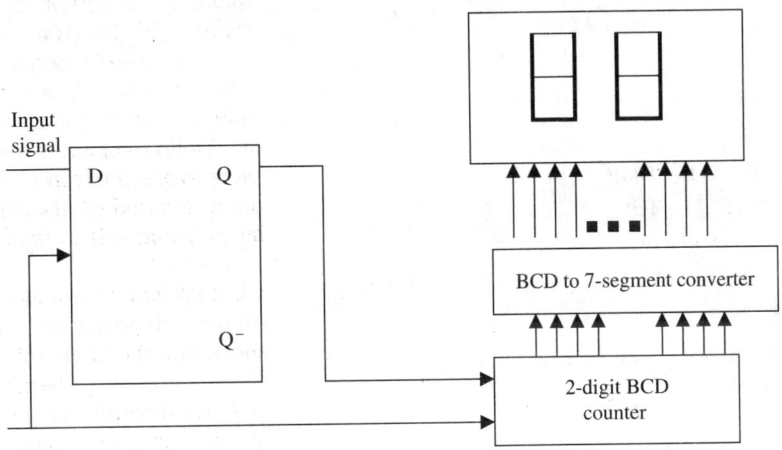

Figure C.1 Block Diagram of Digital Timer

The proposed digital timer consists of a D flip-flop, a 2-digit BCD counter to count pulses, a BCD to 7-segment display and a 2-digit 7-segment display. The input signal which acts as enable input to start the counter operation to count the clock pulses is given through a D flip-flop. A clock signal is given as a clock to both the flip-flop and the counter, and serves as reference clock whose pulses are counted.The use of flip-flop at the input serves a two fold purpose. Firstly, it acts as divide by two frequency enable signal.

Code for D flip-flop

```
LIBRARY IEEE;
USE IEEE.STD_LOGIC_1164.ALL;

ENTITY dff IS
  PORT (d : IN  STD_LOGIC;
       clk : IN STD_LOGIC;
       q :OUT  STD_LOGIC);
END dff;

ARCHITECTURE behave OF dff IS                -- behavioural architecture model of the entity
BEGIN
  P1: PROCESS(clk)                           -- declarative part of process
```

```
        BEGIN
            IF clk'EVENT AND clk='1' THEN
                q<=d;
            END IF;     -
        END PROCESS P1;
END behave;
```

Code for BCD counter

```
LIBRARY IEEE;
USE IEEE.STD_LOGIC_1164.ALL;
USE IEEE.STD_LOGIC_ARITH.ALL;
USE IEEE.STD_LOGIC_UNSIGNED.ALL;
ENTITY bcd_ctr IS
  PORT (clk : IN  STD_LOGIC;
        clr : IN STD_LOGIC;
        en : IN  STD_LOGIC;
  Bcd1, Bcd2:BUFFER std_logic_vector(3 DOWNTO 0));
END bcd_ctr;
ARCHITECTURE Behavioral OF bcd_ctr IS
BEGIN
PROCESS(clk)
BEGIN
IF clk'event AND clk='1' THEN
 IF clr='1' THEN
  bcd1<="0000";
   bcd2<="0000";
ELSIF en='1' THEN
 IF bcd1="1001" THEN
    bcd1<="0000";
   IF bcd2<="1001" THEN
        bcd2<="0000";
   ELSE
        bcd2<= bcd2 + '1';
  END IF;
 ELSE
 bcd1<= bcd1 + '1';
END IF;
END IF;
END IF;
END process;
END behavioral;
```

Code for BCD to 7-segment converter

```
LIBRARY IEEE;
USE IEEE.STD_LOGIC_1164.ALL;
ENTITY bcd_7seg IS
  PORT (in1 : IN  STD_LOGIC_vector(3 DOWNTO 0);
       out1 : OUT  STD_LOGIC_vector(1 TO 7));
END bcd_7seg;
ARCHITECTURE Behavioral OF bcd_7seg IS
BEGIN
PROCESS(in1)
BEGIN
 CASE in1 IS
 WHEN "0000"=> out1<="1111110";
 WHEN "0001"=> out1<="0110000";
 WHEN "0010"=> out1<="1101101";
 WHEN "0011"=> out1<="1111001";
 WHEN "0100"=> out1<="0110011";
 WHEN "0101"=> out1<="1011011";
 WHEN "0110"=> out1<="1011111";
 WHEN "1000"=> out1<="1111111";
 WHEN "1001"=> out1<="1111011";
 WHEN OTHERS=> out1<="-------";
 END CASE;
 END PROCESS;
END behavioral;
```

Main code for digital timer

```
Library IEEE;
use IEEE.STD_LOGIC_1164.ALL;
use IEEE.STD_LOGIC_ARITH.ALL;
use IEEE.STD_LOGIC_UNSIGNED.ALL;
                    -- uncomment the following library declaration if instantiating
                    -- any Xilinx primitives in this code.
                    -- library UNISIM;
                    -- use UNISIM.VCOMPONENTs.all;

ENTITY digi_timer IS
  PORT (clk1 : IN  STD_LOGIC;
       clr1 : IN  STD_LOGIC;
```

```
        INp : IN  STD_LOGIC;
        o1 : Buffer  STD_LOGIC_VECTOR(1 TO 7);
        o2 : Buffer  STD_LOGIC_VECTOR(1 TO 7));
END digi_timer;

ARCHITECTURE Behavioral OF digi_timer IS
COMPONENT bcd_ctr IS
   PORT (clk : IN  STD_LOGIC;
        clr : IN  STD_LOGIC;
        en : IN  STD_LOGIC;
   Bcd1, Bcd2:BUFFER STD_LOGIC_VECTOR(3 DOWNTO 0));
END COMPONENT;
COMPONENT dff IS
   PORT (d : IN  STD_LOGIC;
        clk : IN  STD_LOGIC;
        q : OUT  STD_LOGIC);
END COMPONENT;

COMPONENT bcd_7seg IS
   PORT (IN1 : IN  STD_LOGIC_VECTOR(3 DOWNTO 0);
        OUT1 : OUT  STD_LOGIC_VECTOR(1 TO 7));
END COMPONENT bcd_7seg;

SIGNAL en1:STD_LOGIC;
SIGNAL bcd_OUT1, bcd_OUT2:STD_LOGIC_VECTOR(3 DOWNTO 0);
begIN
ff1: dff PORT MAP(INp,clk1,en1);
ctr: bcd_ctr PORT MAP(clk1, clr1,en1,bcd_OUT1, bcd_OUT2);
seg1:bcd_7seg PORT MAP(bcd_OUT1,o1);
seg2:bcd_7seg PORT MAP(bcd_OUT2,o2);
END behavioral;
```

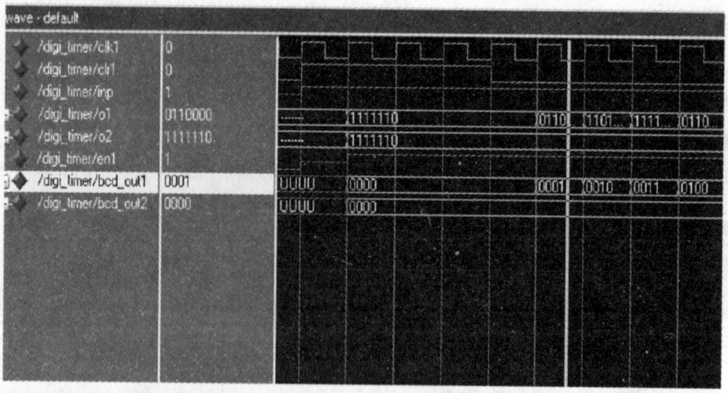

Figure C.2 Simulation Waveforms for Digital Timer

Sequence Detector

A sequence generator circuit is a circuit which gives output high when a particular sequence occurs at the input. Let's say, if a circuit has one input and one output the output goes high only when the sequence of "0", "1", "0", "1" occurs at input. This circuit is acting as a sequence detector which detects the presence of sequence 0101 at input in consecutive clock cycles.

A sequence detector circuit can be modeled as a finite state machine. Since the output depends on the input and the present state, it is modeled as Mealy Machine.

A case study of designing a synchronous sequence detector which has one input and one output is given. The data at input is sampled three times and the output of the circuit goes high if the input is high for odd number of times. (Courtesy An Engg. Approach to Digital Design—Fletcher).

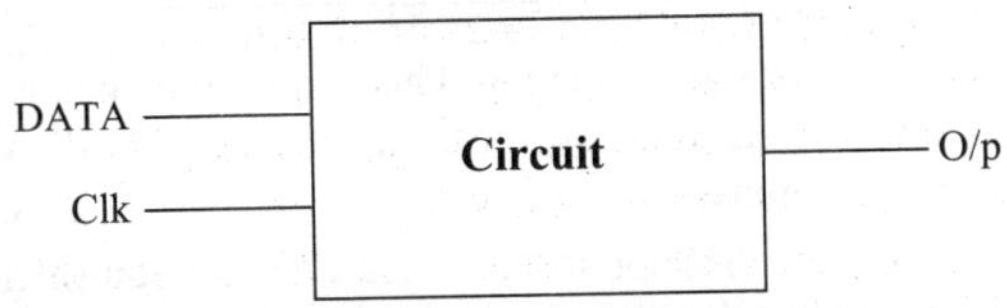

Figure C.3 Block Diagram of Sequence Detector Circuit

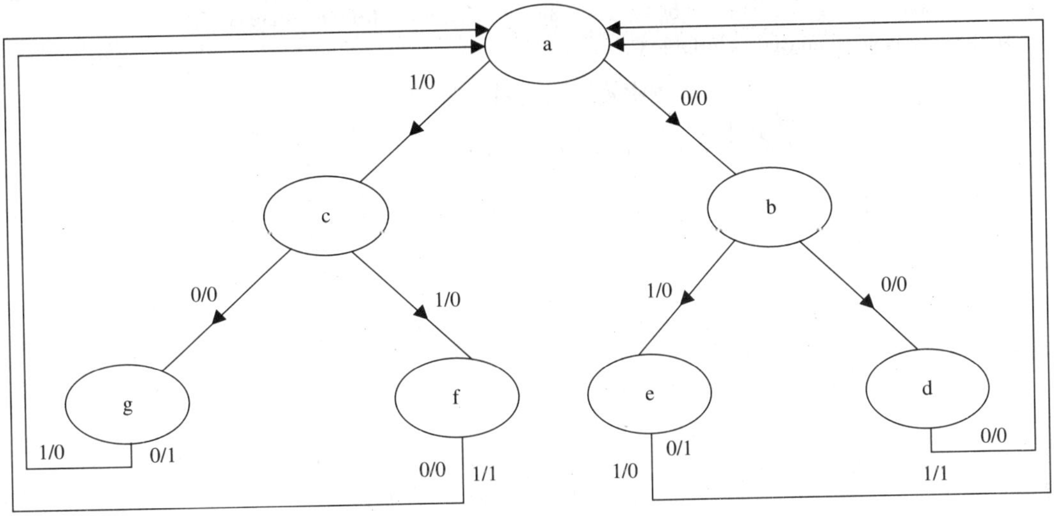

Figure C.4 Primitive State Diagram for the Sequence Detector Circuit

Table C.1 Primitive State Table

Present state	Input (inp)	Next state	Output
a	0	b	0
	1	c	0
b	0	d	0
	1	e	0
c	0	g	0

(*Continued*)

Present state	Input (inp)	Next state	Output
	1	f	0
d	0	a	0
	1	a	1
e	0	a	1
	1	a	0
f	0	a	0
	1	a	1
g	0	a	1
	1	a	0

Next step in design process is to see for state reduction. This helps eliminate the redundant states by combining together the equivalent states. Any two states are said to be equivalent iff they have same output and same next state for every possible combination of inputs.

For example in our design of sequence detector states "d" and "f" are equivalent as they have same outputs and same next states for given input conditions. So all instances of "f" are replaced by "d" and state "f" is removed from the state table.

On similar lines, states "e" and "g" are also equivalent. As any one state is to be retained from the equivalent states, all instances of "g" are replaced by "e" and state "g" is removed from the state table.

So the state table is reduced as in Table C.2.

Table C.2 Reduced State Table

Present state	Input (inp)	Next state	Output
a	0	b	0
	1	c	0
b	0	d	0
	1	e	0
c	0	e	0
	1	d	0
d	0	a	0
	1	a	1
e	0	a	1
	1	a	0

Once the reduced table is obtained, FSM can be modeled in VHDL for implementing the sequential detector.

```
LIBRARY IEEE;
USE IEEE.STD_LOGIC_1164.ALL;
USE IEEE.STD_LOGIC_ARITH.ALL;
USE IEEE.STD_LOGIC_UNSIGNED.ALL;
ENTITY seq_det IS
  PORT ( clk : IN STD_LOGIC;
      inp : IN STD_LOGIC;
      outp : OUT  STD_LOGIC);
END seq_det;
```

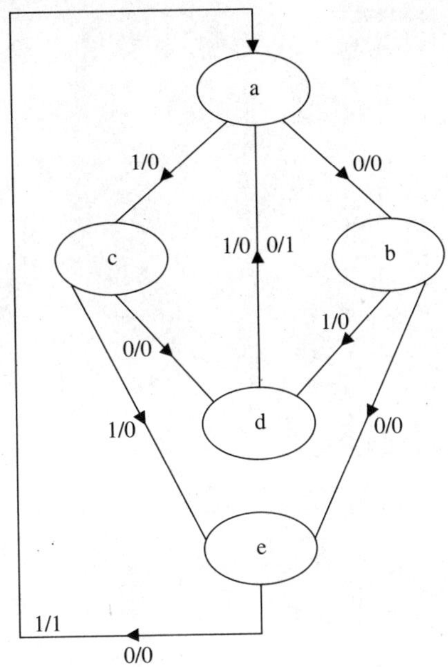

Figure C.5 Reduced State Diagram for Sequence Detector

Code for sequential detector

```
ARCHITECTURE Behavioral OF seq_det IS
TYPE state IS(a,b,c,d,e);
SIGNAL p_state:state:=a;
BEGIN
PROCESS(clk)
BEGIN
IF clk'EVENT AND clk='1' THEN
CASE p_state IS
WHEN a =>
 IF inp='0' THEN
 outp<= '0';
 p_state<=b;
 ELSE
  outp<= '0';
   p_state<= c;
 END IF;
WHEN b =>
 IF inp='0' THEN
  outp<= '0';
  p_state<=e;
 ELSE
```

```
outp<= '0';
p_state<= d;
END IF;
WHEN c =>
IF inp='0' THEN
outp<= '0';
p_state<=d;
ELSE
outp<= '0';
p_state<= e;
END IF;
WHEN d =>
IF inp='0' THEN
outp<= '1';
p_state<=a;
ELSE
outp<= '0';
p_state<= a;
END IF;
WHEN e =>
IF inp='0' THEN
outp<= '0';
p_state<=a;
ELSE
outp<= '1';
p_state<= a;
END IF;
END CASE;
END IF;
END PROCESS;
END Behavioral;
```

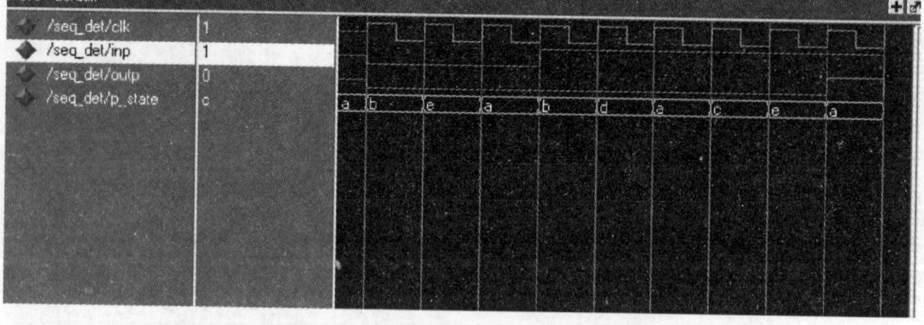

Figure C.6 Simulation Waveform for Sequential Detector Unit

Floating Point Numbers

Floating point numbers are decimal numbers of the form:

1. 23.45754312
2. 2.45×10^{12}
3. 45.33478…

In general, a floating point number can be written as:.

$$\pm M \times B^E$$

where

M is the mantissa

B is the base of number system

E is the exponent

Thefloating point numbers are generally represented by their scientific representation.

For example, number 34.56×10^{-8} can be expressed as

34.56 E–8 is its scientific representation.

In our design of floating point multiplier and floating point adder, we are dealing with binary numbers 32-bit representation of floating numbers is used as

S	E7	. …	E0	M22	….	M0

S is the Sign Bit

E7-E0 are 8 bits representing the

M22-M0 are bits representing mantissa

Decimal point is implied between the exponent and the mantissa parts.

Effectively there are 24 bits in mantissa part because in any binary number, the leftmost bit is always "1". This is taken as implied "1" without occupying the actual bit position.

The 8 bits in exponent represent a biased exponent, which is obtained by adding 127 to the actual exponent. The biasing accommodates large or very small values without requiring a separate sign bit for exponents. So the actual exponents range from –126 to + 128.

Let's say there is binary number 1011100101. Now it is a 10-bit number which is moved nine places to left with "1" implied in leftmost position and shifting expressed as exponent of 2.

So, $10111000101 = 1.011100101 \times 2^9$

Now the exponent is 9, so by adding 127 to it, the value obtained is 127 + 9= 136.

Now, 136 in binary is 10001000. So, complete 32-bit floating point representation of a given binary number is

0 10001000 01110010100000000000000

Sign Exponent

Mantissa

Floating point representation is very important in digital processing circuits. Considering the significance of floating point arithmetic and addition as the basic operation, floating point adder has been designed in VHDL.The floating point adder performs the addition operation for two floating point numbers.The numbers are represented in scientific form with mantissa and exponent parts. Floating point adder is one of the very important circuits used in complex large digital systems, in particular, for the signal processing part of digital systems. In the design given here, a floating point number is represented as 32-bit std_logic _vector with upper 8 bits from 30 down to 23 representing the mantissa part and lower bits from 22 down to 0 representing the exponent part.

Floating Point Multiplier

A 32-bit floating point multiplier is designed, which multiplies two 32-bit numbers. One of the numbers(multiplier) is stored in ROM while the other is the multiplicand which is given as input to the multiplier.

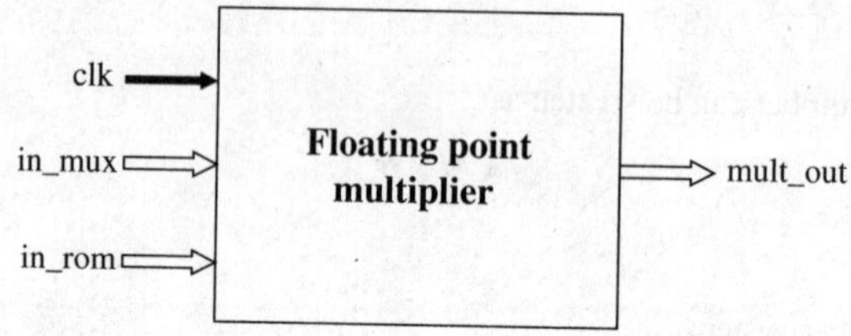

Figure C.7 Block Diagram of Floating Point Multiplier

The flowchart for floating point multiplication is given below:

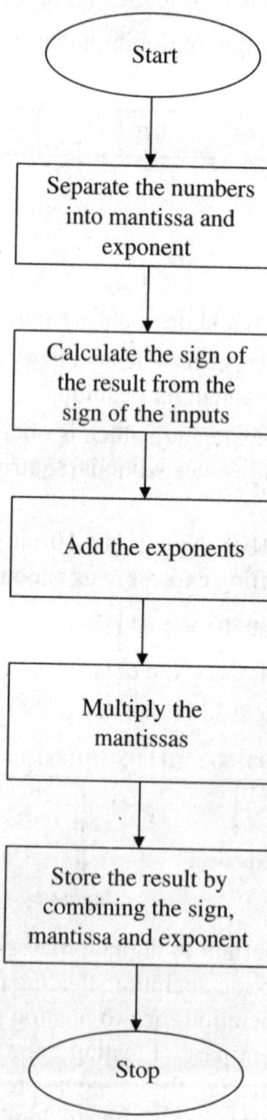

Figure C.8 Flowchart for Floating Point Multiplication

VHDL code for floating point multiplier

```
LIBRARY IEEE;
USE IEEE.STD_LOGIC_1164.ALL;
USE IEEE.STD_LOGIC_ARITH.ALL;
USE IEEE.STD_LOGIC_UNSIGNED.ALL;

ENTITY fp_mul IS
PORT(
in_mux, in_rom : IN std_logic_vector(31 DOWNTO 0) ;
clk : IN std_logic ;
mult_out : OUT std_logic_vector(31 DOWNTO 0)
 );
END ENTITY fp_mul ;
ARCHITECTURE behav OF fp_mul IS
BEGIN

PROCESS(IN_mux, IN_rom , clk)

VARIABLE sign,flag: std_logic := '0' ;
VARIABLE x, y : std_logic_vector(22 DOWNTO 0) ;
VARIABLE exp_mux, exp_rom : std_logic_vector(7 DOWNTO 0) ;
VARIABLE mult_mant : std_logic_vector(45 DOWNTO 0) ;
VARIABLE mult_exp, mux_temp, rom_temp : std_logic_vector(8 DOWNTO 0) ;
VARIABLE result : std_logic_vector(31 DOWNTO 0) ;

BEGIN
x := '1' & IN_mux(22 DOWNTO 1) ;                                -- '1' for implicit '1'
y := '1' & IN_rom(22 DOWNTO 1) ;
IF (IN_mux(31) = '1' AND IN_rom(31) = '1' AND clk = '1') THEN   -- sign of results
sign := '0' ;
ELSIF (IN_mux(31) = '0' AND IN_rom(31) = '0' AND clk = '1') THEN
sign := '0' ;
ELSIF(clk = '1') THEN
sign := '1' ;
END IF ;

IF (IN_mux = 0 AND clk = '1') THEN                              -- i.e., the number is zero
flag := '1' ;
ELSIF (IN_rom = 0 AND clk = '1') THEN
flag := '1' ;
ELSIF (clk = '1') THEN
flag := '0' ;
```

```
END IF ;

IF (flag = '0' AND clk = '1') THEN                    -- separation of mantissa AND exponent
exp_mux := IN_mux (30 DOWNTO 23) ;
exp_rom := IN_rom (30 DOWNTO 23) ;
mux_temp := '0' & exp_mux(7 DOWNTO 0) ;
rom_temp := '0' & exp_rom(7 DOWNTO 0) ;
mult_exp := mux_temp + rom_temp ;
mult_exp := mult_exp – 127 ;

mult_mant := x * y ;

IF(mult_mant(45) = '1') THEN                          -- normalisation
mult_exp := mult_exp + 1 ;
result := sign & mult_exp(7 DOWNTO 0) & mult_mant(44 DOWNTO 22) ;
mult_out <= result(31 DOWNTO 0) ;
ELSIF(mult_mant(45) = '0') THEN
result := sign & mult_exp(7 DOWNTO 0) & mult_mant(43 DOWNTO 21) ;
mult_out <= result(31 DOWNTO 0) ;
END IF ;

ELSIF (flag = '1' AND clk = '1') THEN                 -- number zero
mult_out <= X"00000000" ;
flag := '0' ;
END IF ;
END PROCESS ;
END behav ;
```

Figure C.9 Simulation Waveform for Floating Point Multiplier

Floating Point Adder

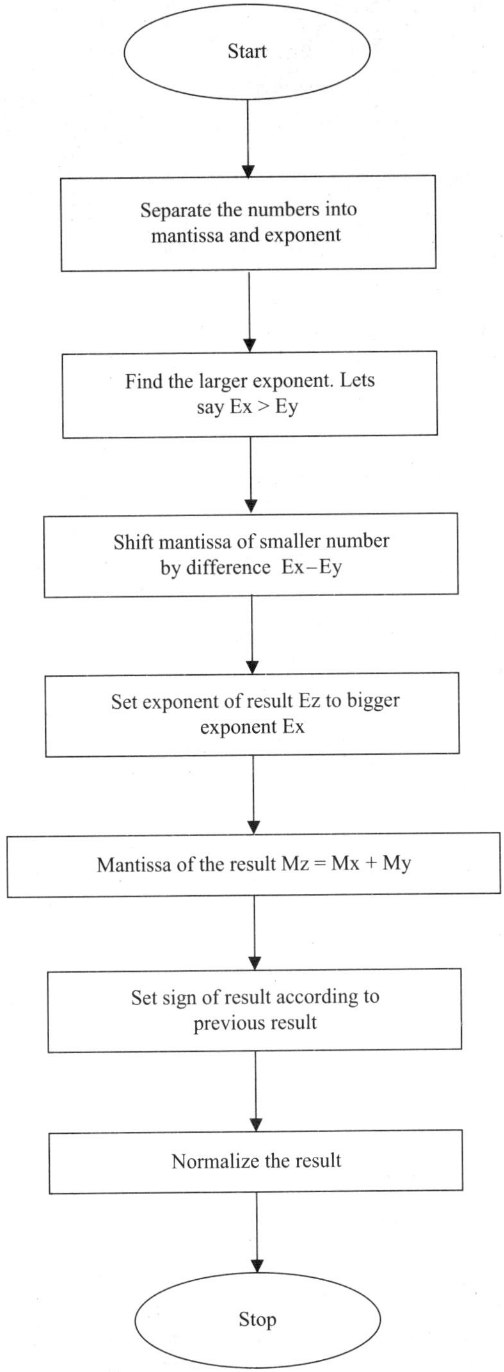

Figure C.10 Flowchart for Floating Point Addition

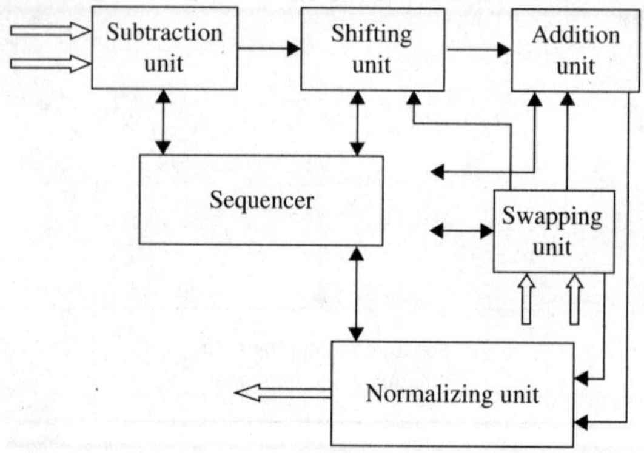

Figure C.11 Block Diagram of Floating Point Adder

It consists of the following sub-units:

- Subtraction sub-unit
- Swap sub-unit
- Shifter sub-unit
- Summer sub-unit
- Normalizing sub-unit
- Sequencer sub-unit

Each sub-unit of the floating point adder is separately explained and coded in VHDL. These sub-units are used as components for top level entity—floating point adder.

Subtraction Unit

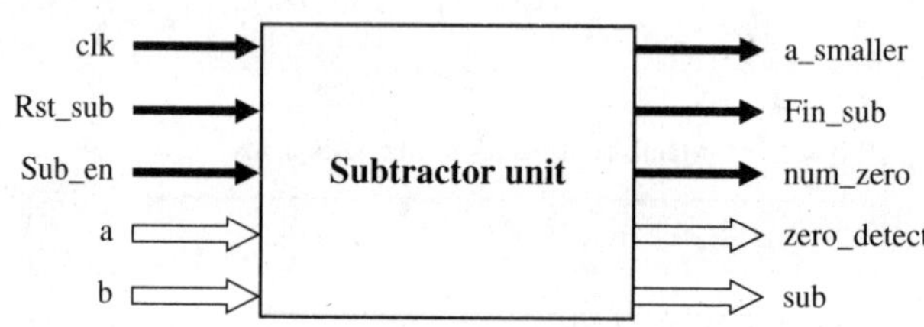

Figure C.12 Block Diagram of Subtraction Sub-unit

The function of subtractor unit is to give the difference between the mantissas of the two floating point numbers. This information is given to the shifter, which shifts the smaller number by the difference between the mantissas. Apart from this, the subtractor gives information as to which number is smaller and if any number is zero. The VHDL code for subtraction unit is given in figure.

Inputs "rst_sub" signal resets the subrator unit, "sub_en" is the enable signal to enable the subtraction operation. "clk" input is used as a synchronizing signal and "a" and "b" are two floating point numbers to be subtracted.

The first process begins with the "if(rst_sub='0')" statement. That is, if the reset port to the subtractor is high, then set the outputs to zero. Note that the ninth bit is an extra bit. This bit is set to "1" whenever there is a valid output number. Since the involvement of zero in calculations need to be treated separately, the presence of zero in any one of the numbers is detected by the statements "if(c=0)" and "if(d=0)". When one of the numbers is zero, the num_zero signal is set high. If "a" is zero then the output of "zero_detect" is "01". If "b" is zero then this signal is set to "10".

First the exponent and the mantissa are separated and written into separate variables. If the exponents of the two numbers are different, then the smaller one is to be found out and corresponding subtractions made. If the exponents are same, then the smaller of the mantissas is to be found out. In certain cases the numbers are the same. All these cases need to be treated separately. The signal "a_smaller" is used to give information to the control unit as to which number is smaller. When the calculations are finished, the "fin_sub" signal goes high. All these signals are reset at the start of the next set of calculations.

Code for subtractor sub-unit for floating point adder

```
LIBRARY IEEE;
USE IEEE.STD_LOGIC_1164.ALL;
USE IEEE.STD_LOGIC_ARITH.ALL;
USE IEEE.STD_LOGIC_UNSIGNED.ALL;

entity subtractor IS
PORT (
a : IN std_logic_vector (31 DOWNTO 0);
b : IN std_logic_vector (31 DOWNTO 0);
clk , rst_sub , sub_en : IN std_logic;
a_smaller , fIN_sub , num_zero : OUT std_logic ;
zero_detect : OUT std_logic_vector(1 DOWNTO 0) ;
sub : OUT std_logic_vector (8 DOWNTO 0)) ;
END subtractor ;

ARCHITECTURE behave OF subtractor IS

BEGIN
PROCESS (a , b , clk , rst_sub , sub_en)
VARIABLE temp ,c , d : std_logic_vector (7 DOWNTO 0) ;
VARIABLE e , f : std_logic_vector (22 DOWNTO 0) ;
BEGIN
   IF (rst_sub = '0') THEN
      c := a (30 DOWNTO 23) ;
      d := b (30 DOWNTO 23) ;
      e := a (22 DOWNTO 0) ;
      f := b (22 DOWNTO 0) ;
         IF(sub_en = '1') THEN
            IF (clk = '1') THEN
               IF ((c=0)) THEN
                     zero_detect <= "01" ;
                     num_zero <= '1' ;
               ELSIF ((d=0)) THEN
                     zero_detect <= "10" ;
                     num_zero <= '1' ;
               ELSIF (c < d ) THEN
                     temp := d – c ;
                     a_smaller <= '1' ;
                     sub <= '1' & temp (7 DOWNTO 0) ;
```

```
                fin_sub <= '1' ;
                zero_detect <= "00" ;
                num_zero <= '0' ;
            ELSIF (d < c) THEN
                temp := c – d ;
                a_smaller <= '0' ;
                sub <= '1' & temp (7 DOWNTO 0) ;
                fin_sub <= '1' ;
                zero_detect <= "00" ;
                num_zero <= '0' ;
            ELSIF((c=d) AND e < f) THEN
                a_smaller <= '1' ;
                temp:= c – d ;
                sub <= '1' & temp (7 DOWNTO 0) ;
                fin_sub <= '1' ;
                zero_detect <= "00" ;
                num_zero <= '0' ;
            ELSIF ((c=d) AND e > f) THEN
                a_smaller <= '0' ;
                temp := c – d ;
                sub <= '1' & temp (7 DOWNTO 0) ;
                zero_detect <= "00" ;
                num_zero <= '0' ;
                fin_sub <= '1' ;
            ELSIF ((c=d) AND (e = f)) THEN
                temp := c – d ;
                a_smaller <= '0' ;
                sub <= '1' & "00000000" ;
                fin_sub <= '1' ;
                zero_detect <= "00" ;
                num_zero <= '0' ;
            END IF ;
        END IF ;
    END IF ;
ELSIF(rst_sub = '1') THEN
    fin_sub <= '0' ;
    sub <= "000000000" ;
    num_zero <= '0' ;
    zero_detect <= "00" ;
END IF ;
END PROCESS ;
END behave ;
```

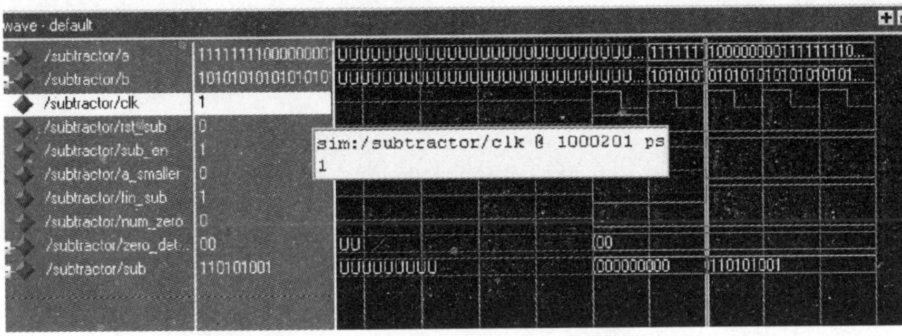

Figure C.13 Simulation Waveforms for Subtraction Unit

Shifter Sub-unit

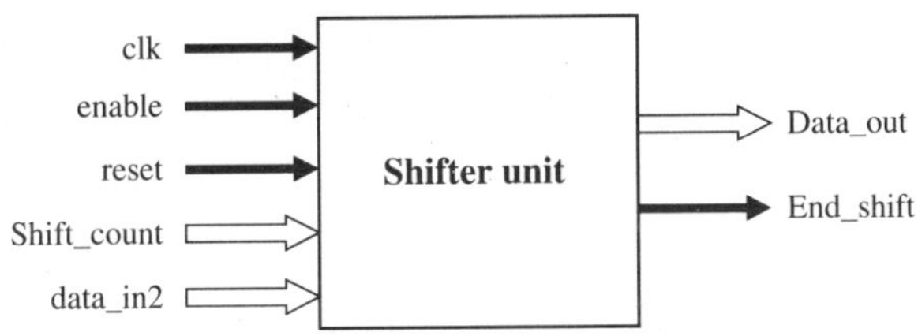

Figure C.14 Block Diagram of Shifter Sub-unit

The function of the shifter unit is to shift one of the two numbers in order to equalize exponents of both numbers. The VHDL code for this unit is given in figure.

Input "enable" is used to enable shift operation while "reset" is used to rest the shifter unit. The data to be shifted is given as input "data_in2" and "shift_count" is used to keep the count of bits shifted in mantissa part of the floating point number. The shifted number is given as output "data_out".

When any one of the signals in the sensitivity list changes, process is executed. Initially the shifter unit is reset if "reset" = "1". If reset is "0" then, shift operation is to be started if "enable" = "1". "tmp" and "flag" signals are used for controlling the sequence of operations to be performed. If "tmp" = "1" then the number of times shift operation to be performed is loaded in a temporary variable temp_cnt. Once done, "tmp" is set to zero. Then the actual shifting of data bits is done only if the "flag" variable is high. The shifting operation is done till "tmp_cnt" is zero. The value of number of times shift operation has to be performed comes from subtraction sub-unit. Once the shift operation is completed, "flag" is set zero and will be set high only when shifter unit will be reset for a new floating point addition operation. If flag = "1" then process first checks whether "tmp_cnt" is zero (last 8 bits only). If so, no more shifting is required and the number is outputted and end_shift is set to "1". If "tmp_cnt" is not zero then the mantissa is shifted to the right. At the same time "tmp_cnt" is decremented. This is given by the lines:

```
elsif (clk = '1') then
tmp_datain := '0' & tmp_datain (31 downto 1) ;
tmp_cnt := tmp_cnt - "00000001" ;
```

In the next clock cycle, the unit will first check if "tmp_cnt" is zero. If so, it outputs the shifted mantissa to the summer, else the number is shifted again.

VHDL code for shifter unit

```
LIBRARY IEEE;
USE IEEE.STD_LOGIC_1164.ALL;
USE IEEE.STD_LOGIC_ARITH.ALL;
USE IEEE.STD_LOGIC_UNSIGNED.ALL;
ENTITY shifter is
PORT (
shift_count : IN STD_LOGIC_VECTOR (8 DOWNTO 0) ;
data_in2 : IN STD_LOGIC_VECTOR (32 DOWNTO 0) ;
data_out : OUT STD_LOGIC_VECTOR (31 DOWNTO 0) ;
clk, enable, reset : IN STD_LOGIC ;
end_shift : OUT STD_LOGIC
) ;
END ENTITY shifter ;
ARCHITECTURE behav OF shifter IS
BEGIN
PROCESS(clk)
VARIABLE tmp_cnt : STD_LOGIC_VECTOR(7 DOWNTO 0) ;
VARIABLE tmp_datain : STD_LOGIC_VECTOR(31 DOWNTO 0) ;
VARIABLE tmp , flag : STD_LOGIC ;
BEGIN
   IF(reset='1') THEN
      tmp := '1' ;
      end_shift <= '0' ;
      flag := '1' ;
   ELSIF(enable = '1') THEN
      IF(tmp = '1') THEN
         IF(shift_count(8) = '1') THEN
            tmp_cnt := shift_count (7 DOWNTO 0) ;
            tmp_datain := '1' & data_in2 (31 DOWNTO 1) ;          -- '1' for implicit one
            tmp := '0' ;
         END IF ;
      END IF ;
   IF(flag = '1') THEN
      IF (shift_count(8) = '1') THEN
         IF (conv_integer(tmp_cnt(7 DOWNTO 0)) = 0) THEN
            data_out <= tmp_datain ;
            end_shift <= '1' ;
            flag := '0' ;
         ELSIF ( clk = '1') THEN
            tmp_datain := '0' & tmp_datain (31 DOWNTO 1) ;
```

```
                tmp_cnt := tmp_cnt - "00000001" ;
            END IF ;
        END IF ;
    END IF ;
END IF ;
END PROCESS ;
END behave ;
```

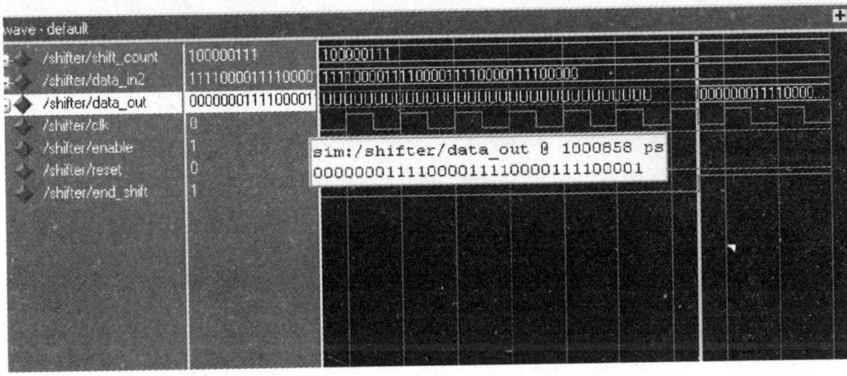

Figure C.15 Simulation Waveform for Shifter Unit

Swap Sub-unit

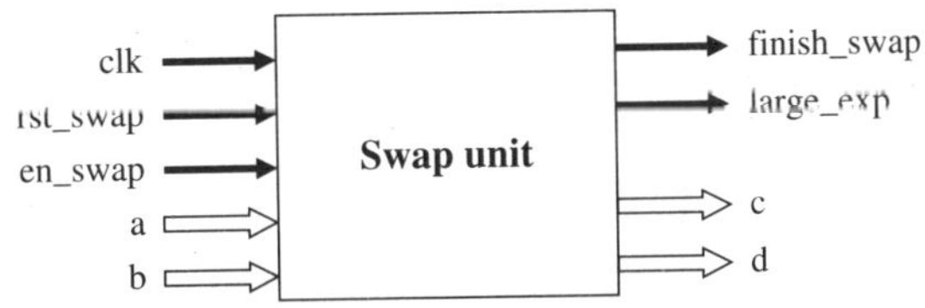

Figure C.16 Block Diagram of Swap Sub-unit

The function of the swap unit is to input the mantissa of the smaller number to the shifter, so that it can shift it by the difference in the exponents of the two numbers. The implicit "1" in the IEEE standard format is concatenated with the mantissa of the larger number and inputted to the summer. Also, the least significant bits (last 8 bits) of the mantissa of larger number are set to zero. The mantissa of the smaller number is given to the shifter. In this case also a "1" is concatenated with the mantissa (for checking in the shifter for a valid number) and the last nine bits are set to zero. After the swapping process the signal finish_swap is set high. When the rst_swap signal is high, the signals are reset. The complete VHDL code is given.

VHDL code for swap unit

```
LIBRARY IEEE;
USE IEEE.STD_LOGIC_1164.ALL;
USE IEEE.STD_LOGIC_ARITH.ALL;
USE IEEE.STD_LOGIC_UNSIGNED.ALL;
```

```
ENTITY swap is
   PORT (
      a : IN STD_LOGIC_VECTOR (31 DOWNTO 0) ;
      b : IN STD_LOGIC_VECTOR (31 DOWNTO 0) ;
      clk : IN STD_LOGIC ;
      rst_swap , en_swap : IN STD_LOGIC ;
      finish_swap : OUT std_logic ;
      d : OUT STD_LOGIC_VECTOR (31 DOWNTO 0) ;
      large_exp : OUT STD_LOGIC_VECTOR (7 DOWNTO 0) ;
      c : OUT STD_LOGIC_VECTOR (32 DOWNTO 0 ) ) ;
END swap ;

ARCHITECTURE rtl OF swap is
BEGIN
   PROCESS (a , b , clk , rst_swap , en_swap)
      VARIABLE x , y : STD_LOGIC_VECTOR (7 DOWNTO 0) ;
      VARIABLE p , q : STD_LOGIC_VECTOR (22 DOWNTO 0) ;
   BEGIN
      IF(rst_swap = '1' ) THEN
         c <= '0' & a(22 DOWNTO 0) & "000000000" ;
         finish_swap <= '0' ;
      ELSIF(rst_swap = '0') THEN
         IF(en_swap = '1') THEN
            x := a (30 DOWNTO 23) ;
            y := b (30 DOWNTO 23) ;
            p := a (22 DOWNTO 0) ;
            q := b (22 DOWNTO 0) ;
            IF (clk = '1') THEN
               IF (x < y) THEN
                  c <= '1' & a (22 DOWNTO 0) & "000000000" ;          -- '1' for checking
                  d <= '1' & b (22 DOWNTO 0) & "00000000" ;           -- '1' for implicit one
                  large_exp <= b (30 DOWNTO 23) ;
                  finish_swap <= '1' ;
               ELSIF (y < x) THEN
                  c <= '1' & b (22 DOWNTO 0) & "000000000" ;
                  d <= '1' & a (22 DOWNTO 0) & "00000000" ;           -- '1' for implicit 1
                  large_exp <= a (30 DOWNTO 23) ;
                  finish_swap <= '1' ;
               ELSIF ((x=y) and (p < q)) THEN
                  c <= '1' & a (22 DOWNTO 0) & "000000000" ;          -- '1' for checking
                  d <= '1' & b (22 DOWNTO 0) & "00000000" ;           -- '1' for implicit one
                  large_exp <= b (30 DOWNTO 23) ;
                  finish_swap <= '1' ;
```

```
            ELSE
                c <= '1' & b (22 DOWNTO 0) & "000000000" ;
                d <= '1' & a (22 DOWNTO 0) & "00000000" ;          -- '1' for implicit 1
                large_exp <= a (30 DOWNTO 23) ;
                finish_swap <= '1' ;
            END IF ;
        END IF ;
    END IF ;
  END IF ;
END PROCESS;
END rtl;
```

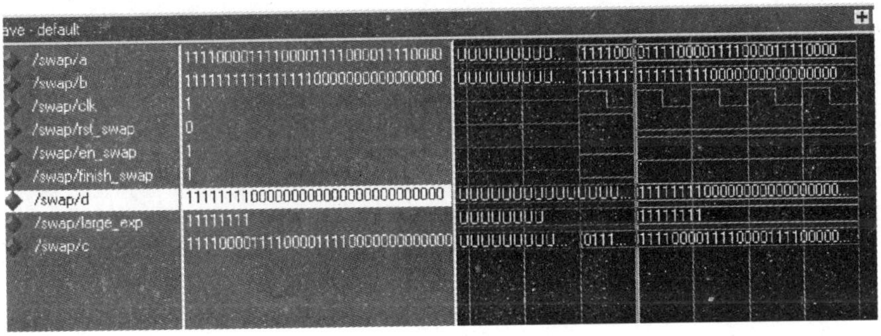

Figure C.17 Simulation Waveforms for Swap Sub-unit

Summer Sub-unit

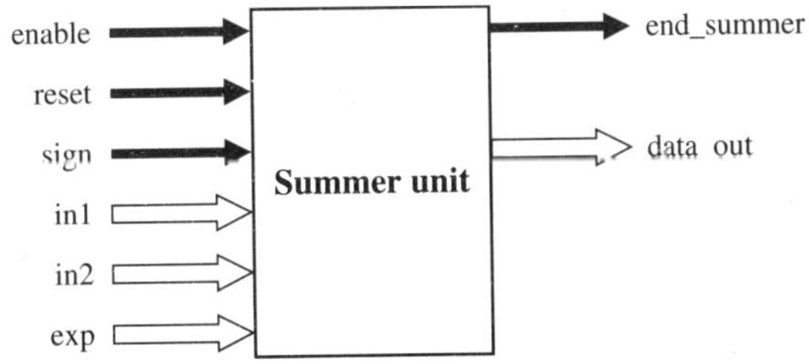

Figure C.18 Block Diagram of Summer Sub-unit

The summer unit sums/subtracts the shifted mantissa of the smaller number to/from the mantissa of the larger number. The summer adds an extra bit "0" and then sums/subtracts. This is to find out whether normalization is required or not. If normalization (that is converting the result to IEEE format) is required, the extra added bit will be set to "1" else it is "0". The information as to whether addition or subtraction is to be done is received from the signal "sign". After the addition process, the "end_summer" signal is set. "reset" signal reset the summer unit while "enable" signal enables the summer unit. "in1" and "in2" are the inputs to the units which are numbers to be added subtracted.

```
LIBRARY IEEE;
USE IEEE.STD_LOGIC_1164.ALL;
USE IEEE.STD_LOGIC_ARITH.ALL;
USE IEEE.STD_LOGIC_UNSIGNED.ALL;ED.ALL;

ENTITY summer IS
PORT (
    in1 , in2 : IN STD_LOGIC_VECTOR (31 DOWNTO 0) ;
    exp : IN STD_LOGIC_VECTOR (7 DOWNTO 0) ;
  enable , sign , reset : IN std_logic ;
  end_summer : OUT std_logic ;
  data_out : OUT STD_LOGIC_VECTOR ( 32 DOWNTO 0)
) ;
END summer ;

ARCHITECTURE behav OF summer IS
BEGIN
PROCESS (in1, in2, enable, reset)
    VARIABLE temp_in1, temp_sum, temp_in2, temp_sum2 : STD_LOGIC_VECTOR (32 DOWNTO 0);
    VARIABLE temp_exp : STD_LOGIC_VECTOR (7 DOWNTO 0) ;
BEGIN
    IF (reset = '1') THEN
        end_summer <= '0';
    ELSIF (enable = '1') THEN
        temp_in1 := '0' & in1 (31 DOWNTO 0) ;        -- 0 to find whether normalisation is required
        temp_in2 := '0' & in2 (31 DOWNTO 0) ;        -- if required MSB will be 1 after addition
        IF (sign = '1') THEN
            temp_sum := temp_in1 + temp_in2 ;
            data_out <= temp_sum ;
            end_summer <= '1' ;
        ELSE
            temp_sum := temp_in2 – temp_in1 ;
            data_out <= temp_sum ;
          end_summer <= '1' ;
        END IF ;
    END IF ;
END PROCESS ;
END behav ;
```

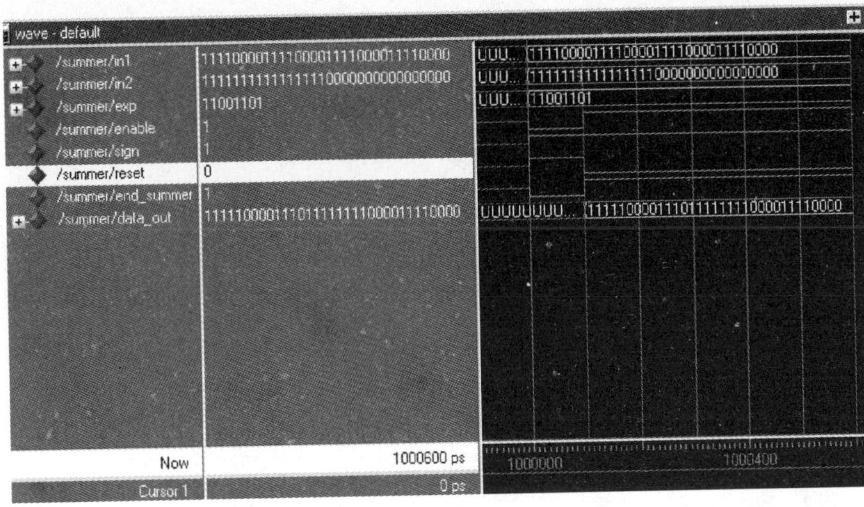

Figure C.19 Simulation Waveforms for Summer Unit

Normalizing Sub-unit

The function of this unit is to normalize the output of the adder unit on similar lines as done in case of scientific numbers. The complete VHDL code for this unit is given. The normalization processes when one number is zero and when addition or subtraction is used are all different from one another. The block under "if (operation='0')" gives the normalization procedure for the difference of two numbers. The first statement under this section is "if(temp=0)". Such a condition occurs only when both the numbers are same and they have been subtracted (or they are of opposite sign and they have been added). Obviously, the result is zero. If a number is normalized then temp(31) is zero. In that case, the final difference can be outputted. The normalization process of addition is different. Here, if normalization is required, the bit "temp(32)" will be "1". In that case, the exponent has to be incremented by one. If temp(32) = "0", the bits of the sign, exponent and the mantissa just have to be concatenated. When one of the numbers is zero, the sum is the other number. The subtractor gives the information as to which number is zero. If "a" is zero, "zero_number" is "01" and the output is "b". If "b" is zero, "zero_number" is "10" and the output is "a".

VHDL code for normalizer unit

```
LIBRARY IEEE;
USE IEEE.STD_LOGIC_1164.ALL;
USE IEEE.STD_LOGIC_ARITH.ALL;
USE IEEE.STD_LOGIC_UNSIGNED.ALL;

ENTITY normaliser is
PORT (
    a , b : IN STD_LOGIC_VECTOR (31 DOWNTO 0) ;
  data_in2 : IN STD_LOGIC_VECTOR (32 DOWNTO 0) ;
  exp : IN STD_LOGIC_VECTOR (7 DOWNTO 0) ;
```

```
  sign, operation, clk, enable, reset : IN STD_LOGIC ;
  zero_number : IN STD_LOGIC_VECTOR(1 DOWNTO 0) ;
 end_norm : OUT STD_LOGIC ;
 normalise_out : OUT STD_LOGIC_VECTOR (31 DOWNTO 0) ) ;
END ENTITY normaliser ;

ARCHITECTURE behav OF normaliser IS
BEGIN
PROCESS (clk)
 VARIABLE temp : STD_LOGIC_VECTOR (31 DOWNTO 0) ;
 VARIABLE texp : STD_LOGIC_VECTOR (7 DOWNTO 0) ;
 VARIABLE t , t2 : STD_LOGIC := '1' ;
BEGIN
 IF (reset = '0') THEN
   IF (enable = '1') THEN
     IF (t = '1') THEN
          temp := data_in2(31 DOWNTO 0) ;
          texp := exp (7 DOWNTO 0) ;
          t := '0';
    END IF ;
     IF (t2 = '1') THEN
          IF (zero_number = 0) THEN
               IF (operation = '0') THEN
                    IF (temp = 0) THEN
                         normalise_out <= temp(31 DOWNTO 0) ;
                         end_norm <= '1' ;
                         t2 := '0' ;
                ELSIF (temp(31) = '1' AND clk = '1') THEN
               normalise_out <= sign & texp(7 DOWNTO 0) & temp(30  DOWNTO 8) ;
                         end_norm <= '1' ;
                       t2 := '0' ;
               ELSIF (clk = '1') THEN
                       temp := temp(30 DOWNTO 0) & '0' ; texp := texp – "00000001" ;
               END IF ;
          ELSIF (operation = '1' AND data_IN2(32) = '1' AND clk = '1') THEN
               texp := texp + "00000001" ;
       normalise_out<=sign & texp(7 DOWNTO 0) & temp(31 DOWNTO 9) ;
       end_norm <= '1' ;
               t2 := '0' ;
```

```
            ELSIF (clk = '1') THEN
                normalise_out <= sign & texp(7 DOWNTO 0) & temp(30 DOWNTO 8) ;
                end_norm <= '1' ;
                 t2 := '0' ;
            END IF;
        ELSIF (zero_number = 1) THEN
              normalise_out <= b;
                end_norm <= '1 ' ; t2 := '0' ;
         ELSIF (zero_number = 2) THEN
                normalise_OUT <= a ;
                end_norm <= '1' ;
                 t2 := '0' ;
        END IF ;
    END IF ;
END IF ;
ELSIF (reset = '1') THEN
    end_norm <= '0' ;
     t := '1' ;
     t2 := '1' ;
END IF ;
END PROCESS ;
END behav ;
```

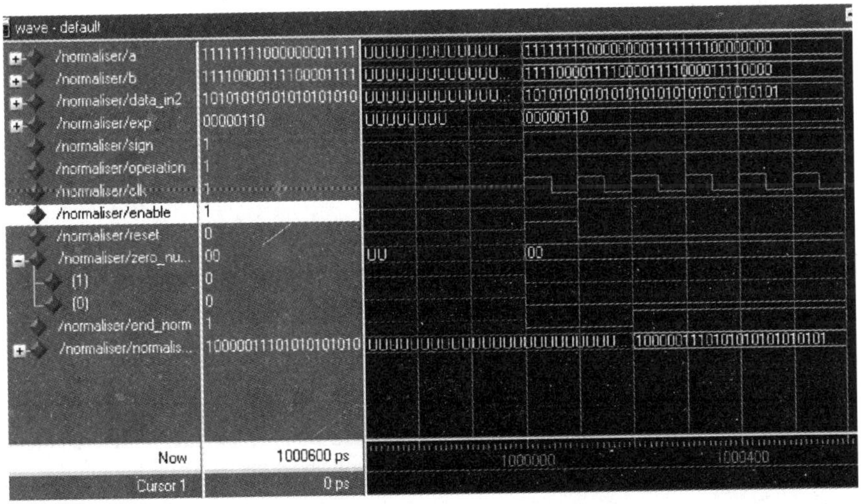

Figure C.20 Simulation Waveform for Normalizer Unit

Sequencer Sub-unit

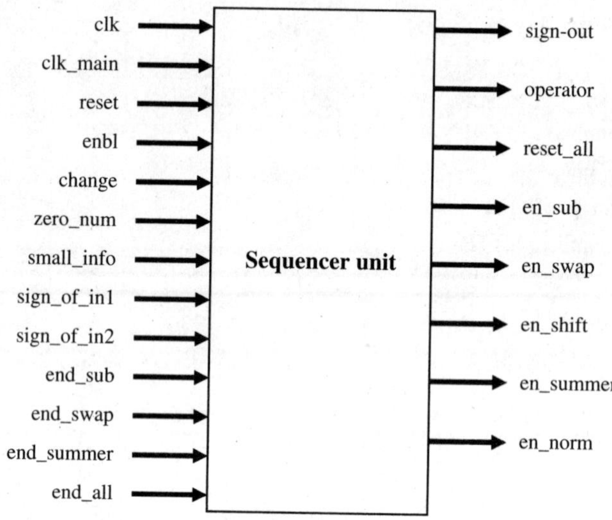

Figure C.21 Block Diagram of Sequencer Unit

The sequencer unit is the controller unit of the floating point adder It controls all the activities of the adder. It is modeled as a finite state machine. The main function of this unit is to synchronize the operation of all other blocks of the adder unit. The transition from one state to another takes place during the rising clock edge of the positive cycle.

(sequence 1). The second process carries out all these actions. The actions performed in each state in the first process are given below:

sequence 1: Here, the various signals are reset.

sequence 2: The reset signal is brought back to "0". (Otherwise the outputs will not change, it will remain as 0). Also the subtractor and swap units are enabled.

sequence 3: If one of the numbers is zero (indicated by "zero_num") the shifting and adding operations can be skipped. Therefore the next state is set as sequence 5. Else, the shifter is enabled when the operations in the subtractor and swap unit are over. Also the subtractor and swap unit are disabled.

sequence 4: When the function of the shifter is over, it is disabled and the adder unit is enabled.

sequence 5: If one of the numbers is zero, the normalize unit is enabled. Else, the normalize unit is enabled when the function of the summer is over. Also, the summer is disabled in this state.

sequence 6: Here, when the normalization process is over, the normalize unit is disabled in the positive cycle. Later, in the negative cycle, the state is transferred to sequence 1.

sequence 7: This is the state into which the controlcomes when the adder is disabled. There is a third process in this unit, which gives information to the summer as to whether addition or subtraction is to be performed. It also gives information to the normalize unit about the sign of the result.

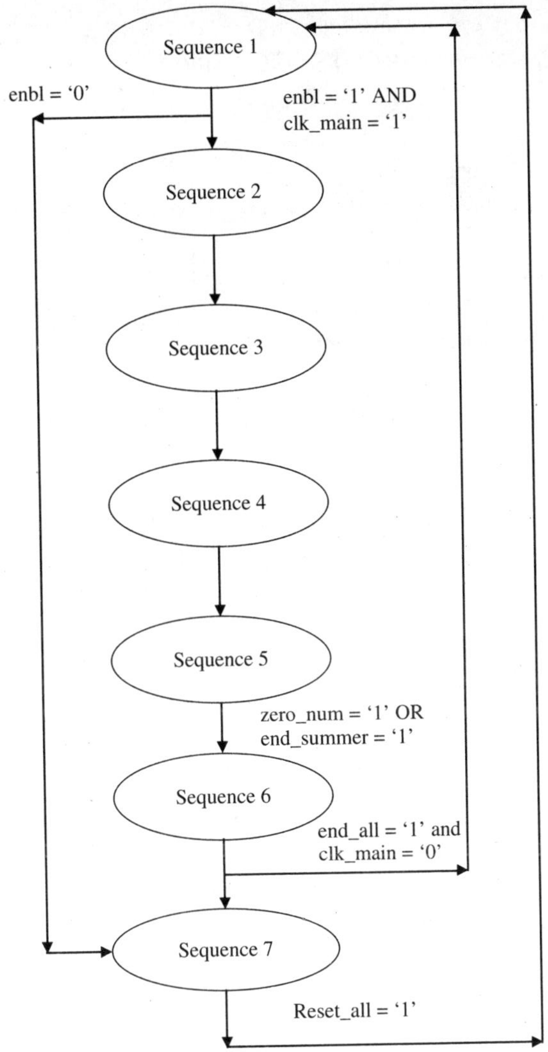

Figure C.22 State Diagram for the Sequencer Unit

VHDL code of sequencer

```
USE IEEE.STD_LOGIC_1164.ALL;
USE IEEE.STD_LOGIC_ARITH.ALL;
USE IEEE.STD_LOGIC_UNSIGNED.ALL;

ENTITY sequencer IS
PORT (small_info, sign_of_in1, sign_of_in2 : IN STD_LOGIC ;
sign_out, operator, reset_all :OUT STD_LOGIC ;
en_sub, en_swap, en_shift, en_summer, en_norm : OUT STD_LOGIC ;
```

```
end_sub, end_swap, end_shift, end_summer, end_all : IN STD_LOGIC ;
clk_main, clk, reset, enbl, zero_num, change : IN STD_LOGIC
) ;
END ENTITY sequencer ;

ARCHITECTURE behav OF sequencer IS
TYPE state IS (sequence1, sequence2, sequence3, sequence4, sequence5, sequence6,
sequence7) ;
SIGNAL present_state : state ;
BEGIN
PROCESS (present_state, end_summer, end_shift, end_all, enbl, clk_main, end_sub,
end_swap, zero_num, change)
BEGIN
IF present_state = sequence1 THEN
   IF(enbl = '1' AND clk_main = '1') THEN
      en_norm <= '0' ;                  -- enable normaliser unit
      en_summer <= '0' ;                -- enable summer unit
      reset_all <= '1' ;                -- reset all the blocks of floating-point adder
      present_state <= sequence2 ;      -- go to next state
   ELSIF ( enbl = '0') THEN
      present_state <= sequence7 ;      -- last state to exit
   ELSE
      present_state <= sequence1 ;
   END IF ;
ELSIF present_state = sequence2 THEN
   reset_all <= '0' ;                   -- disable reset SIGNAL else values will not change
   en_sub <= '1' ;                      -- enable subtractor unit
   en_swap <= '1' ;                     -- enable swap unit
   present_state <= sequence3 ;
ELSIF present_state = sequence3 THEN
   IF (zero_num = '1') THEN             -- if first number is zero skip sequences 3 and 4
      present_state <= sequence5 ;
      en_sub <= '0' ;
      en_swap <= '0' ;
   ELSIF(end_sub= '1') THEN
      IF(end_swap = '1') THEN
      en_shift <= '1' ;                 -- enable shifter unit
      en_sub <= '0' ;
      en_swap <= '0' ;
      present_state <= sequence4 ;
    END IF ;
 ELSE
    present_state <= sequence3 ;
 END IF ;
```

```
ELSIF present_state = sequence4 THEN
  IF (end_shift = '1') THEN
       en_shift <= '0' ;
       en_summer <= '1' ;                          -- enable summer unit
       present_state <= sequence5 ;
    ELSE
       present_state <= sequence4 ;
    END IF ;
ELSIF present_state = sequence5 THEN
    IF (zero_num = '1') THEN en_norm <= '1' ;  -- enable normaliser
       present_state <= sequence6 ;
    ELSIF (end_summer = '1') THEN
       en_norm <= '1' ; en_summer <= '0' ;
       present_state <= sequence6 ;
    ELSE
       present_state <= sequence5 ;
    END IF ;
ELSIF present_state = sequence6 THEN
    IF (end_all = '1' AND clk_main = '1') THEN
       en_norm <= '0' ; present_state <= sequence6 ;
    ELSIF (end_all = '1' and clk_main = '0') THEN
       present_state <= sequence1 ;
    ELSE
       present_state <= sequence6 ;
    END IF ;
ELSIF present_state = sequence7 THEN
    present_state <= sequence7 ;
ELSE
    presentt_state <= sequence1 ;
END IF ;
END PROCESS ;
PROCESS (small_info , sign_of_in1 , sign_of_in2)
BEGIN
IF (sign_of_in1 = '0' AND sign_of_in2 = '0') THEN
    sign_out <= '0' ;
    operator <= '1' ;
ELSIF (sign_of_in1 = '1' AND sign_of_in2 = '1') THEN
    sign_out <= '1' ;
    operator <= '1' ;
ELSIF (small_info = '1' AND sign_of_in1 = '0') THEN
    sign_out <= '1' ;
    operator <= '0' ;
```

```
ELSIF (small_info = '0' AND sign_of_in1 = '1') THEN
    sign_out <= '1' ;
    operator <= '0' ;
ELSE
    sign_out <= '0' ;
    operator <= '0' ;
END IF ;
END PROCESS ;
END behave ;
```

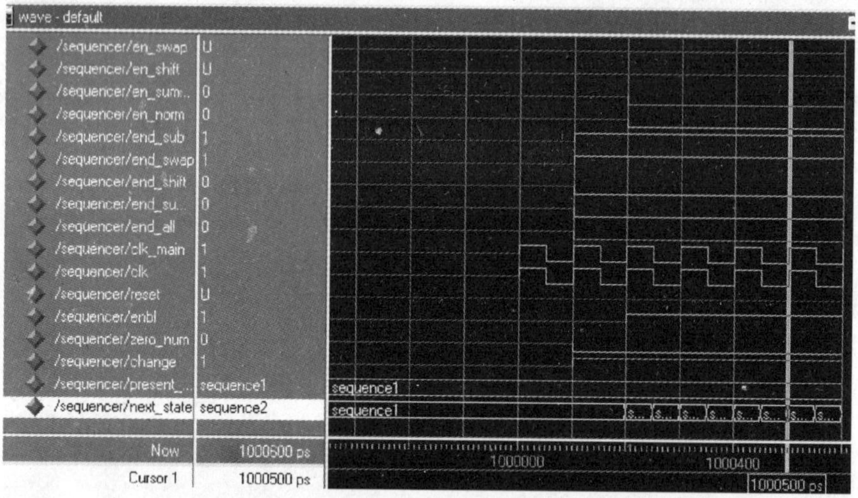

Figure C.23 Simulation Waveforms for Sequencer Unit

Overall high level floating point adder VHDL code is given below:

```
LIBRARY IEEE;
USE IEEE.STD_LOGIC_1164.ALL;
USE IEEE.STD_LOGIC_ARITH.ALL;
USE IEEE.STD_LOGIC_UNSIGNED.ALL;

ENTITY fp_adder IS
PORT (
    a,b : IN STD_LOGIC_VECTOR(31 DOWNTO 0);
    result_out : OUT STD_LOGIC_VECTOR(31 DOWNTO 0) ;
    clk_main, clk, enbl, reset : IN STD_LOGIC) ;
END ENTITY fp_adder ;

ARCHITECTURE behav OF fp_adder IS

COMPONENT subtractor IS
```

```
PORT (
a : IN std_logic_vector (31 DOWNTO 0) ;
b : IN std_logic_vector (31 DOWNTO 0) ;
clk, rst_sub, sub_en : IN std_logic ;
a_smaller, fin_sub, num_zero : OUT std_logic ;
zero_detect : OUT std_logic_vector(1 DOWNTO 0) ;
sub : OUT std_logic_vector (8 DOWNTO 0)) ;
END COMPONENT subtractor ;

COMPONENT swap is
  PORT (
    a : IN STD_LOGIC_VECTOR (31 DOWNTO 0) ;
    b : IN STD_LOGIC_VECTOR (31 DOWNTO 0) ;
    clk : IN STD_LOGIC ;
    rst_swap, en_swap : IN STD_LOGIC ;
    finish_swap : OUT std_logic ;
    d : OUT STD_LOGIC_VECTOR (31 DOWNTO 0) ;
    large_exp : OUT STD_LOGIC_VECTOR (7 DOWNTO 0) ;
    c : OUT STD_LOGIC_VECTOR (32 DOWNTO 0 ) ) ;
END COMPONENT swap ;

COMPONENT shifter is
PORT (
shift_count : IN STD_LOGIC_VECTOR (8 DOWNTO 0) ;
data_in2 : IN STD_LOGIC_VECTOR (32 DOWNTO 0) ;
data_out : OUT STD_LOGIC_VECTOR (31 DOWNTO 0) ;
clk , enable , reset : IN STD_LOGIC ;
end_shift : OUT STD_LOGIC
 ) ;
END COMPONENT shifter ;

COMPONENT sequencer IS
PORT (small_info, sign_of_in1, sign_of_in2 : IN STD_LOGIC ;
sign_out, operator, reset_all :OUT STD_LOGIC ;
en_sub, en_swap, en_shift, en_summer, en_norm : OUT STD_LOGIC ;
end_sub, end_swap, end_shift, end_summer, end_all : IN STD_LOGIC ;
clk_main, clk, reset, enbl, zero_num, change: IN STD_LOGIC
 ) ;
END COMPONENT sequencer ;

COMPONENT summer IS
PORT (
  in1, in2 : IN STD_LOGIC_VECTOR (31 DOWNTO 0) ;
  exp : IN STD_LOGIC_VECTOR (7 DOWNTO 0) ;
 enable , sign , reset : IN std_logic ;
```

```
  end_summer : OUT std_logic ;
  data_out : OUT STD_LOGIC_VECTOR ( 32 DOWNTO 0)
 ) ;
 END COMPONENT summer ;

 COMPONENT normaliser is
 PORT (
     a , b : IN STD_LOGIC_VECTOR (31 DOWNTO 0) ;
        data_in2 : IN STD_LOGIC_VECTOR (32 DOWNTO 0) ;
        exp : IN STD_LOGIC_VECTOR (7 DOWNTO 0) ;
        sign, operation, clk, enable, reset : IN STD_LOGIC ;
        zero_number : IN STD_LOGIC_VECTOR(1 DOWNTO 0) ;
        end_norm : OUT STD_LOGIC ;
        normalise_out : OUT STD_LOGIC_VECTOR (31 DOWNTO 0) ) ;
 END COMPONENT normaliser ;

SIGNAL shift0 , waves : STD_LOGIC_VECTOR(3 DOWNTO 0) ;
type state is (reset1, reset2, reset3, reset4, reset5, reset6, reset7) ;
SIGNAL shift, changeb, finish_sum, signbit, normalise, END_all, a_small, addsub, sum_out2, shift_done,
done, num_rec, setbit, addpulse : STD_LOGIC := '0' ;
SIGNAL link16, final_sum, link7, final_sumb, shift_outa, swap_num2 : STD_LOGIC_VECTOR ( 31
DOWNTO 0 ) := (OTHERS => '0') ;
SIGNAL swap_num1, sum_out : STD_LOGIC_VECTOR (32 DOWNTO 0) := (OTHERS => '0') ;
SIGNAL sub2, suba: STD_LOGIC_VECTOR (8 DOWNTO 0) := (OTHERS => '0') ;
SIGNAL exp : STD_LOGIC_VECTOR (7 DOWNTO 0) := (OTHERS => '0') ;
SIGNAL rst, enswap, ensub, enshift, finsub, finswap, finshift, numzero : STD_LOGIC := '0' ;
SIGNAL zerodetect : STD_LOGIC_VECTOR(1 DOWNTO 0) ;
SIGNAL swap_num1b, sum_outb : STD_LOGIC_VECTOR (32 DOWNTO 0) := (OTHERS => '0') ;
SIGNAL subb, sub2b : STD_LOGIC_VECTOR (8 DOWNTO 0) := (OTHERS => '0') ;
SIGNAL expb: STD_LOGIC_VECTOR (7 DOWNTO 0) := (OTHERS => '0') ;
SIGNAL shift_outb, swap_num2b : STD_LOGIC_VECTOR ( 31 DOWNTO 0 ) := (OTHERS => '0') ;
SIGNAL rstb, enswapb, ensubb, enshiftb, finsubb, finswapb, finshiftb, numzerob, clk_mainb, resetb, enblb :
STD_LOGIC := '0' ;
SIGNAL zerodetectb : STD_LOGIC_VECTOR(1 DOWNTO 0) ;
type state_values is (st0, st1, st2, st3) ;
SIGNAL pres_state1, present_state1 : state_values ;
SIGNAL shiftb, finish_sumb, signbitb, normaliseb, end_allb, a_smallb, addsubb, sum_out2b, shift_doneb,
doneb, num_recb, setbitb, addpulseb, clkb : STD_LOGIC := '0' ;
BEGIN
sub0 : subtractor PORT MAP (a=> a , b=>b , clk=>clk , rst_sub=>reset, sub_en=>ensubb, a_smaller=>a_
smallb , fin_sub=>finsubb, num_zero=>numzerob, zero_detect=>zerodetectb, sub=>subb ) ;
swap0 : swap PORT MAP (a=>a, b=>b, clk=>clk, rst_swap=>rstb, en_swap=>enswapb, finish_
swap=>finswapb, d=>swap_num2b, large_exp=>expb, c=>swap_num1b) ;
shifter0 : shifter PORT MAP (subb, swap_num1b, shift_outb, clk, enshiftb, rstb, finshiftb) ;
```

```
seq0 : sequencer PORT MAP ( a_smallb, a(31), b(31), signbitb, addsubb, rstb, ensubb,
enswapb, enshiftb, addpulseb, normaliseb, finsubb, finswapb, finshiftb, finish_sumb, end_allb,
clk_main, clk, reset, enbl, numzerob, changeb) ;
sum0 : summer PORT MAP (shift_outb, swap_num2b, expb, addpulseb, addsubb, rstb, finish_sumb, sum_
outb) ;
norm0 : normaliser PORT MAP (a , b, sum_outb, expb, signbitb, addsubb, clk, normaliseb, rstb, zerodetectb,
end_allb, result_out) ;
END behave ;
```

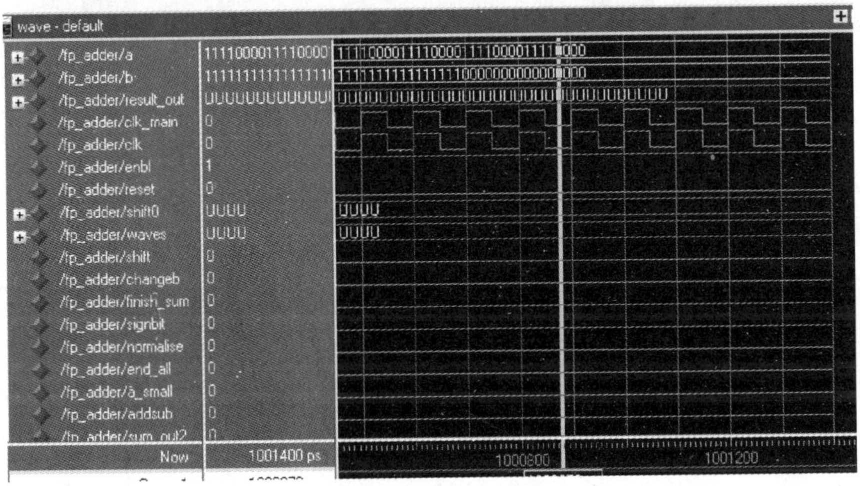

Figure C.24 Simulation Result of Complete Floating Point Adder Unit

Appendix A

Xilinx 9.1 Tutorial

This tutorial gives an overview of how to create a new Xilinx project, add source files, synthesize, and implement your design.

(i) Open up Xilinx project navigator.

(ii) The open window has four panes as follows:

- A source pane that shows the organization of the source files that make up your design. There are three tabs so that you can view the functional modules or HDL libraries for your project, or look at various snapshots of the project.
- A process pane that lists the various operations you can perform on a given object in the source pane.
- A log pane that displays the various messages from the currently running process.
- An editor pane where you can enter the HDL code.

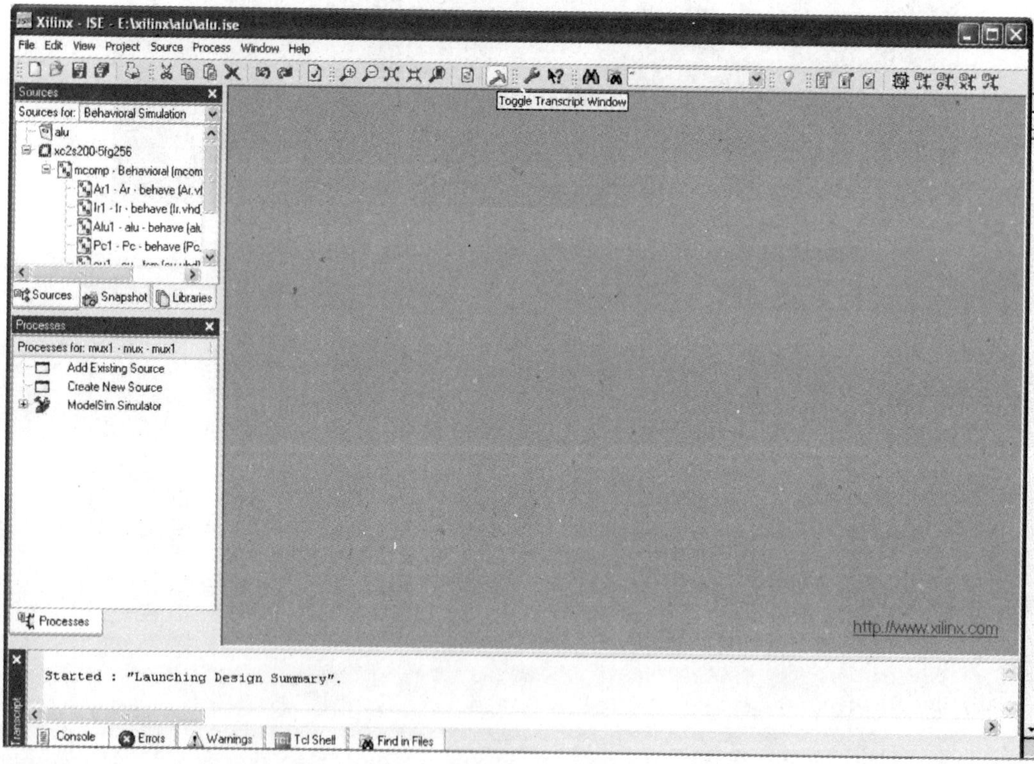

Creating a New Design

Create a new project by selecting the File → New Project from the menu.

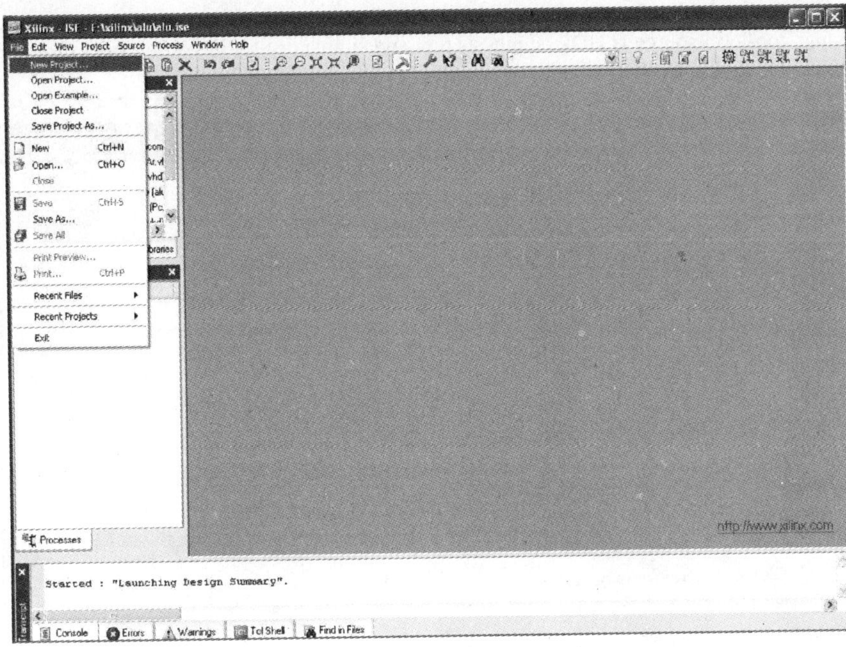

A new window pops up asking for the project name, its location, and the type of source—HDL, EDF, and so on.

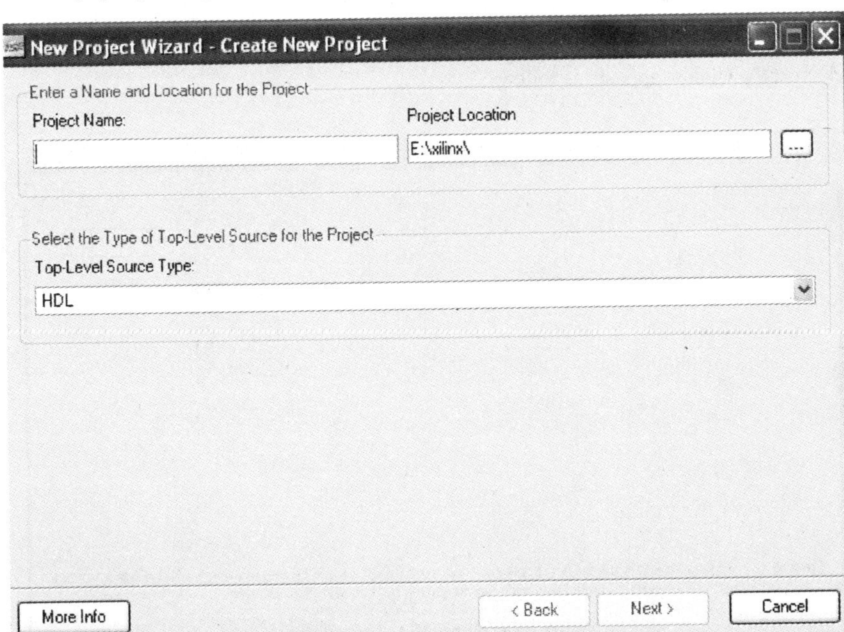

Enter a valid name for the project. Then choose its location which is generally XILINX folder in your drive. Choose the type of source as HDL and click on NEXT. Once done, a new window pops.

Up asking for device (FPGA/CPLD) to be used for implementation and its characteristics as speed, package. For HDL coding, top-level source is HDL, synthesis tool and simulator used is chosen from drop-down menu. The language used is chosen as VHDL.

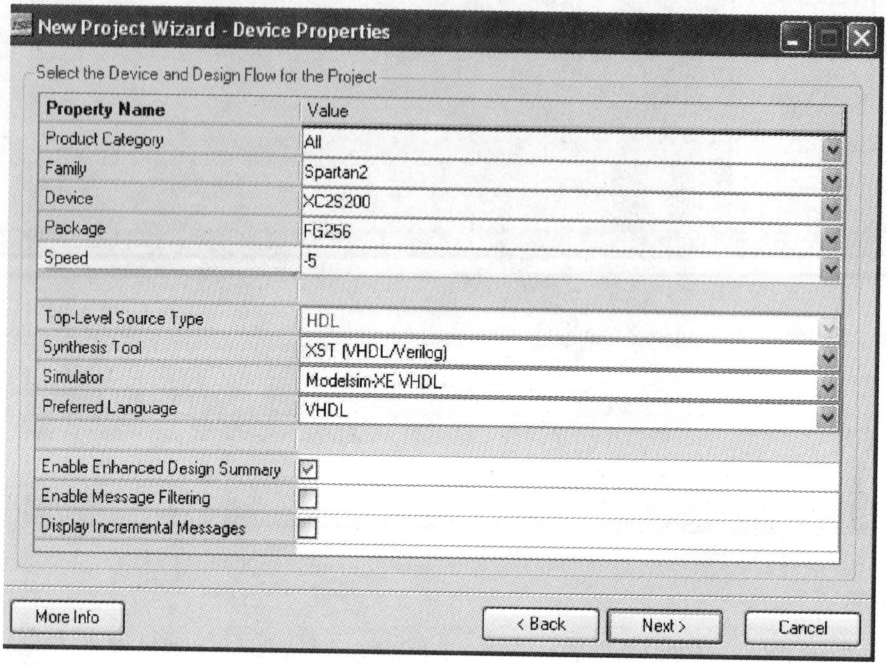

Then click Next. A window appears as follows:

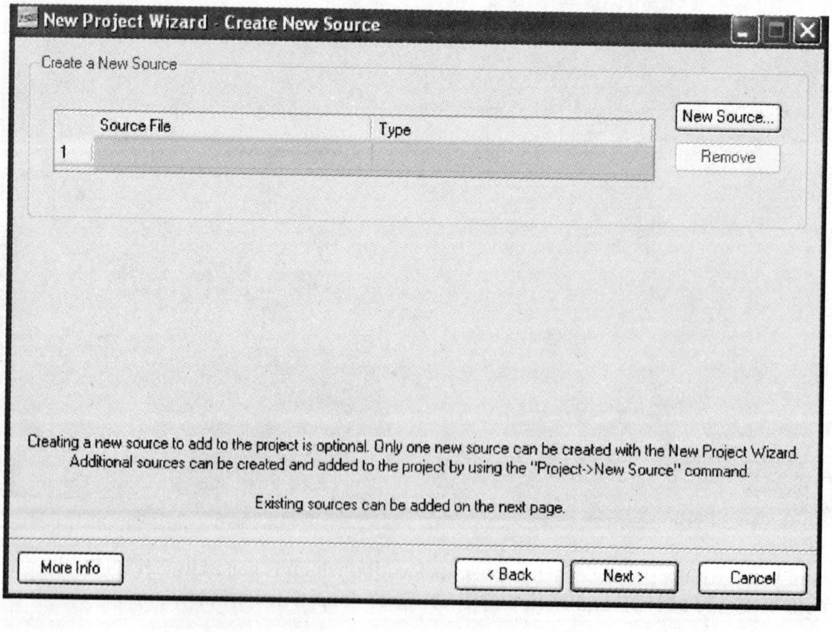

Then Click on New Source. The following window appears:

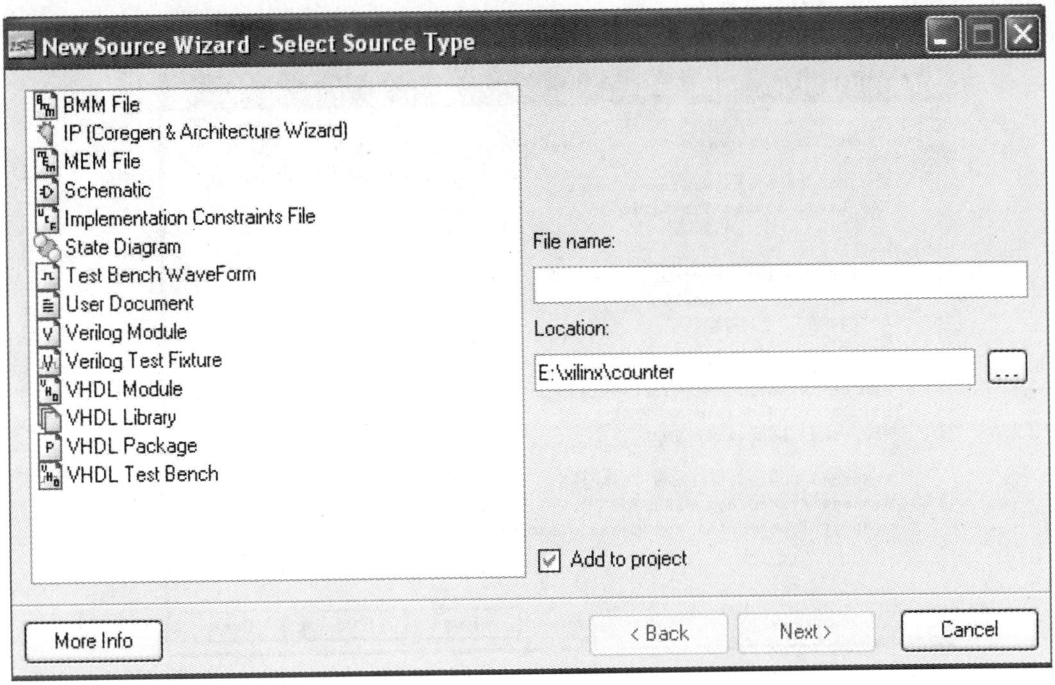

Select the source type as VHDL module and give a name to the design (this is the name of the entity to be designed).

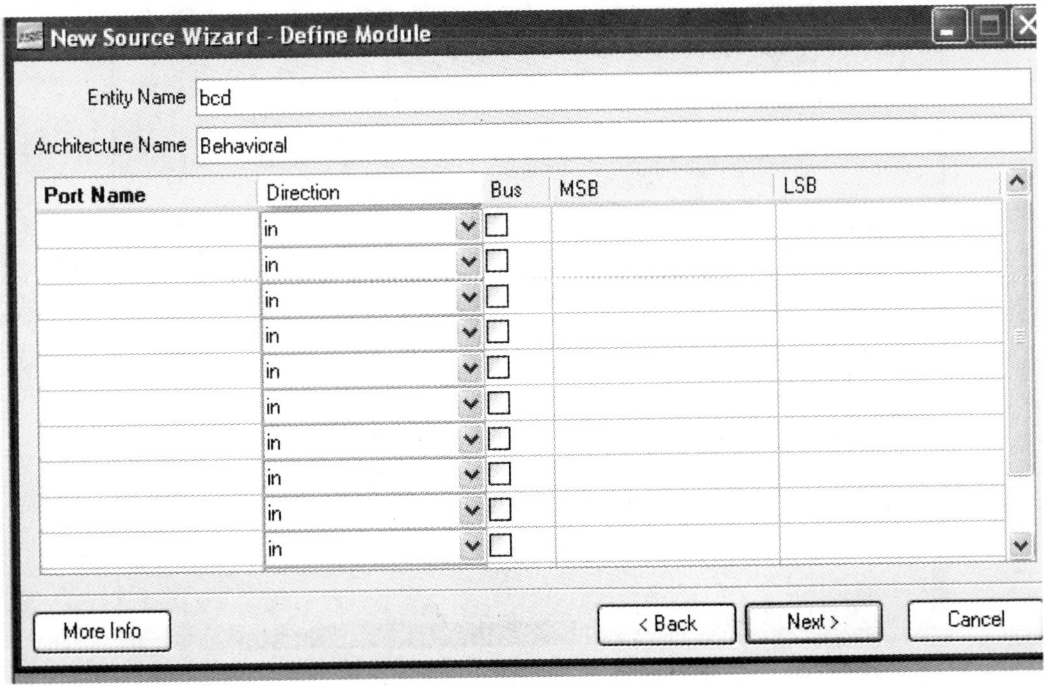

In the previous window, define the ports of the entity under design. By default all ports are of type Std_Logic.

If the already written VHDL code has to be used, click on Add Source and select the VHDL design file stored on disk; else, for a new VHDL source to be created, click on Next.

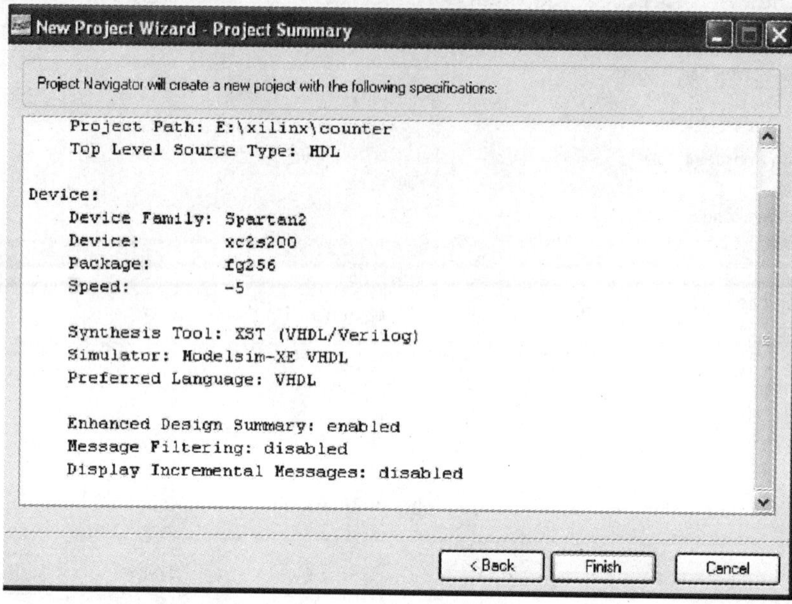

This summarizes all the options selected for the project to be designed. Click on Finish and editor window appears as follows:

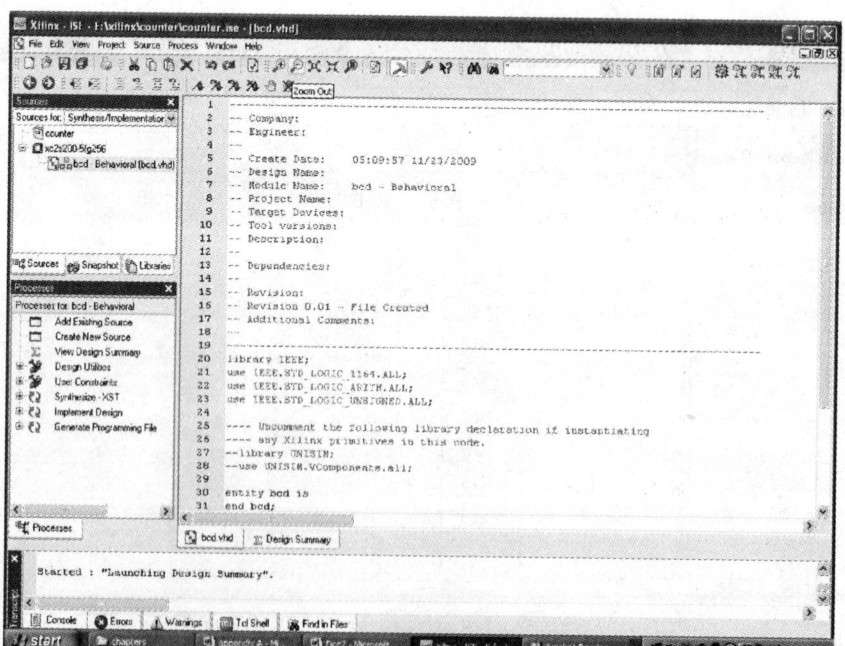

The code is entered in the editor window.

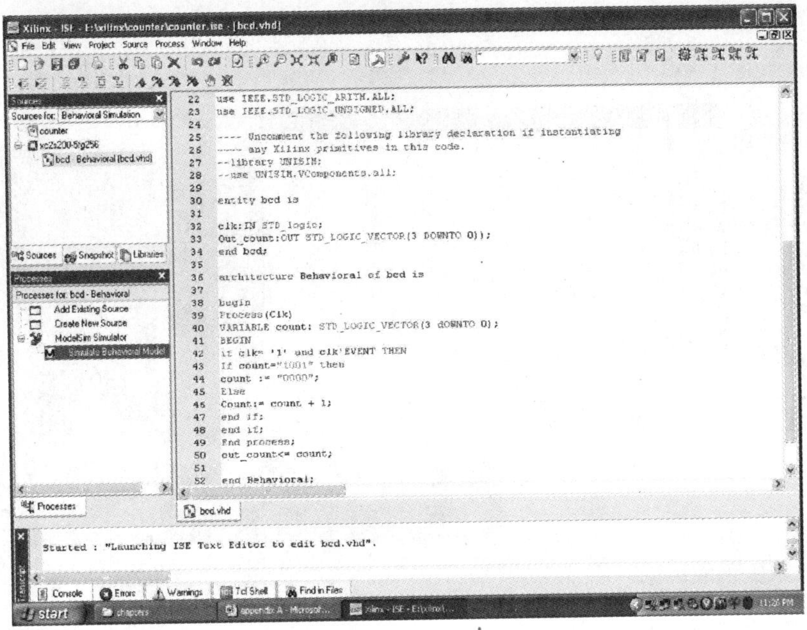

Simulating the Design

Once the code is entered in the sources window, click the entity to be simulated and in tab Sources For, select behavioral simulation from the drop-down menu. The process window will show the simulator that is to be used for simulation (i.e., simulator used will be the one selected in the beginning in device properties). In this case, Modelsim 6.2 SE is used as the simulator.

Double click on the simulator and a new window of modelsim appears as shown:

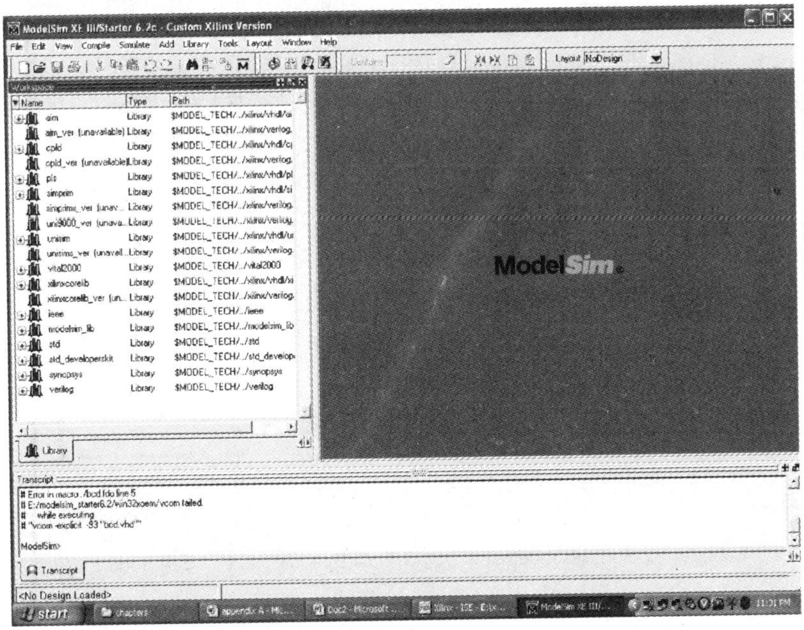

In the previous window, go to file menu → Open → and the following window pops up:

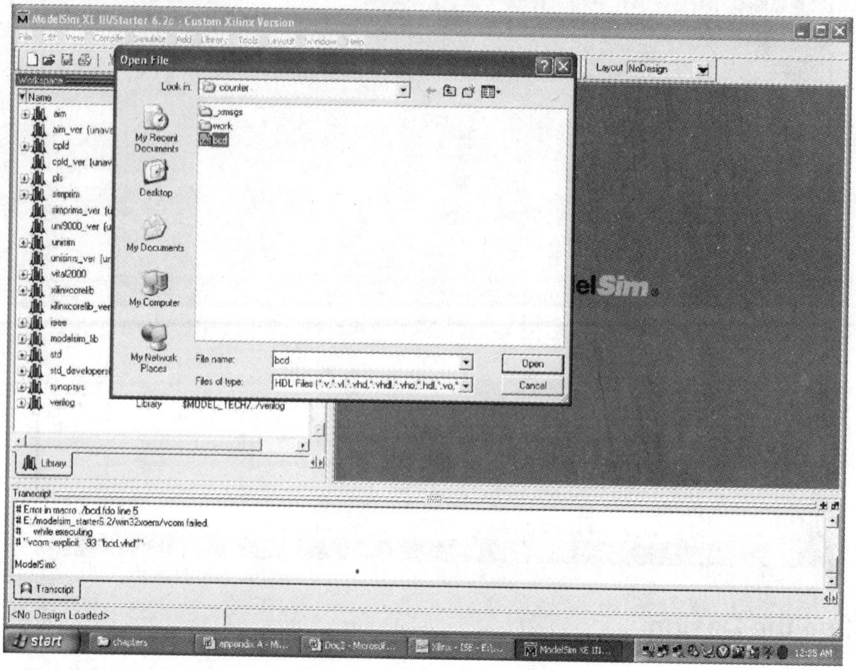

In this open the VHD file BCD.

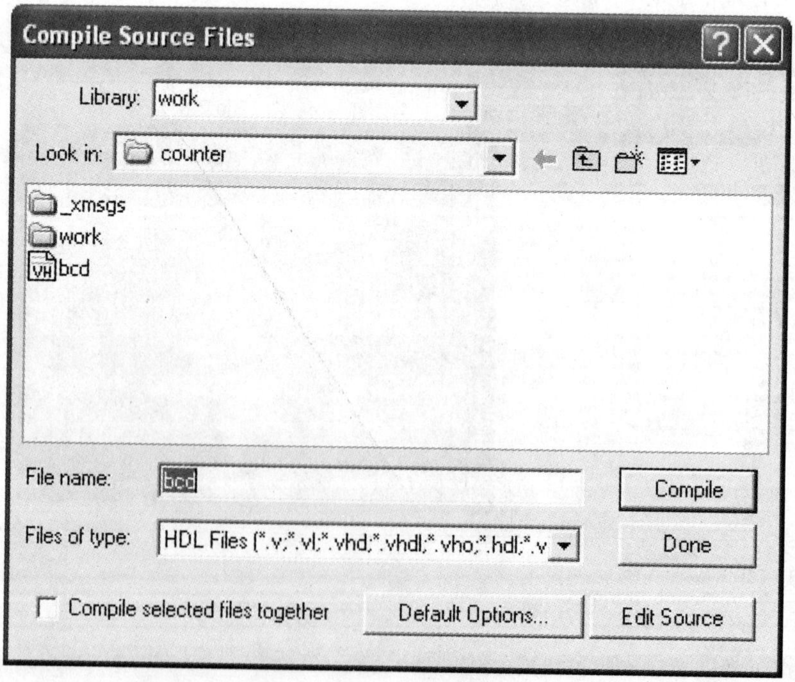

The compiler will show errors and warnings in the log window in red colour.

Remove errors if any. Then recompile the design. If there are no errors, then it means no syntax errors in the design. Next, go to View tab, and click on Add Wave.The wave window opens.

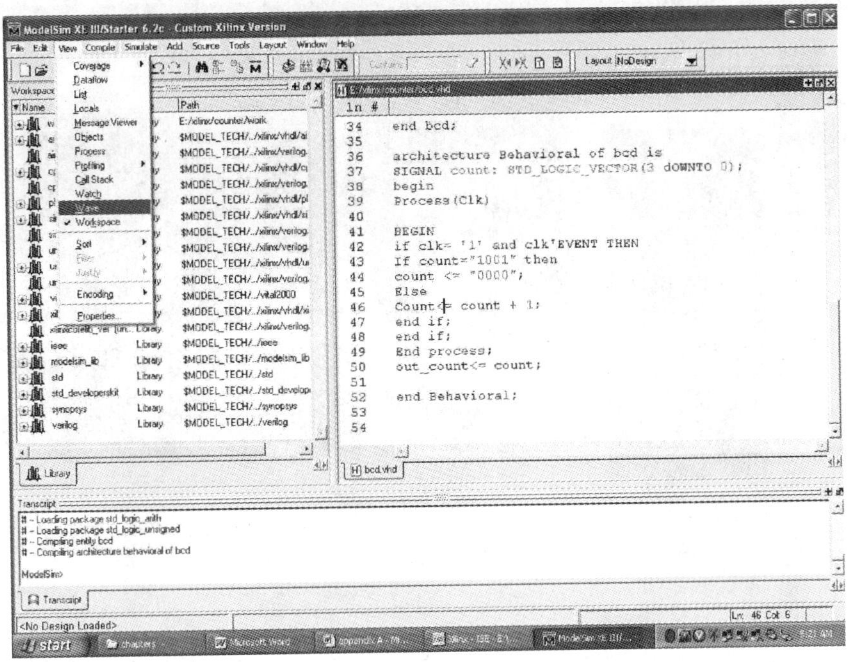

Click on Simulate tab → When you start simulation, a new window pops up, Then open the work library and choose the design that is to be simulated.

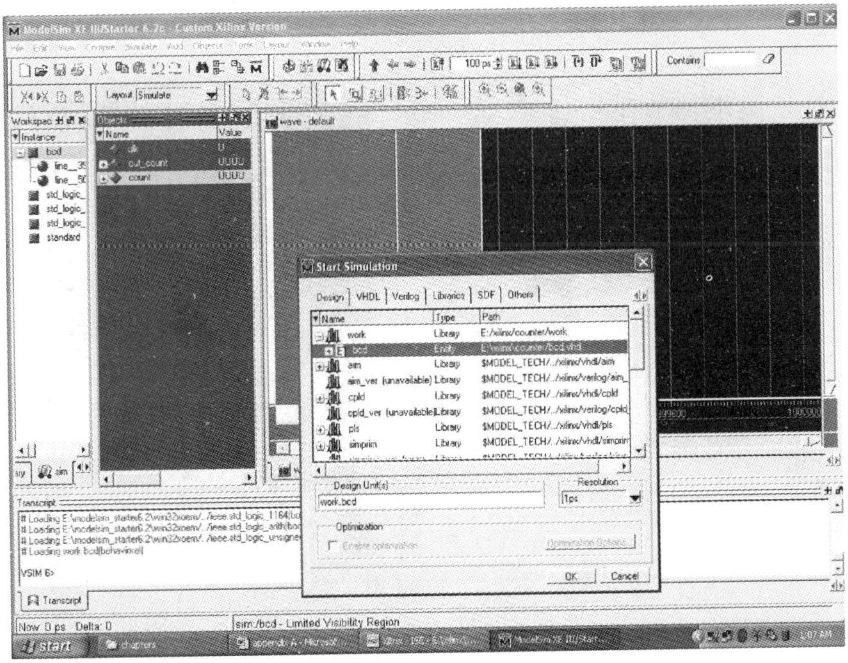

Add signals from the object area to the wave window, by selecting a signal and right clicking on it. Choose the option Add to Wave → Selected Signals. If all signals are to be selected for simulation, then choose the option Add to Wave → Selected Signals in the region, and all signals will be added to the wave window.

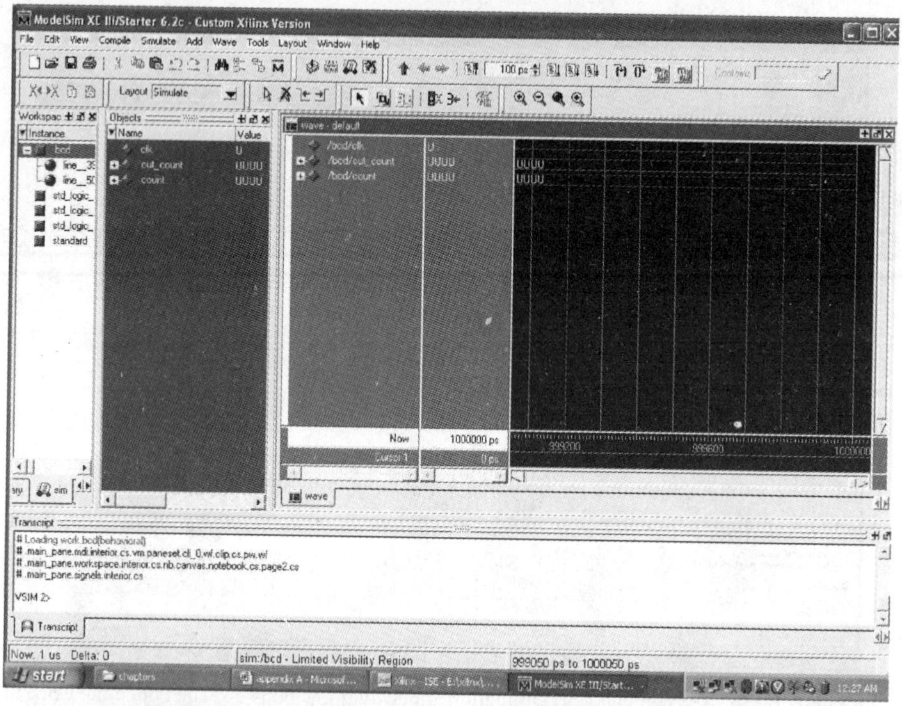

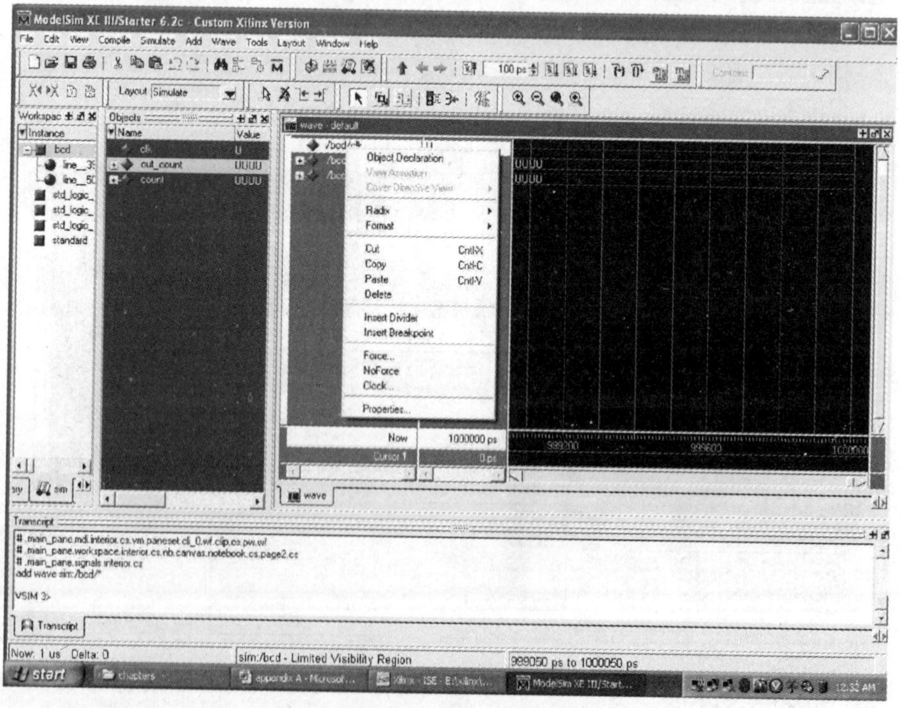

To start the simulation signal drivers need to be stimulated. For that, right click on the signal in the wave window as shown in the previous window. If the signal is clock signal as in our current example, then click on option Clock.

For stimulation of input to binary values, choose the option Force and fill in the value to be forced on the signal driver. If any other radix is to be used except binary, then choose radix and change the radix number too. If hex or decimal values are to be displayed, then the same can be changed using option Format as found in the following window.

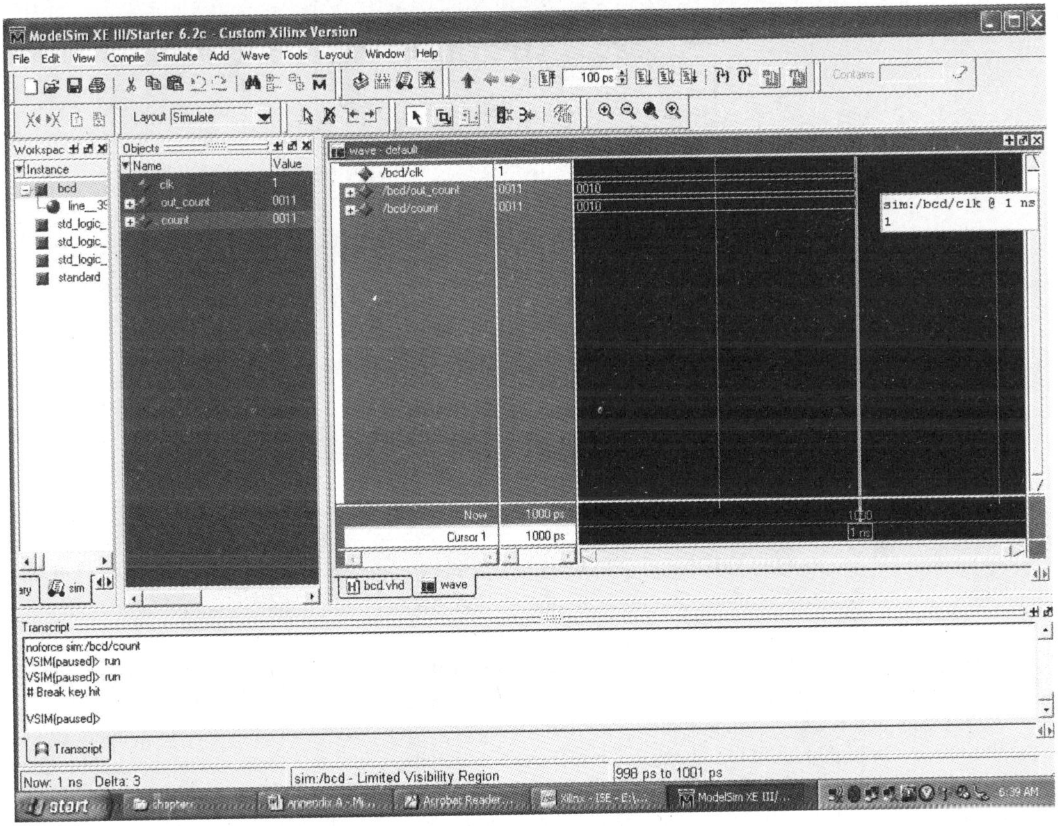

Synthesizing the Circuit

In the sources window, select synthesis/implementation from the drop-down menu. Also select the design that is to be synthesized and in the process window, open the option of synthesis—XST. The first step to synthesis is checking syntax which checks for design if all constructs used are synthesizable. In case of any semantic/syntax error or use of any non-synthesizable construct, then errors/warnings are shown in log window—else, message says Check Syntax Successful.

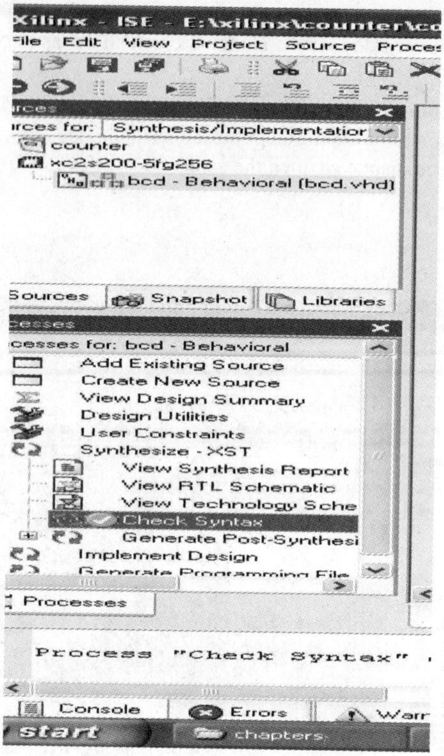

The synthesizer will read the VHDL code and transform it into a netlist of gates as in the previous window.

Double click on the View Synthesis Report to see the various synthesizer options that were enabled and some device utilization and timing statistics for the synthesized design.

Double-click on View RTL Schematic to see the schematic that was derived from the VHDL source code.

After synthesis, the next step is hardware implementation of the design which requires the designer to translate, map, place, and route the same into the logic resources of the FPGA/CPLD in order to actually use it. Start this process by highlighting the design that is to be implemented in the source window, and then double click on the Implement Design process.

Implementation is dependent on the chip to be chosen (FPGA/CPLD ands its family) and hence, left to the user.

Appendix B

Synthesizable Constructs of VHDL-93 and VHDL-2008

Synthesis of Data Types

Type	Synthesizable
Bit	Synthesizable
Bit_vector	Synthesizable
Boolean	Synthesizable
Character	Synthesizable
Std_logic	Synthesizable
Std_ulogic	Synthesizable
Integer	Synthesized as bus
Real	Not synthesizable
Enumerated	Synthesizes to a bus of bits with 1st value as binary 0, the 2nd as binary 1, the third as binary 10 and so on.
Physical	Not synthesizable
Array	Some synthesis tools do not support multi-dimensional arrays; mostly, arrays of bits or bit vectors are synthesizable but not ports which are arrays of arrays.
Access	Not synthesizable
Record	Synthesized as wires but all synthesis tool may not support records
File	Not synthesizable

Concurrent Statements

Statement	Synthesizable
Aggregates	Generally not synthesizable
Alias	Not synthesizable
Assert	Not synthesizable
Block	Synthesizable
Conditional select statements (with–when)	Synthesized as a combinational logic
Generate	Synthesizable

Sequential Statements

Statement	Synthesizable
Assert	Not synthesizable
Case	Synthesized as multiplexer; unassigned conditions if in synchronized process (with clock) synthesize to registers else as latches
Exit	Synthesizable (no hardware generation; so ignored by synthesis tools but no errors caused)
FOR loop	Synthesizable for static ranges
IF	Same as case statement; sometimes, nested if statements synthesize to multiple logic levels which can be avoided by using a case statement
Loop	Not synthesizable
Next	No hardware generation
Null	No hardware generation
Report	No hardware generation
Variable	A wire or a register
Wait	Wait for, wait on and wait; are not synthesizable. Wait until is synthesizable only when used to synchronize a process to a clock edge
While loop	Not synthesizable

Attributes

Statement	Synthesizable
Attributes of enumeration types are not synthesizable	

VHDL Constructs

Statement	Synthesizable
Configurations	Not synthesizable
Resolution functions	Not synthesizable
Functions	Each call to a function is synthesized as a separate block of combinational logic
Generics	Synthesizable but only as integer values
Procedures	Synthesizable, provided that they do not detect clock edges or contain 'wait' statements
Shared variables	Not synthesizable

Operators

Operator	Synthesizable
+	Synthesizable as adder
–	Synthesizable as subtractor
=, /=, >, < , <=, >=	Synthesizable as comparator
Shift operators	Synthesizable
Logical opeartors	Synthesizable
Mod, rem, **, /	Not synthesizable
Multiplication	Synthesizable as adder

In VHDL-2008, new constructs have been added which are synthesizable along with all synthesizable constructs of VHDl-93. Two new packages have been added—IEEE-fixed_generic_pkg and IEEE-floating_generic_pkg. These packages make all floating- and fixed-point arithemetic operations on STD_logic types synthesizable. Also Generic types, packages and subprograms are synthesizable in VHDL 2008.

Appendix C

VHDL-2008: New Features

VHDL 2008 has come as a major modification to IEEE VHDL after its 1076-93 version. Significant changes are made making VHDL more designer friendly.

■ PSL is integrated into VHDL

PSL is a language tool developed for verification using assertions regarding properties of design under a set of assumptions. VHDL already includes the assert statement, which is used for adding simple check on violations in VHDL models.

VHDL extends this idea by integrating PSL as a part of VHDL, providing a syntax for describing the expected behaviour of a circuit over time and for checking that the VHDL code implements that behaviour.

VHDL 2008 includes the basic subset of PSL as a part of its standard syntax. Therefore, all the keywords of PSL are reserved words in VHDL.

■ Package and subprogram generics

In VHDL 2008 the idea of generics has been extended to packages and subprograms too. Earlier, only generic entities were allowed in VHDL, but now VHDL-2008 allows generics on packages and subprograms also. This has added to reusability of code with VHDL.

■ Generic types, subtypes and packages

A major improvement in VHDL 2008 is generic types, subtypes and packages against the earlier allowable only generic constants. This has added to flexibility and extensibility of VHDL. Even the inbuilt features of VHDL have used this feature. Two new packages have been included in library IEEE—fixed_generic_pkg and fixed_float_types for fixed- and floating-type arithmetic, respectively.

■ Inclusion of new synthesizable generic packages

The new generic packages for fixed- and floating-point arithmetic defined in library IEEE are fully synthesizable.

■ Support for hierarchical identifiers

The concept of external names has been introduced in VHDL 2008 to support hierarchical naming convention. An external name refers to a (shared) variable, signal, or constant which is in another part of the design hierarchy. External names are embedded in double angle brackets << >>. Special characters may be used

- ^ to move up the hierarchy
- @ to root the path in a package

External names may also be used

■ New condition operator

VHDL-2008 introduces a condition operator, which converts a STD_LOGIC or BIT expression to a BOOLEAN: '1' and 'H' are considered as TRUE and everything else FALSE.

Enhanced bit string literals

In VHDL 2008, bit string literals can have an explicit width and can be either signed or unsigned type—say a hexadecimal string length can be 11 bits instead of the previously allowed only multiple of 4. Further use of meta values such as U, X, and _ is allowed in VHDL 2008.

Use of vectors in aggregates

In VHDL 2008, aggregates can be assigned values using a slice or a vector as
A:= ("00", others => '1') – which implies for a bit vector as 00111111.
B:= (3 DOWNTO 0 = > '0', others => '1') – which implies a vector 11110000.

- Usage of conditional select statements in sequential programming

The conditional select statements (WHEN ....ELSE and WITH ..SELECT.....WHEN) can be used inside a process as sequential statements in VHDL 2008.

New form of generate statement

The generate statement is allowed to use Else part and Elsif for If generate. Even a case statement can be used for a generate statement. This has made the separation between the combinational and sequential design statement quite thin. The designer has flexibility to follow a uniform style of programming (combinational and sequential).

Easy updation of sensitivity list

A new keyword 'all' has been added to VHDL 2008 which represents the sensitivity list of a process and implies that all the signals read inside the process are included in the sensitivity list. This eases the task of a designer as any modification inside the process to signals does not require the designer to update the sensitivity list. All signals read inside the process will automatically be treated as a part of sensitivity list.

New packages for arithmetic operations on signed and unsigned STD_LOGIC

Two new packages—IEEE.Numeric_Std_Unsigned and IEEE.Numeric_Std_Signed, have been included in the VHDL-2008 to facilitate arithmetic operations on STD_logic_vectors.

All the important packages have been standardized

VHDL 2008 standard includes the following packages:

- IEEE.Numeric_Std
- IEEE.Numeric_Bit
- IEEE.Math_Real
- IEEE.Math_Complex
- IEEE.Std_logic_1164

Apart from the above new packages, IEEE.numeric_std_signed and IEEE.numeric_std_unsigned have been included to replace the non-standard packages— std_logic_signed and std_logic_unsigned, respectively.

IEEE.std_logic_1164 package too has been improved. All the features of non-standard package std_logic_textio have been included in std_logic_1164. Also std_logic has been included as a subtype of std_ulogic, making it easier to work with the two types. Further, new operators and functions have been included in std_logic_1164.

The new operators defined in VHDL 2008 are

- Reduction operator
- Matching operator
- Shift operator

New functions defined in VHDL 2008 are

- Minimum and maximum functions (for array and scalar types).
- String conversion functions_(to_string is defined for scalar and array types; to_bstring, to_binarystring, to_ostring, to_octalstring, to_hstring and to_hexstring for arrays).
- Function rising edge is defined for type Boolean.

Use of context to include packages

A new VHDL construct called 'context' has been introduced in VHDL 2008 which is user-defined or pre-defined and includes set of library and USE statements to include packages. Once defined, the context can be included in the design file instead of including individual libraries and the packages. Moreover, context once declared can be used again and again in different design files.

- **New array types** are defined as boolean_vector, integer_vector, real_vector and time_vector.
- /* */ **block comments** have been introduced instead of single-line comments in the previous versions.

IP Encryption

VHDL 2008 has IP encryption enabled, that is, VHDL-2008 has a means of specifying that a block of data is encrypted. This uses an additional feature—the *tool directive*. Tool directives are arbitrary words preceded by a backtick character. The idea of tool directives is that they are interpreted by tools and they do not have any meaning to a VHDL compiler.

For an IP encryption, a set of predefined tool directives is defined as follows:

```
`protect begin
`protect end
`protect begin_protected
`protect end_protected
`protect author
`protect author_info
`protect encrypt_agent
`protect encrypt_agent_info
`protect key_keyowner
`protect key_keyname
`protect key_method
`protect key_block
`protect data_keyowner
`protect data_keyname
`protect data_method
`protect data_block
`protect digest_keyowner
`protect digest_keyname
`protect digest_key_method
```

```
`protect digest_method
`protect digest_block
`protect encoding
`protect viewport
`protect decrypt_license
`protect runtime_license
`protect comment
```

All these features have made VHDL a very flexible and designer-friendly language.

Appendix D

VHDL-AMS

VHDL-AMS is an extension of VHDL to design analogue systems. It was introduced in 1999 as IEEE 1076.1–1999 standard as Superset of VHDL IEEE 1076–1993. It supports modelling of continuous time-driven systems, discrete time-driven systems and mixed signal systems. VHDL-AMS has added features to digital VHDL standard IEEE1076–1993 in terms of object interfaces, behaviour, types, and attributes. It supports modelling of continuous time systems based on differential equations. The simulation kernel has a dedicated part called analogue solver, which does the computation part for differential equations.

VHDL supports the following modelling approaches:

- Modelling of conservative systems (kirchoff equation-based systems)
- Modelling of non-conservative systems (using signal-flow graphs)
- Modelling of digital systems
- Modelling of mixed systems (analogue and digital)

It supports time domain, frequency domain and Quiescent domain representation of simulation.

Conservative Systems Modelling

For solving conservative systems KVL and KCL equations are used. A new object called 'quantity' is defined in VHDL-AMS, which has value of any user-defined type such as displacement, current, and so on, or a predefined type as Real. For object quantity new attributes have been defined:

- Q'Dot: is defined as a derivative of quantity with reference to (wrt) time.
- Q'Integ: is defined as an integral of quantity Q over time from zero to current time.
- Q'Slew (max_rising_slope, max_falling_slope): Follows Q, but its derivative wrt is limited by the specified slopes. While the default for max_falling_slope is max_rising_slope, the default for max_rising_slope is infinity.
- Q'Delayed (T) is defined as the quantity Q delayed by T (ideal delay, T >= 0).
- Q'Ltf (num, den) defines Laplace transfer function whose input is Q.
- Q'ZOH (T, initial_delay) defines a sampled version of quantity Q (zero-order hold).
- Q'Ztf (num, den, T, initial_delay) defines Z-domain transfer function whose input is Q.

A new set of statements is defined in VHDL-AMS—simple simultaneous statements. These statements are used to express relationships between quantities related by operator "= =". The left- (LHS) and right-hand side (RHS) must be expressions with scalar sub elements of a floating-point type. Expressions may involve quantities, constants, literals, signals and functions. The solution of simultaneous statements is done by the analogue solver. Simultaneous statement is just like a concurrent statement where the order is not important.

Other forms for simultaneous statements are as follow:

- Simultaneous IF statement
- Simultaneous CASE statement
- Simultaneous PROCEDURAL statement.

Analogue solver gives approximate results which may not be true to exact precision. So, every solution is subject to some tolerance band. Tolerance band is an indicator of exactness of the solution. Each and every quantity and simultaneous statement belongs to a tolerance band indicated by a string expression. A simple simultaneous statement whose LHS or RHS is a quantity gets its tolerance group from the quantity; otherwise, the tolerance group must be specified.

For modelling of electrical systems, a new port TERMINAL has been added which has attributes reference and contribution. Every terminal is associated with type 'NATURE' which is a user-defined energy domain representation of the terminal. Nature has two aspects related to physical quantities—across the quantity and through the quantity. Nature can be a scalar or a composite type. Reference defines the terminal as positive reference and reference terminal of its associated nature as negative terminal. Contribution defines through the quantity aspect with a value equal to the sum of all through quantities incident at terminal T.

In VHDL-AMS, analogue waveforms are treated as piecewise linear with each part represented by IF statement. Each IF part further contains different simultaneous equations. To represent constitutive relation of branches, free quantities are supported which do not have an across and through aspect. Conditions after discontinuity are represented using a break statement. Even the initial conditions are specified using break statements. Initial conditions are used for boundary conditions.

Signal Flow Graphs Modelling

For signal flow modeling, ports of entity are quantities defined in the modes IN or OUT. A quantity port with mode IN can be associated with any kind of quantity as an actual, whereas a quantity port with mode OUT can be associated only with an actual which is a quantity port with mode OUT or a free quantity or a branch quantity.

For signal flow graphs, quantities cannot be directly connected to terminals. Conversion models are used through simultaneous equations.

Solving Differential Equations

Although analogue solver is responsible for solving differential equations, a necessary condition is imposed for solving set of equations in VHDL-AMS. Any equation to be solved is represented by a scalar simultaneous statement, and the number of equations to be solved should be equal to the number of through quantities and free quantities and number of port quantities with mode OUT. If equations are less than the above, they cannot be solved.

Mixed System Modelling

It includes features of both VHDL and VHDL-AMS. In the entity interface for digital design by default, every port is treated as SIGNAL; whereas in VHDL-AMS for analogue, systems 'terminals' are used. So, in mixed system modelling both can be used in entity interface, and keyword SIGNAL is explicitly used for port which is not terminal.

Process sensitivity list can include quantity attributes.

Appendix E

Unsolved Design Problems

(i) Design a 4-bit adder using instances of full adder. How does the code differ for a 16-bit full adder with representation as integers.

(ii) Design a 16-bit adder for bit_vector inputs. Compare it with a 16-bit adder for std_logic_vector inputs.

(iii) Design a 4-bit full adder with overflow and carry signals.

(iv) Design a 4-bit BCD adder circuit.

(v) Design a 2×2 multiplier circuit in VHDL.

(vi) Implement the Boolean function $y = \Sigma\ (3,5,6,7)$ using 2:1 Multiplexer.

(vii) Design a 8:1 multiplexer using generate statements with 2:1 multiplexer as a component.

(viii) Design a VHDL entity d2t4 (2 to 4 decoder). Using a 2t4 decoder design a 4t16 decoder.

(ix) Design a 4-bit bidirectional shift register from a generic n-bit register.

(x) Design a mod-8 bidirectional counter. How many flip-flops are needed?

(xi) Design a mod-8 presettable counter with presettable count as 101.

(xii) Write a VHDL code to implement a bus structure.

(xiii) Implement tri-state buffers in VHDL.

(xiv) Design a 3-digit BCD counter. Further, extend the VHDL code to convert this BCD count to 7-segment display.

(xv) Design an 8-bit Johnson counter in VHDL.

(xvi) Design a 2-bit sequence detector circuit in VHDL. What type of FSM is it?

(xvii) Design a serial adder circuit in VHDl as FSM.

(xviii) For the following FSM write the VHDL code.

(xix) Design a VHDL circuit for finding the average of given inputs.

(xx) Design a 4-bit priority encoder circuit with don't-care inputs. Which predefined library function is to be used for such don't-care conditions?

(xxi) Design a BCD-to-binary converter?

(xxii) Design a comparator circuit which gives output high for an input that is greater than binary 6. Design using concurrent modelling.

(xxiii) Design a 2-bit binary to reflected-code converter circuit.

(xxiv) Implement a circuit with the following truth table:

A	B	C	Y
0	0	0	0
0	0	1	1
0	1	0	0
0	1	1	X
1	0	0	1
1	0	1	1
1	1	0	X
1	1	1	1

How are don't-care conditions synthesized in a hardware?

(xxv) Design a 2421-to-binary-code converter. Treat the inputs as std_logic_vectors? Will the synthesis output differ if inputs are bit_vectors instead of std_logic_vectors?

(xxvi) Design a synchronous 4-bit counter using J-K flip-flops.

(xxvii) Design a 2's-complement-based full subtractor circuit. Extend your code for a generic subtractor.

(xxviii) Design the above question by calling a user-defined function sub2comp to subtract numbers in 2's complement form using the overloading of '-' operator.

(xxix) Design a universal 4-bit shift register which can work in all four modes and is bidirectional.

(xxx) Design an 8-bit parallel adder circuit with temporary registers used for storing results and operands.

(xxxi) Write a VHDL code for checking key debounce for keyboard scanner. Use assert statements to design the circuit.

(xxxii) Write a VHDL code for implementing 2 8 × 8 size ROM.

(xxxiii) Write a VHDL code which converts JK flip-flop to D flip-flop.

(xxxiv) Design a timer circuit in VHDL with a 7-segment output.

(xxxv) Design a Fibonacci series generator in VHDL.

(xxxvi) Design a circuit in VHDL using state machine approach which has one input of type bit. It samples data on input three times. If input is one for odd number of times, then the output goes high; else, it is zero. After three samples the system is reset to initial stage.

Index